THE
PHARMA PRODUCT MANAGER

Handbook for Navigating the Digital Frontier

THE
PHARMA PRODUCT MANAGER

Handbook for Navigating the Digital Frontier

Subba Rao Chaganti

PharmaMed Press

An imprint of BSP Books Pvt. Ltd.
4-4-309/316, Giriraj Lane,
Sultan Bazar, Hyderabad - 500 095.

The Pharma Product Manager: Handbook for Navigating the Digital Frontier
by *Subba Rao Chaganti*

© 2025, *by Author,* All rights reserved.

No part of this book or parts thereof may be reproduced, stored in a retrieval system or transmitted in any language or by any means, electronic, mechanical, photocopying, recording or otherwise without the prior written permission of the publishers.

Published by

PharmaMed Press

An imprint of BSP Books Pvt. Ltd.
4-4-309/316, Giriraj Lane, Sultan Bazar, Hyderabad - 500 095.
Phone: 040-23445688, 23445600; Fax: 91+40-23445611
E-mail: info@pharmamedpress.com
www.pharmamedpress.com/pharmamedpress.net

ISBN: 978-9-3487344-0-2 (Paperback)

DEDICATION

This book, *The Pharma Product Managers Handbook for Navigating the Digital Frontier*, is dedicated posthumously to two visionary pioneers who have profoundly shaped India's pharmaceutical marketing and product management fields.

To **Professor Chitta Mitra**, the pioneer of prescription research whose groundbreaking work laid the foundation for evidence-based practices and scientific rigor in our industrt.

And

Professor Tarun Gupta, the pioneer of pharmaceutical marketing and product management, contributed significantly to creating scientific marketing processes and elevated the standards and scope of pharmaceutical marketing.

They have illuminated the path for future generations, demonstrating the power of combining scientific processes with strategic insight. Their legacies continue to inspire and guide us as we navigate the ever-evolving landscape of pharmaceutical product management.

DISCLAIMER

The information provided in this book, *The Pharma Product Manager: Handbook for Navigating the Digital Frontier*, is intended for educational and informational purposes only. While every effort has been made to ensure the accuracy, completeness, and reliability of the information contained herein, the author and publisher make no representations or warranties of any kind, express or implied, about the accuracy, reliability, suitability, or availability of the content, including but not limited to any specific strategies, tactics or recommendations.

The content of this book is based on the author's research, experience, and understanding of the pharmaceutical industry as of the date of publication. However, the pharmaceutical industry is constantly evolving, and new developments, regulations, and technologies may emerge that could affect the relevance or accuracy of the information provided. Readers are encouraged to conduct their research and seek professional advice to ensure their knowledge is current and applicable to their specific circumstances.

The author and the publisher shall not be liable for any loss or damage, including but not limited to direct, indirect, incidental, consequential, or punitive damages, arising out of or in connection with the use of or reliance on the information contained in this book. This includes but is not limited to any errors or omissions in the content, any actions taken or not taken based on the information provided, and any external websites or resources referenced in this book.

The opinions and views expressed in this book are those of the author and do not necessarily reflect the views of any organizations, institutions, or individuals with whom the author may be affiliated.

Including any specific company, product, or service in this book does not imply endorsement or recommendation by the author or publisher.

This book is not intended to replace professional advice, including but not limited to medical, legal, regulatory, financial, or business advice. Readers should seek the services of qualified professionals for any specific issues or concerns.

By using this book, the readers acknowledge and agree to the terms of this disclaimer.

ACKNOWLEDGMENTS

Writing *The Pharma Product Manager: Handbook for Navigating the Digital Frontier* has been an enriching journey, and I am grateful to many individuals who contributed to its realization. Their support, insights, and encouragement have been invaluable.

First and foremost, I would like to express my deepest gratitude to my family—my better half Mahalakshmi, my children and their spouses—Srinivasa Phanindra, Geetha, Lavanya, Aditya, Sowmya, and Chaitanya—and my grandchildren Aditi, Eesha, Surya, and Shriya. Your unwavering support, patience, and understanding have been my anchor throughout this project. Thank you for believing in me and giving me the space and time to bring this book to life.

I am profoundly grateful to my colleagues and mentors in the pharmaceutical industry. Your shared experiences, knowledge, and guidance have significantly shaped the content of this book.

Thanks to my publishers—the ebullient Anil Shah, whose enthusiasm is contagious, and Nikunjesh Shah, who made the digital editions possible. A special thanks to Mr. Naresh Davergave and his team at PharmaMed Press for designing this book so well.

I also want to acknowledge the many pharmaceutical product managers and marketers who shared their views, experiences, and challenges through formal and informal conversations. Your insights have brought practical perspectives to the theories and concepts discussed in this book.

To my friends and peers in the pharmaceutical marketing community, your camaraderie and intellectual exchange have inspired and motivated me. Thank you for being a sounding board and encouraging me throughout this journey.

Finally, to the readers of this book, thank you for your interest and trust. I hope this handbook is a valuable resource in your professional journey, equipping you with the knowledge and skills to navigate pharmaceutical product management's dynamic and ever-evolving landscape.

PREFACE

The product manager's role has become increasingly complex and dynamic in the constantly evolving pharmaceutical industry. As we stand at the cusp of the digital revolution, pharmaceutical product managers must navigate a frontier where science, technology, and patient care converge. This book, *The Pharma Product Manager: Handbook for Navigating the Digital Frontier*, is a comprehensive guide for current and aspiring product managers who aim to thrive in this transformative era.

The genesis of this book lies in the recognition that the pharmaceutical industry is undergoing a profound shift. Advances in artificial intelligence, big data analytics, telemedicine, and personalized medicine are reshaping how drugs are developed, marketed, and delivered. Product managers are no longer just brand custodians; they are strategic leaders who must harness these technologies to drive innovation, enhance patient outcomes, and achieve commercial success.

Throughout my career in pharmaceutical marketing, I have witnessed firsthand the challenges and opportunities that product managers face. This book is a culmination of years of experience, extensive research, and insights gathered from industry experts. It is designed to equip you with the knowledge and skills needed to excel in your role, whether developing a new drug, managing an existing product portfolio, or leading cross-functional teams.

The structure of this book is deliberate and comprehensive. It begins with an introduction to the digital frontier, setting the stage for the profound changes we are witnessing. We then delve into a pharmaceutical product manager's core responsibilities, providing a realistic portrayal of the role. From there, we explore the required knowledge and skills, covering everything from drug development and regulatory affairs to business acumen and financial management.

Understanding the importance of social sciences, we decode how sociology, anthropology, and behavioral economics can be leveraged to understand patient populations and influence healthcare behaviors. The scientific foundation section provides a deep dive into critical areas such as pharmacology, clinical research, epidemiology, and pharmacoeconomics, ensuring a robust understanding of the science that underpins your products.

Marketing research and analysis are crucial for strategic decision-making, and this book offers practical tools and techniques for conducting market research, analyzing sentiment, and performing competitive analysis. Leadership and communication skills are also highlighted, emphasizing the importance of effective team management, conflict resolution, and stakeholder engagement.

The book covers the latest trends and technologies in digital marketing, email marketing, HCP marketing, and patient education as we move into the digital age. We explore how artificial intelligence, big data, and emerging technologies such as telemedicine and blockchain are transforming product management.

Finally, we address career development and advancement, providing guidance on career pathways, continuing education, networking, and mentorship. The book concludes with a look towards the future, offering insights into emerging trends and encouraging continuous learning and adaptation.

This handbook is a valuable resource and a trusted companion in your journey as a pharmaceutical product manager. The digital frontier is vast and filled with opportunities. With the right knowledge and skills, you can navigate it successfully and make a meaningful impact in healthcare.

Welcome to the future of pharmaceutical product management.

Subba Rao Chaganti

CONTENTS

Part I. Introduction to Pharmaceutical Product Management

Pharmaceutical product management is a specialized and dynamic field that plays a crucial role in a drug's lifecycle, from its development to its entry into the market and beyond. It sits at the intersection of science, business, and strategy, requiring a unique blend of knowledge, skills, and experience.

At its core, pharmaceutical product management ensures that a drug reaches the right patients at the right time and in the right way. This involves a deep understanding of the drug's therapeutic benefits, competitive landscape, and potential market. But it also requires a keen awareness of the broader business environment, including regulatory requirements, market dynamics, and the ever-evolving needs of healthcare providers and patients.

Pharmaceutical product managers are the architects of a drug's market strategy. They are responsible for planning and executing the product's development and launch, managing its lifecycle, and driving its commercial success. This role demands a thorough understanding of the drug development process and strong leadership and communication skills, as product managers must collaborate with cross-functional teams that include R&D, marketing sales, regulatory affairs, and more.

In today's digital age, the role of pharmaceutical product managers is evolving rapidly. The rise of digital health, the increasing importance of data analytics, and the growing influence of artificial intelligence are all reshaping how drugs are brought to market and managed throughout their lifecycle. As a result, modern product managers must be agile and adaptable, staying abreast of emerging trends and technologies to remain competitive.

This introduction sets the stage for understanding the pivotal role of pharmaceutical product management in ensuring that innovative therapies reach the patients who need them, driving both clinical and commercial success in a highly regulated and competitive industry. Whether you are a seasoned professional or new to the field, this exploration of pharmaceutical product management will provide you with the insights and tools needed to navigate the challenges and opportunities that lie ahead.

Welcome to the Digital Frontier

The pharmaceutical industry is undergoing a profound transformation driven by the rapid advancement of digital technologies. As we step into the digital frontier, pharmaceutical product management is being redefined, creating challenges and opportunities for those responsible for managing drug portfolios.

The digital age has brought a suite of tools and technologies that are reshaping every aspect of pharmaceutical product management, from the early stages of drug development to post-market surveillance; digital innovations enable more efficient processes, more effective communication, and more personalized patient care. These advancements are not just enhancing traditional practices but creating entirely new ways of thinking about and managing pharmaceutical products.

At the heart of this transformation is data — big data, to be precise. The ability to collect, analyze, and leverage vast amounts of data empowers product managers to make more informed decisions, predict market trends, and tailor strategies to meet the specific needs of patients and healthcare providers. Predictive analytics, machine learning, and artificial intelligence are no longer futuristic concepts but integral components of the modern pharmaceutical product manager's toolkit.

However, the digital frontier is not just about technology but the mindset required to thrive in this new environment. Product managers must be agile, adaptable, and forward-thinking. They must be willing to embrace change, continuously learn, and stay ahead of the curve in a rapidly evolving landscape. The traditional roles and responsibilities of product management are expanding, requiring a deeper understanding of digital marketing, patient engagement, and regulatory compliance in a digital context.

This chapter, "Welcome to the Digital Frontier," sets the stage for your journey into the new era of pharmaceutical product management. It explores the key digital trends shaping the industry, the challenges of digital transformation, and the opportunities that await those ready to lead in this brave new world. As you delve deeper into the content, you'll discover how to leverage digital tools and strategies to manage your products more effectively and deliver better patient and stakeholder outcomes.

The digital frontier is not just a phase; it's the future of pharmaceutical product management. Welcome to this exciting journey, where innovation meets tradition and where the possibilities are as vast as the data that drives them.

1.1 Introduction to the Digital Age in Pharma

The pharmaceutical industry, long known for its rigorous processes and systematic approach to drug development, is now standing at the crossroads of a digital revolution. The digital age is not merely an era of technological advancements; it is a fundamental shift in how the industry operates, interacts with stakeholders, and delivers value to patients.

In the past, pharmaceutical companies relied on traditional methods for everything from research and development (R&D) to marketing and sales. However, the advent of digital technologies has introduced new paradigms reshaping the industry. Artificial intelligence (AI), big data analytics, machine learning, blockchain, digital therapeutics, and telemedicine are just a few of the innovations driving this transformation.

The digital age in pharma is characterized by unprecedented access to data. Pharmaceutical companies can now collect and analyze vast amounts of information from clinical trials, electronic health records, social media, and wearable devices. This data-driven approach enables more precise decision-making, leading to the development of personalized medicines, improved patient outcomes, and more efficient drug development processes.

Moreover, digital technologies are breaking down traditional barriers in healthcare. Patients increasingly manage their health through mobile apps, wearable devices, and telehealth services. This shift empowers patients and gives pharmaceutical companies direct channels to engage with their end-users, offering a more personalized and responsive approach to patient care.

The digital age brings both challenges and opportunities for pharmaceutical product managers. On one hand, they must navigate the complexities of integrating digital technologies into their existing workflows. On the other hand, they can leverage these technologies to innovate, enhance product offerings, and create more value for patients and healthcare providers.

The transition to the digital age is also reshaping the skills and knowledge required for success in pharmaceutical product management. Traditional competencies, such as understanding drug development and regulatory affairs, are now complemented by new areas of expertise, including digital marketing, data analytics, and cybersecurity. Product managers must be adaptable, continuously learning, and ready to embrace new technologies that can drive their products and companies forward.

This introduction to the digital age in pharma serves as a gateway to understanding the profound changes underway in the industry. It sets the stage for exploring how digital technologies transform pharmaceutical product management and highlights the importance of staying ahead in a rapidly evolving landscape. As you delve into this new era, you will discover the tools, strategies, and mindsets needed to thrive in the digital age of pharma.

1.2 The Transformation of Pharmaceutical Product Management

Pharmaceutical product management has undergone a remarkable transformation over the past few decades, shaped by evolving market dynamics, technological advancements, regulatory changes, and shifting patient expectations. This transformation reflects the industry's response to modern healthcare's growing complexity and demands, requiring product managers to continuously adapt and redefine their roles.

From Traditional to Modern Product Management

In its earliest form, pharmaceutical product management was primarily focused on the tactical aspects of marketing. Product managers managed drugs' lifecycles, from launch to patent expiry, strongly emphasizing sales support, promotional activities, and relationships with healthcare professionals (HCPs). The role was largely reactive, centered around traditional marketing channels like print advertising, medical conferences, and sales reps.

However, as the industry evolved, so did the responsibilities of product managers. The increasing pressure to bring innovative therapies to market faster and the rise of generics forced companies to think more strategically. Product management expanded beyond marketing to encompass broader responsibilities, including market access, pricing strategies, and competitive intelligence.

The Impact of Digitalization

The digital revolution has been one of the most significant drivers of change in pharmaceutical product management. Digital technologies have introduced new tools, platforms, and channels that fundamentally alter how pharmaceutical products are developed, marketed, and managed.

For instance, digital marketing has transformed the way product managers reach and engage with HCPs and patients. Social media, online communities, and digital content have replaced traditional marketing tactics, enabling more personalized and targeted communication strategies. Additionally, the rise of telemedicine and digital therapeutics has opened up new avenues for product managers to deliver to patients beyond the pill.

Data analytics and artificial intelligence (AI) have also become integral to modern product management. Today's product managers have access to vast amounts of data from various sources, including clinical trials, real-world evidence, and social media. This data-driven approach allows for more informed decision-making, enabling product managers to anticipate market trends, optimize marketing strategies, and improve patient outcomes.

The Shift Toward Patient-Centricity

One of the most profound shifts in pharmaceutical product management is the move toward patient-centricity. In the past, the focus was primarily on HCPs as the key decision-makers. However, with the rise of empowered patients and the growing emphasis on outcomes-based healthcare, product managers now place patients at the center of their strategies.

Patient-centricity means understanding patients' needs, preferences, and behaviors and using this insight to guide product development, marketing, and communication strategies. It involves engaging with patients throughout the product lifecycle, from early-stage clinical development to post-market surveillance. This shift has led to the development of more personalized treatments, improved patient adherence, and better health outcomes.

The Role of Regulatory and Market Access Challenges

The regulatory landscape has also become more complex, with increasing scrutiny from regulatory agencies and a greater emphasis on demonstrating value for money. Product managers must now navigate these challenges while ensuring compliance with global and local regulations. This requires a deep understanding of market access strategies, health economics, and outcomes research (HEOR).

Market access has become a critical component of pharmaceutical product management. With healthcare systems worldwide under financial pressure, payers demand more evidence of a product's value before agreeing to reimburse it. Product managers must work closely with cross-functional teams to develop pricing strategies, negotiate with payers, and ensure their products are accessible to patients.

Adapting to Emerging Technologies and Trends

The rapid pace of technological advancement continues to shape the future of pharmaceutical product management. Emerging technologies such as AI, machine learning, blockchain, and digital health tools create new opportunities and challenges for product managers. These technologies have the potential to streamline operations, enhance decision-making, and improve patient outcomes. Still, they also require product managers to upskill and adapt to new working methods continuously.

For example, AI can predict market trends, identify potential patient populations, and optimize marketing strategies. Digital health tools like mobile apps and wearables offer new ways to engage with patients and collect real-world data. Blockchain technology promises to improve transparency and security in the pharmaceutical supply chain, which could have significant implications for product management.

The Future of Pharmaceutical Product Management

Looking ahead, the role of pharmaceutical product managers will continue to evolve as the industry faces new challenges and opportunities. The increasing complexity of healthcare, the rise of personalized medicine, and the growing importance of digital health will require product managers to be more agile, data-driven, and patient-focused.

In this new era, successful product managers will be those who can balance strategic thinking with a deep understanding of the science behind their products, leverage emerging technologies to stay ahead of the competition and build strong relationships with both internal and external stakeholders. As the pharmaceutical industry transforms, so will the product manager role, making it one of the field's most dynamic and rewarding careers.

1.3 Opportunities and Challenges in the Digital Era

The digital era has ushered in a transformative period for the pharmaceutical industry, offering unprecedented opportunities while presenting significant challenges. For pharmaceutical product managers, navigating this landscape requires a keen understanding of digital technologies and a strategic approach to leveraging them for competitive advantage.

Opportunities in the Digital Era

1. **Enhanced Patient Engagement and Empowerment**

   - Digital tools like mobile health apps, wearables, and telemedicine platforms have revolutionized patient engagement. These technologies empower patients to manage their health actively, providing them with real-time access to information, personalized treatment plans, and continuous monitoring.

   - Pharmaceutical companies can use these tools to offer value-added services that improve patient adherence, enhance outcomes, and build stronger relationships with patients. For example, digital therapeutics can be integrated with drug treatments to provide comprehensive care solutions.

2. **Data-Driven Decision-Making**

   - The proliferation of digital data sources, including electronic health records (EHRs), social media, and wearable devices, has created vast amounts of data that can be harnessed for insights. Advanced analytics and artificial intelligence (AI) can analyze this data to identify trends, predict outcomes, and optimize marketing strategies.

   - Product managers can use these insights to tailor marketing campaigns, identify potential patient populations, and refine product development strategies. Data-driven decision-making enables more precise targeting and better resource allocation.

3. Personalized Medicine

- The digital era has facilitated the rise of personalized medicine, where treatments are tailored to each patient's characteristics. Advances in genomics, proteomics, and AI-driven analysis allow for the development of targeted therapies that address specific genetic profiles or disease markers.

- Product managers have the opportunity to develop and market these personalized treatments, offering more effective solutions that meet patients' unique needs. This shift towards precision medicine opens new avenues for product differentiation and market segmentation.

4. Efficient Clinical Trials and Research

- Digital technologies are streamlining clinical trials, making them more efficient and cost-effective. Virtual trials, remote patient monitoring, and real-world evidence (RWE) collection are reducing the time and expense of bringing new drugs to market.

- These innovations enable product managers to accelerate the development timeline, reduce risks, and gather more comprehensive data on the safety and efficacy of new treatments. Additionally, digital tools facilitate better patient recruitment and retention in clinical trials.

5. Global Reach and Market Expansion

- Digital platforms allow pharmaceutical companies to reach global audiences with relative ease. Online marketing campaigns, telemedicine, and e-commerce platforms enable product managers to expand into new markets and engage with healthcare professionals (HCPs) and patients worldwide.

- The ability to quickly adapt and localize marketing strategies for different regions is a significant advantage in today's interconnected world. Digital tools also support compliance with local regulations and cultural nuances.

Challenges in the Digital Era

1. Data Privacy and Security Concerns

- Data privacy and security have become critical challenges for pharmaceutical companies with the increased collection and use of digital health data. Major concerns include cybersecurity threats, data breaches, and compliance with regulations like the General Data Protection Regulation (GDPR).

- Product managers must navigate these challenges by implementing robust data protection measures, ensuring compliance with global and local regulations, and maintaining patient trust. Failure to address these concerns can lead to reputational damage and legal consequences.

2. Regulatory Compliance

- The rapid pace of digital innovation often outstrips regulatory frameworks, creating uncertainty and complexity in compliance. Regulatory bodies are still adapting to digital health technologies, leading to a need for clear guidelines on digital marketing, telemedicine, and AI-driven diagnostics.

- Product managers must stay informed about evolving regulations and work closely with legal and regulatory teams to ensure compliance. This requires a proactive approach to monitoring regulatory changes and anticipating potential challenges.

3. Integration of Digital and Traditional Marketing

- Balancing digital and traditional marketing strategies is a significant challenge. While digital tools offer new opportunities for engagement, traditional methods like sales rep visits and medical conferences remain important in many markets.

- Product managers must develop integrated marketing strategies that leverage the strengths of both digital and traditional channels. This requires a deep understanding of customer preferences and behaviors across different segments.

4. Digital Literacy and Skills Gap

- The rapid adoption of digital technologies has created a skills gap in the pharmaceutical industry. Not all product managers or HCPs have the digital literacy required to fully utilize these tools, leading to challenges in implementation and adoption.

- Pharmaceutical companies must invest in training and development programs to upskill their workforce. To remain competitive in the digital era, Product managers must be proficient in digital marketing, data analytics, and emerging technologies.

5. Adapting to Rapid Technological Change

- The fast pace of technological innovation presents both opportunities and challenges. Staying ahead of the curve requires continuous learning and adaptation, which can be resource-intensive.

- Product managers must be agile, constantly updating their knowledge and skills to keep up with the latest developments. This includes understanding new digital tools, platforms, and trends and their implications for product management.

6. Patient Expectations and Experience

- Patients' expectations for personalized, seamless, and responsive healthcare experiences have increased as they become more digitally savvy. Meeting these expectations requires a patient-centric approach that leverages digital tools to enhance the patient journey.

- Product managers must focus on creating value beyond the pill by offering digital solutions that improve patient outcomes and satisfaction. This involves understanding patient needs, preferences, and behaviors and integrating these insights into product development and marketing strategies.

Conclusion

The digital era offers immense opportunities for pharmaceutical managers to innovate, differentiate, and deliver value to patients and HCPs. However, it also presents significant challenges that require strategic foresight, continuous learning, and adaptability. By embracing digital tools and overcoming these challenges, product managers can lead their organizations to success in a rapidly evolving healthcare landscape.

The Role of the Pharmaceutical Product Manager

The pharmaceutical industry, a critical pillar of global healthcare, is marked by its complexity, innovation, and high stakes. At the head of this dynamic sector lies the role of the pharmaceutical product manager — a pivotal figure responsible for guiding a product from conception through to its life in the market. The product manager's operational and strategic role requires a deep understanding of science, market dynamics, regulatory landscapes, and customer needs.

In this chapter, we delve into the multifaceted role of the pharmaceutical product manager, exploring the skills, knowledge, and expertise required to navigate the intricacies of drug development, marketing, and lifecycle management. We will examine how product managers balance scientific innovation with commercial viability and act as the linchpin connecting various functions within a pharmaceutical company—from research and development (R&D) to marketing, sales, and regulatory affairs.

As the pharmaceutical landscape evolves, the role of the product manager has expanded beyond traditional responsibilities. Today's product managers must be adept at leveraging digital tools, interpreting vast amounts of data, and adapting to the rapid pace of technological change. They orchestrate a product's success, ensuring that every aspect of its journey — from clinical trials to post-market surveillance — is executed precisely and aligned with the company's strategic goals.

This chapter provides a comprehensive overview of the role, offering insights into the day-to-day responsibilities of a product manager, the challenges they face, and the strategies they employ to ensure that their products not only reach the market but also thrive within it. Whether you are an aspiring product manager or a seasoned professional, this chapter will equip you with a deeper understanding of the critical importance of this role in the pharmaceutical industry.

2.1 Defining the Role and Responsibilities

The role of a pharmaceutical product manager is multifaceted and pivotal to a pharmaceutical company's success. Acting as the central hub for a product's lifecycle, from development to commercialization, the product manager orchestrates a wide range of activities, ensuring that a product meets market needs, regulatory requirements, and company objectives. The role requires scientific knowledge, business acumen, and strategic vision. Here is a detailed breakdown of the role and responsibilities:

Key Responsibilities

1. **Product Lifecycle Management**
   - **Development:** Product managers are involved from the early stages of product development, often contributing to the design of the clinical development plan. They work closely with research and development (R&D) teams to ensure that the product aligns with market demands and has a clear value proposition.
   - **Launch**: A critical phase, the product launch requires careful planning and execution. The product manager coordinates marketing, sales, regulatory, and supply chain teams to ensure a successful launch. This includes developing marketing strategies, pricing models, and sales training programs.
   - **Growth and Maturity**: Post-launch, the product manager monitors market performance, adjusts strategies, and works to maximize the product's market share. This phase involves lifecycle management strategies such as line extensions, market expansion, and formulation improvements.
   - **Decline and Exit**: As a product approaches the end of its lifecycle, the product manager plans for a managed decline, which might include exploring product repositioning or exit strategies to minimize losses and optimize the portfolio.

2. Market Research and Analysis

- Product managers must deeply understand market dynamics, the competitive landscape, and customer needs. This involves conducting market research, analyzing data, and translating insights into actionable strategies. They must also identify opportunities and threats, understand the patient journey, and anticipate trends.

3. Strategic Planning

- Developing and implementing a strategic plan for the product is a core responsibility. This includes setting objectives, defining the product positioning, creating a unique selling proposition (USP), and aligning the strategy with the overall business goals. Product managers must also create long-term forecasts and financial models to ensure the product's profitability and sustainability.

4. Cross-Functional Leadership

- The product manager leads cross-functional teams, including marketing, sales, medical affairs, regulatory, manufacturing, and supply chain. They act as a bridge between these departments, ensuring alignment and collaboration. Effective communication and leadership skills are essential to guide these teams toward common goals.

5. Regulatory and Compliance Management

- In the highly regulated pharmaceutical industry, compliance with regulatory standards is non-negotiable. Product managers must ensure that all promotional materials, product claims, and marketing activities comply with regulatory guidelines. They work closely with regulatory affairs teams to navigate the complex approval processes and ensure that the product adheres to local and international regulations.

6. Financial Management

- Managing the product's budget, including marketing and sales expenditures, is a key responsibility. Product managers must ensure that resources are allocated effectively to maximize return on investment (ROI). This includes monitoring financial performance, conducting cost-benefit analyses, and adjusting the budget as needed.

7. Customer and Stakeholder Engagement

- It is crucial to build and maintain relationships with key stakeholders, including healthcare professionals (HCPs), key opinion leaders (KOLs), patients, and payers. Product managers must understand these stakeholders' needs and expectations and develop strategies to meet them. They also gather feedback to inform product development and improvement.

8. Innovation and Digital Integration

- Product managers must be at the forefront of innovation in the digital age. This includes leveraging digital tools and technologies to enhance product marketing, patient engagement, and data analytics. Whether through digital marketing campaigns, telemedicine, or AI-driven insights, product managers must integrate digital solutions into their strategies to remain competitive.

9. Risk Management

- Identifying and mitigating risks associated with the product is a vital function. This includes market, regulatory, competitive, and operational risks. Product managers must develop contingency plans and be prepared to adapt to changing circumstances.

Skills and Competencies

- **Scientific Knowledge**: A strong understanding of pharmacology, clinical research, and the science behind the product is essential. This knowledge enables the product manager to make informed decisions, communicate effectively with R&D teams, and create scientifically accurate marketing materials.

- **Business Acumen**: Product managers must understand the business side of pharmaceutical product management, including finance, marketing, and operations. This allows them to balance scientific considerations with commercial objectives, ensuring the product's success in the market.

- **Leadership and Communication**: Leading cross-functional teams requires strong leadership and communication skills. Product managers must be able to motivate and guide their teams, resolve conflicts, and communicate strategies clearly to all stakeholders.

- **Analytical Thinking**: Analyzing data, understanding market trends, and making data-driven decisions is critical. Product managers must be adept at using analytical tools and methodologies to derive insights that drive strategy.

- **Adaptability and Resilience**: The pharmaceutical industry is dynamic, with frequent changes in market conditions, regulations, and competitive landscapes. Product managers must be adaptable and resilient and able to pivot strategies quickly in response to new challenges.

Conclusion

The role of a pharmaceutical product manager is both challenging and rewarding. It requires a unique blend of scientific knowledge, business acumen, strategic thinking, and leadership skills. As the pharmaceutical industry continues to evolve, the product manager's role will only become more critical in driving innovation, ensuring compliance, and delivering products that meet the needs of patients and healthcare providers. The ability to navigate this complex landscape and bring successful products to market makes pharmaceutical product management a vital function within the industry.

2.2 Core Competencies and Skills Required

Pharmaceutical product managers need diverse competencies and skills to navigate the industry's complexities and drive product success effectively. Here is a detailed breakdown of the core competencies and skills required:

1. **Scientific and Technical Knowledge**
   - **Pharmacology**: Understanding the science of drugs, including their mechanisms of action, side effects, and interactions.
   - **Drug Development**: Knowledge of the drug development process, from preclinical research through clinical trials and regulatory approval.
   - **Medical Affairs**: Awareness of medical and clinical aspects related to the product, including therapeutic areas and disease states.

2. **Business Acumen**
   - **Market Analysis**: Ability to analyze market trends, competitive landscape, and customer needs to inform strategic decisions.
   - **Financial Management**: Budgeting, forecasting, and financial analysis skills to manage product profitability and resource allocation.
   - **Pricing Strategy**: Expertise in developing pricing strategies that align with market conditions and company objectives.

3. **Project Management**
   - **Planning and Execution**: Competence in developing project plans, setting milestones, and managing timelines to ensure successful product launches and lifecycle management.
   - **Cross-functional Coordination**: Ability to lead and coordinate efforts across different departments, such as R&D, marketing, sales, and regulatory affairs.

4. Strategic Planning

- **Product Strategy**: Skills in developing and implementing strategic plans for product positioning, market entry, and growth.
- **Scenario Planning**: Ability to anticipate market changes and develop contingency plans to address potential challenges.

5. Marketing and Sales Expertise

- **Market Segmentation**: Knowledge of segmenting the market to identify target audiences and tailor marketing efforts.
- **Promotional Strategies**: Expertise in creating and executing marketing campaigns, including digital marketing, content creation, and branding.
- **Sales Support**: Skills supporting sales teams with training, tools, and strategies to drive product adoption and sales performance.

6. Regulatory and Compliance Knowledge

- **Regulatory Affairs**: Understanding regulatory requirements and processes for product approvals and compliance.
- **Compliance Management**: Ability to ensure that all marketing materials and promotional activities adhere to industry regulations and guidelines.

7. Leadership and Team Management

- **Team Leadership**: Skills in leading and motivating cross-functional teams to achieve common goals.
- **Conflict Resolution**: Ability to address and resolve conflicts within teams and with external stakeholders.

8. Communication Skills

- **Interpersonal Communication**: Strong skills in interacting with internal teams, healthcare professionals, and other stakeholders.

- **Presentation Skills**: Ability to present information clearly and persuasively to various audiences, including executives and stakeholders.

9. Analytical and Problem-Solving Abilities

- **Data Analysis**: Skills in analyzing market data, clinical trial results, and financial performance to make informed decisions.
- **Problem-Solving**: Ability to identify issues, develop solutions, and implement strategies to overcome challenges.

10. Digital and Technological Proficiency

- **Digital Tools**: Familiarity with digital marketing tools, CRM systems, and data analytics platforms.
- **Emerging Technologies**: Awareness of emerging technologies such as AI, machine learning, and digital therapeutics and their applications in pharma.

11. Customer and Stakeholder Engagement

- **Patient Engagement**: Skills in developing programs and strategies to engage and support patients throughout their treatment journey.
- **Stakeholder Engagement**: Ability to build and maintain relationships with key stakeholders, including healthcare professionals, payers, and advocacy groups.

12. Ethical and Legal Awareness

- **Ethical Standards**: Knowledge of ethical standards and practices in pharmaceutical marketing and product management.
- **Legal Compliance**: Understanding legal requirements for drug promotion, advertising, and patient privacy.

Conclusion

The role of a pharmaceutical product manager is multifaceted and requires a blend of scientific expertise, business acumen, strategic vision, and leadership skills. By developing and honing these core competencies, pharma product managers can effectively manage products through their lifecycle, drive successful market outcomes, and contribute to the overall success of their organizations.

2.3 The Product Manager's Impact on Business Success

A product manager plays a pivotal role in shaping the success of a pharmaceutical company. Their responsibilities extend across various aspects of the product lifecycle, from inception to market delivery and beyond. Here's a detailed look at how a product manager impacts business success:

1. Driving Product Strategy

- **Market Positioning**: The product manager develops a strategic vision for the product, positioning it effectively within the market to meet unmet needs and capitalize on opportunities.

- **Competitive Advantage**: By understanding the competitive landscape, the product manager identifies unique selling propositions (USPs) and differentiates the product from competitors.

2. Optimizing Product Development

- **Strategic Planning**: The product manager outlines the development roadmap, aligning it with the company's goals and market needs. They ensure the product's development is on track and within the budget.

- **Cross-Functional Coordination**: Effective coordination with research and development, clinical trials, regulatory affairs, and other departments.

3. Enhancing Market Launch

- **Go-to-Market Strategy**: The product manager crafts and executes comprehensive go-to-market strategies, including pricing, distribution, and promotional plans. This ensures a successful product launch and market entry.

- **Launch Execution**: They manage the launch process, coordinating activities across teams to ensure the product is introduced effectively to the market.

4. Driving Revenue and Profitability

- **Sales Performance**: The product manager helps drive sales performance and revenue growth by providing sales teams with the necessary tools, training, and strategies.

- **Financial Management**: The product manager oversees budgeting and financial forecasting, optimizing resources to maximize return on investment (ROI) and profitability.

5. Managing Product Lifecycle

- **Lifecycle Management**: The product manager oversees the product throughout its lifecycle, from introduction to growth, maturity, and eventual decline. They implement strategies to extend the product's market presence and maximize its lifetime value.

- **Continuous Improvement**: They gather feedback from stakeholders and market data to make informed decisions about product enhancements, modifications, or updates.

6. Enhancing Customer and Patient Outcomes

- **Patient Engagement**: The product manager develops programs and strategies to improve patient adherence, engagement, and overall outcomes, ensuring that the product delivers value to patients and healthcare providers.

- **Customer Support**: They ensure that customer support mechanisms are in place to address issues or questions related to the product, enhancing customer satisfaction.

7. Navigating Regulatory and Compliance Challenges

- **Regulatory Compliance**: The product manager ensures that the product meets all regulatory requirements, navigating complex approval processes and maintaining compliance with industry standards.

- **Risk Management**: They identify potential risks and implement mitigation strategies to address regulations and market dynamics challenges.

8. Fostering Innovation

- **Innovation Management**: The product manager promotes innovation by staying abreast of industry trends, technological advancements, and emerging market needs. They drive the development of new features or products that address evolving demands.

- **Technology Integration**: They leverage new technologies and digital tools to enhance product development, marketing, and customer engagement.

9. Building and Leading Teams

- **Team Leadership**: The product manager leads cross-functional teams, fostering collaboration and ensuring all team members are aligned with the product's goals and objectives.

- **Motivation and Mentoring**: They motivate and mentor team members, promoting a positive work environment and driving collective success.

10. Strategic Decision-Making

- **Data-Driven Decisions**: The product manager utilizes market research, clinical data, and financial analysis to make informed decisions that impact the product's success.

- **Strategic Insights**: They provide valuable insights to senior management, contributing to overall business strategy and direction.

Conclusion

The impact of a product manager on business success is profound and multifaceted. By driving product strategy, optimizing development, managing the lifecycle, and enhancing market performance, the product manager plays a crucial role in achieving business goals and ensuring the product's success in the market. Their contributions are integral to delivering value to patients, generating revenue, and maintaining a competitive edge in the pharmaceutical industry.

The Evolution of Pharmaceutical Product Management

The role of the pharmaceutical product manager has undergone significant transformation over the past few decades. Initially confined to managing a product's marketing and sales aspects, the role has expanded to encompass various responsibilities, including strategic planning, cross-functional leadership, and integrating digital tools and technologies. Various factors, such as advancements in medical science, changes in regulatory environments, the rise of digital technologies, and the increasing complexity of the pharmaceutical market, have driven this evolution.

Understanding the evolution of pharmaceutical product management is essential for grasping the current landscape and anticipating future trends. The traditional focus on product-centric strategies has shifted towards a more holistic approach that considers the entire product lifecycle, patient-centricity, and the integration of innovative technologies. This chapter explores the key milestones in the field's development, the shift from traditional to digital product management, and the expanding knowledge and skills required for success in this dynamic field.

By tracking the historical progression and examining the factors that have shaped modern pharmaceutical product management, this chapter provides a comprehensive overview of how the role has adapted to meet the demands of an ever-evolving industry. Whether you are an aspiring product manager or a seasoned professional, understanding this evolution will offer valuable insights into the complexities and opportunities that define pharmaceutical product management today.

3.1 Historical Overview of Product Management

Pharmaceutical product management has evolved significantly over the decades, influenced by advancements in drug development, market dynamics, regulatory changes, and technological innovations. Here is a detailed historical overview of how product management in the pharmaceutical industry has developed:

1. **Early 20th Century: Initial Stages**

   - **Pre-1970s**: Before the formalization of product management, pharmaceutical companies primarily focused on drug discovery and production. The role of managing individual products was not distinctly defined.

   - **Informal Practices**: Sales representatives and marketing teams often managed marketing efforts without a structured approach to product lifecycle management. There was limited emphasis on strategic product oversight.

2. **Formalization and Structure (1970s - 1980s)**

   - **Introduction of Product Management**: In the 1970s, the pharmaceutical industry began formalizing product management roles. Upjohn, a pioneering pharmaceutical company, is credited with introducing the structured product management system. This marked the beginning of a more systematic approach to managing pharmaceutical products.

   - **Development of Product Management Frameworks**: During this period, companies established frameworks for product management, including defining roles and responsibilities, managing product lifecycles, and integrating product strategies with overall business goals.

   - **Lifecycle Management**: Product managers started overseeing products throughout their entire lifecycle, from development to commercialization and eventual phase-out. This included managing market entry strategies, regulatory approvals, and marketing campaigns.

3. **Expansion of Responsibilities and Integration (1990s - 2000s)**

   - **Cross-Functional Collaboration**: The 1990s shifted towards more integrated and collaborative approaches. Product managers began working closely with R&D, regulatory affairs, marketing, and sales teams to ensure cohesive product strategies.

   - **Regulatory and Compliance Focus**: With increasing regulatory scrutiny, product managers took on compliance and quality assurance responsibilities. This included navigating complex approval processes and ensuring adherence to industry regulations.

   - **Market Research and Customer Insights**: Product management began to emphasize market research and customer insights. Product managers used data to inform market strategies, understand customer needs, and identify opportunities for product differentiation.

4. **Digital Transformation and Advanced Analytics (2000s - 2010's)**

   - **Emergence of Digital Tools**: The early 2000s marked the advent of digital technologies that transformed product management practices. Product managers began using digital tools for market research, data analysis, and customer engagement.

   - **Data-Driven Decision-Making**: The use of data analytics became crucial. Product managers leveraged data to drive decision-making, optimize marketing strategies, and improve operational efficiency.

- **Patient-Centric Approaches**: There was a growing emphasis on patient-centric strategies. Product managers focused on improving patient outcomes, adherence, and engagement through digital platforms and patient support programs.

5. **Current Trends and Future Outlook (2010s - Present)**

- **Integration of AI and Machine Learning**: Recent years have seen the integration of AI and machine learning into product management. Product managers use these technologies for predictive analytics, personalized marketing, and operational optimization.

- **Digital Therapeutics and Innovation**: The rise of digital therapeutics and other innovative treatments has introduced new aspects to product management. Product managers are involved in developing and managing digital health solutions alongside traditional pharmaceutical products.

- **Real-World Evidence (RWE)**: The focus on real-world evidence has increased. Product managers use RWE to support product claims, improve patient outcomes, and inform market strategies.

- **Cross-Functional Expertise**: Modern product managers are expected to have a diverse skill set, including expertise in digital marketing, data analytics, regulatory affairs, and project management.

Key Milestones in Pharmaceutical Product Management

1. **Formalization (1970s)**: The introduction of structured product management roles, notably by Upjohn, established the foundation for modern product management practices.

2. **Cross-Functional Integration (1990s)**: The emphasis on department collaboration led to more comprehensive and strategic product management approaches.

3. **Digital Transformation (2000s)**: Adopting digital tools and data analytics transformed product management practices, enhancing decision-making and operational efficiency.

4. **Advanced Technologies (2010s)**: Integrating AI, machine learning, and digital therapeutics has reshaped product management, introducing new opportunities and challenges.

Conclusion

The historical evolution of pharmaceutical product management reflects broader changes in the industry, from informal practices to a sophisticated technology-driven role. Product managers must adapt to new trends, technologies, and market dynamics as the pharmaceutical landscape evolves to ensure successful product development and commercialization.

3.2 Key Milestones in the Development of the Role

The role of a pharmaceutical product manager has evolved through several key milestones, reflecting broader trends in the industry and the increasing complexity of drug development and marketing. Here is a detailed look at these milestones:

1. **The Birth of Product Management (1970s)**
   - **Upjohn's Innovation (Early 1970s)**: Upjohn is often credited with formalizing the role of product management in the pharmaceutical industry. The company introduced a structured approach to managing pharmaceutical products, which included appointing dedicated product managers responsible for overseeing products from development through commercialization.
   - **Role Definition**: During this period, the role of the product manager began to take shape, focusing on coordinating cross-functional teams, managing product lifecycles, and developing marketing strategies.

2. **Establishment of Product Management Frameworks (1980s)**
   - **Development of Product Management Practices**: The 1980s saw the establishment of formal frameworks and best practices for product management. Companies developed processes for product planning, market analysis, and lifecycle management.
   - **Integration with Business Strategy**: Product managers began aligning product strategies with overall business goals, emphasizing the need for strategic thinking and market orientation.

3. **Cross-Functional Collaboration (1990s)**
   - **Enhanced Collaboration**: The 1990s marked a shift towards greater collaboration across departments. Product managers increasingly worked with R&D, regulatory affairs, marketing, and sales teams to ensure cohesive product strategies.

- **Emphasis on Regulatory Compliance**: As regulatory requirements became more stringent, product managers took on compliance and quality assurance responsibilities. They played a crucial role in navigating complex approval processes and ensuring product quality.

4. **Adoption of Data-Driven Decision-Making (2000s)**

   - **Introduction of Digital Tools**: The early 2000s saw the adoption of digital tools and technologies that transformed product management practices. Product managers began using data analytics, market research tools, and digital marketing platforms to inform decision-making and optimize strategies.

   - **Patient-Centric Strategies**: There was a growing focus on patient-centric approaches. Product managers developed strategies to improve patient outcomes, adherence, and engagement, leveraging digital platforms and patient support programs.

5. **Integration of Advanced Technologies (2010s - to Present)**

   - **AI and Machine Learning**: Integrating artificial intelligence (AI) and machine learning became a significant milestone. Product managers began using these technologies for predictive analytics, personalized marketing, and operational optimization.

   - **Digital Therapeutics and Innovation**: The rise of digital therapeutics and other innovative treatments introduced new dimensions to product management. Product managers had to adapt to managing traditional pharmaceutical products and digital health solutions.

   - **Real-World Evidence (RWE)**: The focus on real-world evidence increased. Product managers utilized RWE to support product claims, improve patient outcomes, and refine marketing strategies.

6. **Emphasis on Cross-Functional Expertise (Present)**

   - **Diverse Skill Set**: Modern product managers are expected to have a diverse skill set that includes expertise in digital marketing, data analytics, regulatory affairs, project management, and strategic planning.

- **Adapting to Market Dynamics**: The role continues to evolve as product managers navigate new market dynamics, technological advancements, and regulatory changes. They must stay abreast of the trends in precision medicine, telemedicine, and digital health innovations.

Summary of Key Milestones

1. **1970s**: Formalization of the role by Upjohn; introduction of structured product management.

2. **1980s**: Development of frameworks and best practices; alignment with business strategy.

3. **1990s**: Enhanced cross-functional collaboration; focus on regulatory compliance.

4. **2000s**: Adoption of digital tools; emphasis on patient-centric strategies.

5. **2010s - to Present**: Integration of AI and machine learning; management of digital therapeutics; focus on real-world evidence and diverse skill sets.

These milestones reflect the ongoing evolution of the pharmaceutical product management role, driven by changes in technology, market demands, and regulatory environments. The role continues to adapt and expand as the industry faces new challenges and opportunities.

3.3. The Shift from Traditional to Digital Product Management

The transition from traditional to digital product management represents a significant evolution in the pharmaceutical industry, driven by technological advancements, changes in market dynamics, and evolving stakeholder expectations. This shift encompasses various aspects, including adopting digital tools, changes in strategy and processes, and integrating new technologies. Here is a detailed description of this transformation:

1. Traditional Product Management

- **Focus**: Traditionally, pharmaceutical product management focused on managing a product's lifecycle from development through launch and post-launch phases. This involved coordinating cross-functional teams, managing regulatory approvals, and executing marketing strategies.

- **Processes**: The processes were largely manual and relied on standard marketing and sales approaches. Product managers used market research reports, sales data, and traditional marketing channels such as print advertising and direct sales meetings.

- **Data Use**: Data was often limited and fragmented. Product managers relied on historical data, sales performance metrics, and physician feedback collected through surveys and focus groups.

- **Communication**: Communication with healthcare professionals (HCPs) and patients was primarily face-to-face or through printed materials. Marketing efforts were centered around traditional media and in-person interactions.

2. The Shift to Digital Product Management

- **Adoption of Digital Tools**:

  - **Digital Marketing**: Digital marketing tools have become central to pharmaceutical product management. Product managers now use online platforms, social media, email marketing, and digital advertising to reach target audiences.

  - **Analytics and Data Management**: Advanced data analytics and customer relationship management (CRM) systems allow for better tracking of customer interactions, sales performance, and market trends. Product managers can analyze real-time data to make informed decisions.

  - **Data-Driven Decision Making**:

    - **Big Data and AI**: The use of big data and artificial intelligence (AI) has revolutionized decision-making processes. AI algorithms analyze large datasets to identify trends, predict outcomes, and optimize marketing strategies.

    - **Real-World Evidence**: Product managers leverage RWE to understand how products perform in real-world settings. This evidence supports product claims, improves patient outcomes, and refines strategies.

- **Enhanced Communication and Engagement**:

  - **Digital Channels**: Communication with HCPs and patients has expanded to include digital channels such as websites, virtual conferences, and mobile apps. These channels offer more interactive and accessible ways to share information.

  - **Patient Engagement Tools**: Digital tools like patient portals and mobile health apps enhance patient engagement and adherence. Product managers develop and implement these tools to improve patient outcomes and support product usage.

- **Agile and Adaptive Strategies:**
  - **Agile Methodologies**: Adopting agile methodologies allows product managers to respond more quickly to market changes and adjust strategies based on real-time feedback.
  - **Continuous Learning**: Digital product management requires continuous learning and adaptation. Product managers must stay updated with digital trends, technologies, and regulatory changes.
- **Integration of Emerging Technologies**:
  - **Telemedicine**: Integrating telemedicine and remote patient monitoring has become increasingly important. Product managers must adapt strategies to include these technologies and address new opportunities and challenges.
  - **Digital Therapeutics**: The rise of digital therapeutics introduces new dimensions to product management. Product managers must manage traditional pharmaceutical products and digital health solutions, requiring a broader skill set and knowledge base.

3. **Implications for Pharmaceutical Product Managers**

- **Skill Set Expansion**: Product managers must develop skills in digital marketing, data analytics, and technology integration. This requires ongoing training and adaptation to new tools and methodologies.
- **Strategic Focus**: The focus has shifted from traditional marketing tactics to data-driven, personalized approaches. Product managers must use digital tools to create targeted campaigns and measure their effectiveness.
- **Enhanced Collaboration**: Digital product management often involves collaboration with technology teams, data scientists, and digital marketing experts. Product managers must effectively coordinate with these teams to implement and optimize digital strategies.

Summary of the Shift

1. **Traditional Product Management** focused on manual processes, traditional marketing, and limited data use.

2. **Digital Product Management** embraces digital tools, data-driven decision-making, enhanced communication channels, and the integration of emerging technologies.

3. **Implications**: Expanded skill set, strategic focus on digital and data-driven approaches, and enhanced collaboration with technology and marketing experts.

This shift reflects the broader transformation in the pharmaceutical industry, where digital innovation and technological advancements are reshaping how products are managed, marketed, and delivered.

Part II. Knowledge and Skills Required

Pharmaceutical product managers play a crucial role in managing a product's lifecycle, from development through commercialization and beyond. They need a diverse set of knowledge and skills to navigate the complexities of the pharmaceutical industry effectively. Here's a comprehensive description of the essential knowledge and skills required:

1. Scientific Knowledge

- **Pharmacology** involves understanding drug mechanisms, interactions, side effects, and therapeutic effects. This knowledge helps product managers appreciate the product's benefits and risks, crucial for creating accurate and compelling marketing messages.

- **Clinical Research:** Knowledge of clinical trial design, methodology, and regulatory requirements. Product managers must understand how clinical data is generated and interpreted to support product claims and navigate regulatory submissions.

- **Biostatistics**: Ability to interpret statistical data from clinical trials and other research studies. This includes understanding efficacy, safety profiles, and statistical significance.

- **Pharmacoeconomics**: Understanding the drug's cost-effectiveness and value compared to alternatives. This includes knowledge for developing pricing strategies and demonstrating the product's value to payers and healthcare providers.

2. Business Acumen

- **Market Analysis**: Skills in assessing market trends, competitor analysis, and identifying opportunities and threats. This involves understanding market dynamics, patient demographics, and competitive positioning.

- **Financial Management**: Proficiency in budgeting, forecasting, and financial analysis. Product managers must manage product budgets, evaluate financial performance, and make strategic decisions based on financial data.

- **Strategic Planning**: Ability to develop and implement strategic plans for product launch and lifecycle management. This includes setting goals, defining strategies, and measuring performance against objectives.

3. Project Management Skills

- **Planning and Execution**: Expertise in planning and executing projects, including timelines, resource allocation, and risk management. Product managers must oversee cross-functional teams and ensure that project milestones are met.

- **Agile Methodologies**: Familiarity with agile project management practices to adapt quickly to changing market conditions and feedback.

- **Cross-Functional Collaboration**: Skills in coordinating and collaborating with departments such as R&D, regulatory affairs, marketing, and sales.

4. Marketing and Communication Skills

- **Digital Marketing**: Proficiency in digital marketing strategies, including online advertising, social media, and content marketing. Product managers must leverage digital channels to reach target audiences effectively.

- **Communication**: Strong verbal and written communication skills are essential for presenting ideas, creating marketing materials, and interacting with stakeholders. This includes crafting clear and persuasive messages for different audiences.

- **Brand Management**: Ability to develop and manage the brand strategy for the product, including positioning, messaging, and visual identity.

5. Regulatory and Compliance Knowledge

- **Regulatory Affairs**: Understanding regulatory requirements and guidelines for product approval and marketing. Product managers must navigate complex regulatory landscapes and ensure compliance with laws and regulations.

- **Ethics and Compliance**: Awareness of ethical considerations and compliance standards in pharmaceutical marketing. This includes adhering to industry codes of conduct and avoiding practices that can be deemed unethical or illegal.

6. Technological Proficiency

- **Data Analytics**: Skills in using analytics tools to gather insights, measure performance, and inform decision-making. This includes analyzing sales data, market trends, and customer feedback.

- **Emerging Technologies**: Familiarity with emerging technologies such as artificial intelligence (AI), machine learning, and digital therapeutics. Product managers should understand how these technologies can be leveraged to enhance product development and marketing strategies.

7. Leadership and Interpersonal Skills

- **Leadership**: Ability to lead and motivate cross-functional teams. Product managers must inspire team members, manage conflicts, and drive project success.

- **Negotiation**: Skills in negotiating with stakeholders, including suppliers, partners, and internal teams. This includes negotiating contracts, terms, and agreements.

- **Adaptability**: Flexibility to adapt to changing market conditions, new technologies, and evolving business needs. Product managers must be agile and responsive to shifts in the industry.

Summary

Pharmaceutical product managers require scientific knowledge, business acumen, project management skills, marketing expertise, regulatory understanding, technological proficiency, and strong leadership abilities. Mastery of these areas enables product managers to effectively manage the lifecycle of pharmaceutical products, from development through commercialization, and to navigate the complexities of the pharmaceutical industry successfully.

Scientific Understanding of Drug Development

Understanding the scientific aspects of drug development is crucial for pharmaceutical product managers. This knowledge helps them navigate the complexities of bringing a new drug to market, from initial research to commercialization. Here is an in-depth look at the scientific understanding required for effective drug development:

1. **Drug Discovery**

   - **Target Identification and Validation**: The process begins with identifying biological targets (such as proteins or genes) implicated in a disease. Validation ensures that targeting these molecules can lead to therapeutic benefits.

   - **Lead Compound Identification**: Researchers screen chemical libraries or use computational methods to identify potential lead compounds interacting with the target.

   - **Preclinical Testing**: Before advancing to human trials, leads undergo preclinical testing in vitro (in test tubes) and in vivo (in animal models) to assess their efficacy, safety, and pharmacokinetics (how the drug is absorbed, distributed, metabolized, and excreted).

2. **Clinical Development**

   - **Phase I Clinical Trials**: Focus on assessing the drug's safety, tolerability, and pharmacokinetics in a small group of healthy volunteers or patients. This phase aims to determine the maximum tolerated dose and identify any adverse effects.

- **Phase II Clinical Trials**: These trials evaluate the drug's efficacy and safety in a larger group of patients with the targeted disease. This phase helps determine the optimal dose and regimen and provides preliminary data on effectiveness.

- **Phase III Clinical Trials** involve large-scale testing in diverse patient populations to confirm the drug's efficacy, monitor side effects, and compare it to standard treatments. This phase generates the primary evidence required for regulatory approval.

- **Phase IV Clinical Trials (Post-Marketing Surveillance)**: Conducted after the drug Is approved and on the market. These trials monitor long-term effects, rare side effects, and the drug's performance in real-world settings.

3. **Regulatory Affairs**

- **Regulatory Submissions**: Product managers must understand the regulatory requirements for submitting a New Drug Application (NDA) or Biologics License Application (BLA) to agencies such as the FDA or EMA. This includes compiling clinical data, manufacturing information, and labeling details.

- **Regulatory Compliance**: Ongoing compliance with regulatory guidelines throughout the product's lifecycle. This includes adhering to Good Clinical Practice (GCP) and Good Manufacturing Practice (GMP) standards.

4. **Pharmacokinetics and Pharmacodynamics**

- **Pharmacokinetics (PK)**: The study of how the drug is absorbed, distributed, metabolized, and excreted in the body. Understanding PK helps in determining dosing regimens and potential drug interactions.

- **Pharmacodynamics (PD)**: The study of the drug's effects on the body, including its mechanism of action and therapeutic effects. PD helps in understanding the relationship between drug concentration and effect.

5. Drug Formulation and Development

- **Formulation Development**: Designing the drug's formulation (e.g., tablets, injections) to ensure stability, efficacy, and patient compliance. This includes selecting appropriate excipients and optimizing the delivery method.

- **Manufacturing Processes**: Understanding the processes involved in large-scale production, including quality control, scaling up from clinical to commercial production, and ensuring consistency and safety.

6. Biostatistics and Data Analysis

- **Statistical Analysis**: Proficiency in analyzing clinical trial data to determine the drug's effectiveness and safety. This includes understanding statistical methods and interpreting results to support regulatory submissions and marketing claims.

- **Data Interpretation**: Ability to interpret complex data from clinical trials and other studies to make informed decisions about the drug's development and marketing strategies.

7. Drug Safety and Risk Management

- **Adverse Event Monitoring**: Understanding how to monitor, report, and manage adverse events and side effects throughout the drug's lifecycle. This includes assessing risk-benefit ratios and implementing risk mitigation strategies.

- **Risk Management Plans**: Developing and implementing plans to minimize risks associated with the drug, including post-marketing surveillance and risk communication strategies.

Summary

A comprehensive scientific understanding of drug development encompasses the knowledge of the entire process, from discovery through clinical testing, regulatory approval, and post-marketing surveillance. For pharmaceutical product managers, this knowledge is crucial for making informed decisions, managing cross-functional teams, and ensuring successful product launches and lifecycle management. Understanding drug discovery, clinical trials, regulatory affairs, pharmacokinetics, formulation development, biostatistics, and drug safety helps product managers navigate the complexities of bringing a new drug to market and ensure the product's success in a competitive marketplace.

4.1 The Drug Development Process: From Discovery to Market

The drug development process is a complex, multi-stage journey that takes a new drug candidate from initial discovery to its market availability. This process ensures that new medications are safe, effective, and meet regulatory standards. Here's a detailed overview of each stage:

1. Drug Discovery

A. Target identification and Validation

- **Objective**: Identify biological targets (such as proteins or genes) implicated in a disease. Validating these targets confirms that modulating them can lead to therapeutic benefits.

- **Methods**: High-throughput screenings, genomics, and proteomics.

B. Lead Compound Identification

- **Objective**: Discover potential drug candidates (leads) that interact with the validated target.

- **Methods**: Screening chemical libraries, computational drug design, and structure-based drug design.

C. Preclinical Testing

- **Objective**: Assess the lead compounds' safety, efficacy, and pharmacokinetics.

- **Methods**: In vitro assays (test tube experiments), in vivo studies (animal models), and toxicology studies.

2. Clinical Development

A. Phase I Clinical Trials

- **Objective**: Evaluate the drug's safety, tolerability, and pharmacokinetics in healthy volunteers or patients.

- **Key Activities**: Determine the maximum tolerated dose (MTD), assess side effects, and understand drug metabolism.

B. Phase II Clinical Trials

- **Objective**: Test the drug's efficacy and safety in a larger group of patients with the targeted disease.
- **Key Activities**: Identify the optimal dose, gather preliminary efficacy data, and monitor adverse effects.

C. Phase III Clinical Trials

- **Objective**: Confirm the drug's efficacy and safety in a large, diverse patient population.
- **Key Activities**: Compare the drug to standard treatments, assess long-term effects, and gather comprehensive data for regulatory submission.

D. Phase IV Clinical Trials (Post-Marketing Surveillance)

- **Objective**: Monitor the drug's performance in the general population after approval.
- **Key Activities**: Detect rare side effects, assess long-term effectiveness, and study the drug in different populations.

3. Regulatory Affairs

A. Regulatory Submissions

- **Objective**: Submit a New Drug Application (NDA) or Biologics License Application (BLA) to regulatory authorities (e.g., FDA, EMA).
- **Key Activities**: Complie clinical trial data, manufacturing information, and labeling details.

B. Regulatory Review

- **Objective**: Regulatory agencies review the submission to ensure the drug meets safety, efficacy, and quality standards.

- **Key Activities**: Address questions and requests for additional information from regulators and prepare for potential advisory committee meetings.

C. Approval

- **Objective**: Obtain approval to market the drug.
- **Key Activities**: Implement post-marketing commitments and prepare for product launch.

4. Manufacturing and Quality Control

A. Scale-Up:

- **Objective**: Transition from small-scale production in the lab to large-scale manufacturing.
- **Key Activities**: Optimize production processes, ensure consistency, and meet quality standards.

B. Quality Control

- **Objective**: Ensure the drug meets production quality, safety, and efficacy standards throughout manufacturing.
- **Key Activities**: Perform routine testing, implement Good Manufacturing Practices (GMP), and conduct batch release testing.

5. Marketing and Launch

A. Market Strategy

- **Objective**: Develop strategies for market entry, including pricing, positioning, and distribution.
- **Key Activities**: Conduct market research, create marketing campaigns, and plan product launches.

B. Sales and Distribution

- **Objective**: Ensure the drug is available to healthcare providers and patients.
- **Key Activities**: Establish distribution channels, train sales representatives, and manage inventory.

6. Post-Market Surveillance and Lifecycle Management

A. Post-Market Surveillance

- **Objective**: Monitor the drug's safety and effectiveness in the real world.
- **Key Activities**: Collect and analyze adverse event reports, conduct additional studies, and update labeling as needed.

B. Lifecycle Management

- **Objective**: Maximize the drug's value throughout its life cycle.
- **Key Activities**: Explore new indications, develop new formulations, and manage patent life and market exclusivity.

Summary

The drug development process is a rigorous, multi-phase journey that begins with discovery and progresses through clinical trials, regulatory approval, and manufacturing, ultimately leading to market launch. Understanding each stage is crucial for pharmaceutical product managers, as it helps them navigate the complexities of bringing a new drug to market and ensures that the product meets safety, efficacy, and quality standards. Effective management of this process requires a deep understanding of scientific principles, regulatory requirements, market dynamics, and strong project management and strategic planning skills.

4.2 Clinical Trials and Regulatory Milestones

Clinical trials and regulatory milestones are critical components of the drug development process. They ensure that new medications are safe, effective, and meet regulatory standards before being approved for public use. Here is a detailed breakdown of these aspects:

Clinical Trials

Clinical trials are research studies conducted on human volunteers to evaluate the safety and efficacy of new drugs. They are divided into several phases, each with specific objectives and requirements.

1. **Phase I Clinical Trials**

   Objective:

   - To assess a drug's safety, tolerability, and pharmacokinetics in a small group of healthy volunteers or patients.

   Key Activities:

   - **Dosing**: Determine the maximum tolerated dose (MTD) and identify the appropriate dosage range.
   - **Safety Monitoring**: Identify and monitor potential side effects or adverse reactions.
   - **Pharmacokinetics**: Study how the drug is absorbed, distributed, metabolized, and excreted in the body.
   - **Study Population**: Typically involves 20 to 100 patients.

   Outcome:

   - Initial safety data and dosage guidelines.

2. **Phase II Clinical Trials**

   Objectives:

   - To evaluate the drug's efficacy and further assess its safety in a larger group of patients with the target disease or condition.

Key Activities:

- **Efficacy Assessment**: Test the drug's effectiveness in treating the condition compared to placebo or standard treatments.

- **Dose-Response Relationship**: Determine the optimal dose based on efficacy and safety data.

- **Safety Monitoring**: Continue to monitor side effects and adverse reactions.

- **Study Population**: Typically involves 100 to 500 participants.

Outcome:

- Data on efficacy, optimal dosage, and safety in the target patient population.

3. Phase III Clinical Trials

Objective:

- To confirm the drug's efficacy and safety in a large, diverse population and compare it to existing treatments.

Key Activities:

- **Large-Scale Testing**: Conduct trials on a large scale to gather comprehensive data on the drug's benefits and risks.

- **Comparison**: Compare the new drug with existing standard treatments or placebo.

- **Long-Term Data**: Collect data on long-term safety and effectiveness.

- **Study Population**: Typically involves 1,000 to 3,000 or more participants.

Outcome:

- Comprehensive data is required for regulatory approval, including evidence of efficacy, safety, and long-term effects.

4. Phase IV Clinical Trials

Objective:

- To monitor the drug's performance in the general population after it has been approved for market use.

Key Activities:

- **Real-World Data**: Collect data on the drug's long-term effects, rare side effects, and overall effectiveness in the general population.
- **Safety Monitoring**: Monitor and report any adverse effects or safety concerns.
- **Additional Studies**: Conduct further studies as required by regulatory agencies or to explore new indications or uses.

Outcome:

- Ongoing safety and efficacy data and potential labeling or usage guidelines updates.

Regulatory Milestones

Regulatory milestones are key points in drug development where regulatory agencies evaluate a drug's progress. These milestones involve submissions, reviews, and approvals crucial for bringing a new drug to market.

1. **Pre-IND (Investigational New Drug) Meeting**

   Objective:

   - To discuss the proposed clinical trial plans with regulatory agencies (e.g., FDA in the US) before submitting the formal IND application.

   Key Activities:

   - **Feedback**: Obtain feedback on trial design, safety considerations, and regulatory requirements.
   - **Preparation**: Prepare for IND submission based on agency feedback.

   Outcome:

   - Guidance on trial design and regulatory expectations.

2. **IND Application Submission**

   Objective:

   - To obtain approval to begin clinical trials.

Key Activities:

- **Documentation**: Submit a comprehensive application with preclinical data, trial proposals, and safety information.
- **Review**: Regulatory agencies review the submission to ensure that trials are designed to address safety and efficacy.

Outcome:

- Approval to initiate clinical trials.

3. **New Drug Application (NDA) or Biologics License Application (BLA)**

Objective:

- To seek approval for marketing the drug based on clinical trial data.

Key Activities:

- **Submission**: Submit an NDA (for small molecules) or BLA (for biologics) that includes all clinical trial results, manufacturing information, and proposed labeling.
- **Review**: Regulatory agencies thoroughly review the submission, including inspecting manufacturing facilities and assessing clinical trial data.

Outcome:

- Regulatory approval or request for additional information.

4. **Advisory Committee Meeting**

Objective:

- To seek independent expert opinions on the drug's safety and efficacy.

Key Activities:

- **Presentation**: Present clinical data to an advisory committee of experts who review and discuss the findings.
- **Recommendations**: The committee provides recommendations or opinions on the drug's approval.

Outcome:

- Guidance on approval and any additional requirements or conditions.

5. Post-Marketing Surveillance (Phase IV)

Objective:

- Monitor the drug's safety and effectiveness after it is on the market.

Key Activities:

- **Adverse Event Reporting**: Collect and analyze reports of adverse events from healthcare providers and patients.
- **Ongoing Studies**: Conduct additional studies or trials as required by regulatory agencies.

Outcome:

- Continuous evaluation of the drug's safety profile and effectiveness.

Summary

Clinical trials and regulatory milestones are integral to drug development, ensuring new medications are safe, effective, and compliant with regulatory standards. Understanding each phase and milestone helps pharmaceutical product managers navigate the complexities of bringing a new drug to market and align their strategies with regulatory requirements and clinical data.

4.3 The Importance of Scientific Knowledge for Product Managers

Scientific knowledge is crucial for pharmaceutical product managers for several reasons. It helps them make informed decisions, navigate complex regulatory environments, and ensure that products meet clinical and market needs. Here's an in-depth look at why scientific understanding is essential:

1. **Understanding Drug Mechanisms and Efficacy**

   Relevance:

   - **Product Development**: Knowledge of pharmacology, drug mechanisms, and therapeutic areas allows product managers to understand how a drug works, its potential benefits, and its limitations. This understanding is crucial for shaping product development strategies and making informed decisions about clinical trial designs and endpoints.

   - **Market Positioning**: It helps correctly position the drug in the market by comparing its mechanisms of action, efficacy, and safety profile with competitors. This understanding supports the development of compelling product claims and marketing messages.

   Implications:

   - **Informed Decision-Making**: Enables product managers to make decisions about drug development, formulation, and marketing based on a solid understanding of scientific principles.

   - **Strategic Planning**: Helps set realistic goals and timelines based on the drug's expected performance and development challenges.

2. **Navigating Clinical Trials and Regulatory Processes**

   Relevance:

   - **Designing Trials**: Scientific knowledge is essential for designing clinical trials that accurately assess a drug's efficacy and safety. Product managers must understand study designs, endpoints, and statistical methods to collaborate effectively with clinical teams.

- **Regulatory Compliance**: Understanding regulatory requirements and scientific guidelines helps ensure that the drug meets all necessary criteria for approval. Knowledge of pharmacokinetics, pharmacodynamics, and adverse event reporting is crucial for preparing regulatory submissions.

Implications:

- **Efficient Execution**: This function facilitates the smoother execution of clinical trials by ensuring that all scientific and regulatory aspects are considered and addressed.

- **Successful Approval**: Increases the likelihood of regulatory approval by ensuring that all necessary scientific data is provided and meets regulatory standards.

3. Communicating with Stakeholders

Relevance:

- **Healthcare Professionals (HCPs)**: Product managers must communicate complex scientific information to HCPs, including physicians and pharmacists. A solid understanding of the drug's scientific background helps explain its benefits, mechanisms, and clinical data effectively.

- **Patients**: Scientific knowledge aids in developing clear, accurate educational materials for patients, helping them understand how the drug works and its potential benefits and risks.

Implications:

- **Effective Communication**: Enhances the ability to convey scientific concepts clearly and persuasively, building trust with HCPs and patients.

- **Educational Materials**: Supports the creation of accurate and informative content that aids in patient understanding and adherence.

4. Driving Innovation and Development

Relevance:

- **Identifying Opportunities**: Scientific knowledge helps identify opportunities for drug development, including novel indications or combination therapies. Understanding emerging scientific trends and technologies can lead to innovative product solutions.

- **Problem-Solving**: Enables product managers to address scientific challenges during development and find solutions to drug formulation, delivery, and efficacy issues.

Implications:

- **Competitive Advantage**: Promotes innovation by leveraging scientific knowledge to develop cutting-edge products that meet unmet medical needs.

- **Problem Resolution**: Provides the expertise to tackle scientific and technical challenges, ensuring successful product development and commercialization.

5. Understanding Market Needs and Trends

Relevance:

- **Therapeutic Areas**: Knowledge of scientific developments in therapeutic areas helps product managers identify market needs, trends, and gaps. This understanding informs product positioning, differentiation, and competitive strategies.

- **Patient Outcomes**: Understanding scientific principles related to patient outcomes, disease mechanisms, and treatment paradigms supports the development of products that address real-world needs.

Implications:

- **Market Relevance**: Ensures the product meets current market needs and aligns with emerging trends and scientific advancements.

- **Strategic Positioning**: Facilitates the development of strategies that effectively address market demands and position the product competitively.

Summary

Scientific knowledge is indispensable for pharmaceutical product managers. It empowers them to make informed decisions, navigate regulatory requirements, communicate effectively with stakeholders, drive innovation, and align products with market needs. By integrating scientific understanding into their roles, product managers can enhance the development and success of pharmaceutical products, leading to improved patient outcomes and business success.

Business Acumen and Financial Management

Business acumen and financial management are critical competencies for pharmaceutical product managers. They enable these professionals to navigate complex business environments, optimize resource allocation, and drive the commercial success of their products. Here's a detailed overview of these areas and their importance:

1. Business Acumen

Definition:

- Business acumen is understanding and applying business principles, including strategic thinking, market analysis, and competitive positioning. For pharmaceutical product managers, it encompasses understanding the broader business environment in which their products are developed and marketed.

Key Components:

- **Market Analysis**: Understanding market dynamics, including demand, competition, and pricing strategies. Product managers must analyze market trends, identify opportunities and threats, and adapt strategies accordingly.

- **Competitive Intelligence** involves monitoring competitors' activities, product offerings, and market positioning. This helps differentiate the product and develop strategies to gain a competitive advantage.

- **Strategic Planning**: Developing and implementing strategies to achieve business objectives, including market entry, growth, and profitability. This involves setting goals, defining key performance indicators (KPIs), and aligning resources with strategic priorities.

- **Customer Insights** involve understanding customer needs and preferences, including those of healthcare professionals (HCPs) and patients. This insight informs product development, marketing, and sales strategies.

Implications:

- **Effective Decision-Making**: Informed decisions regarding product development, marketing, and commercialization based on a thorough understanding of market conditions and business opportunities.

- **Strategic Advantage**: The ability to position the product effectively in the market and anticipate competitive moves, leading to a stronger marketing presence and increased profitability.

2. Financial Management

Definition:

- Financial management involves planning, organizing, controlling, and monitoring financial resources to achieve business objectives. It includes budgeting, forecasting, financial analysis, and performance measurement for pharmaceutical product managers.

Key Components:

- **Budgeting**: Creating and managing budgets for product development, marketing, and sales activities. This involves allocating resources efficiently, controlling costs, and ensuring expenditures align with financial goals.

- **Forecasting**: Predicting future financial performance based on market trends, historical data, and strategic plans. Accurate forecasting helps set realistic sales targets, manage cash flow, and plan investments.

- **Financial Analysis**: Evaluating financial data to assess product performance, profitability, and return on investment (ROI). This includes analyzing profit margins, cost structures, and financial metrics to make data-driven decisions.

- **Cost Management**: identifying and managing costs associated with product development, production, and marketing. Effective cost management ensures that the product remains profitable and competitive.

Implications:

- **Resource Optimization**: Efficient allocation of financial resources to maximize return on investment (ROI) and support strategic objectives. This involves prioritizing investments, controlling costs, and ensuring financial sustainability.

- **Performance Monitoring**: Regularly reviewing financial performance against targets and benchmarks. This enables timely adjustments to strategies and tactics to improve financial outcomes.

Integration of Business Acumen and Financial Management

1. **Aligning Strategies with Financial Goals**:

   Product managers must ensure that business strategies are aligned with financial objectives. This involves integrating market insights and financial data to develop strategies that drive growth and profitability.

2. **Data-Driven Decision-Making**:

   Utilizing financial data and market analysis to inform strategic decisions. This includes evaluating the financial impact of different strategies, assessing potential risks, and making adjustments based on data-driven insights.

3. **Effective Communication**:

   Communicating financial and business insights to senior management, investors, and team members. Clear communication of financial performance and strategic plans helps secure support and drive alignment across the organization.

4. Risk Management:

Identifying and managing financial and business risks associated with product development and commercialization. This includes assessing potential risks, developing mitigation strategies, and monitoring risk factors throughout the product life cycle.

Summary

Business acumen and financial management are essential for pharmaceutical product managers to succeed in a competitive and complex industry. By understanding market dynamics, developing strategic plans, and managing financial resources effectively, product managers can drive the success of their products, optimize performance, and achieve organizational goals. These competencies enable them to make informed decisions, align strategies with financial objectives, and navigate the challenges of the pharmaceutical landscape.

5.1. Understanding the Pharmaceutical Business Model

The pharmaceutical business model is a complex system encompassing pharmaceutical product development, production, marketing, and distribution. It is designed to address the multifaceted nature of the industry where scientific innovation, regulatory compliance, and commercial strategies intersect. Here's a detailed description of the pharmaceutical business model and its key components:

1. **Research and Development (R&D):**

   Overview:

   - R&D is the foundation of the pharmaceutical business model. It involves discovering and developing new drugs or therapies, from initial research to clinical trials and regulatory approval.

   Key Components:

   - **Drug Discovery** involves identifying potential drug candidates through preclinical research, including in vitro studies and animal testing. This phase involves high-risk and high-cost activities to find compounds with therapeutic potential.

   - **Clinical Trials**: Conducting human studies to evaluate new drugs' safety, efficacy, and dosage. Clinical trials are typically divided into phases (I to IV), each with specific objectives and regulatory requirements.

   - **Regulatory Approval**: Obtaining approval from regulatory agencies (e.g., FDA, EMA) to market the drug. This involves submitting extensive data on the drug's safety, efficacy, and manufacturing processes.

   Implications:

   - **Innovation**: Successful R&D leads to new treatments that address unmet medical needs and improve patient outcomes.

- **Investment**: R&D is capital intensive, with significant investments required to develop and test new drugs.

2. Manufacturing and Supply Chain

Overview:

- Once a drug is approved, it must be manufactured and distributed efficiently. This involves scaling production, ensuring quality control, and managing the supply chain.

Key Components:

- **Manufacturing**: Producing the drug at scale while adhering to Good Manufacturing Practices (GMP). This includes sourcing raw materials, managing production processes, and ensuring product quality.
- **Quality Control**: Testing and validating products to meet safety and efficacy standards.
- **Supply Chain Management**: Distributing the drug to healthcare providers, pharmacies, and patients. This involves logistics, inventory management, and coordination with distributors.

Implications:

- **Efficiency**: Streamlined manufacturing and supply chain operations help reduce costs and ensure timely availability of the drug.
- **Compliance**: Adherence to regulatory standards and quality controls is crucial for maintaining product safety and efficacy.

3. Marketing and Sales

Overview:

- Marketing and sales activities aim to promote the drug, educate healthcare professionals (HCPs), and drive market adoption.

Key Components:

- **Market Access**: Gaining approval and reimbursement from health insurers and government programs. This involves demonstrating the drug's value through health economics and outcomes research.

- **Promotion**: Creating awareness and educating HCPs about the drug's benefits, uses, and clinical data. This includes marketing materials, scientific communication, and drives prescriptions.
- **Sales Force**: Employing sales representatives to engage with HCPs, provide information, and drive prescriptions.

Implications:

- **Market Penetration:** Effective marketing and sales strategies enhance product visibility and drive adoption in the market.
- **Reimbursement**: Securing favorable pricing and reimbursement conditions is essential for market success.

4. Lifecycle Management

Overview:

- Lifecycle management involves optimizing the product's performance throughout its market life. This includes extending the product's lifecycle through new indications, formulations, or line extensions.

Key Components:

- **Product Extensions**: Developing new formulations, combinations, or indications to expand the drug's use and reach.
- **Post-Market Surveillance**: Monitoring the drug's performance and safety in the real world, including adverse event reporting and ongoing clinical studies.
- **Patent Management:** Managing intellectual property rights and addressing patent expirations through strategies like generics or reformulations.

Implications:

- **Sustainability**: Effective lifecycle management helps maintain product relevance and profitability.

- **Adaptation**: Adapting to market changes, competitor actions, and emerging scientific evidence ensures continued success.

5. Financial and Strategic Management

Overview:

- Financial and strategic management encompasses budgeting, forecasting, and strategic planning to ensure the business's long-term success.

Key Components:

- **Budgeting and Forecasting**: Planning financial resources and projecting revenues, costs, and profitability.
- **Strategic Planning** involves developing long-term strategies to achieve business goals, including market positioning, investment priorities, and growth initiatives.

Implications:

- **Resource Allocation**: Efficient financial management supports strategic initiatives and operational efficiency.
- **Growth**: Strategic planning helps identify opportunities for expansion, partnerships, and innovation.

Conclusion

Understanding the pharmaceutical business model is essential for product managers to navigate the industry's complexities. By comprehending the components of R&D, manufacturing, marketing, lifecycle management, and financial management, product managers can effectively contribute to the development and success of pharmaceutical products. This holistic understanding enables them to make informed decisions, optimize processes, and drive the commercial success of their products while addressing regulatory and market challenges.

5.2 Financial Statements and Key Metrics

Understanding financial statements and key metrics is crucial for pharmaceutical product managers to make informed decisions, manage resources effectively, and drive product success. Here's a detailed overview:

1. Financial Statements

A. Income Statement

Overview:

- The income statement (the profit and loss statement) summarizes a company's revenues, expenses, and profits over a specific period.

Key Components:

- **Revenue**: Total income generated from sales of goods or services. In pharma, this includes drug sales and licensing fees.
- **Cost of Goods Sold (COGS)**: Direct costs associated with producing the goods sold, such as manufacturing and raw material costs.
- **Gross Profit**: Revenue minus COGS. It reflects the profitability of the core business activities.
- **Operating Expenses**: Costs incurred to run the business, including R&D expenses, marketing, and administrative costs.
- **Operating Income**: Gross profit minus operating expenses. It indicates the profitability of core operations.
- **Net Income**: Operating income minus non-operating expenses (interest, taxes). It represents the company's total profit.

Importance:

- **Profitability Analysis**: Helps assess how efficiently a company converts revenue into profit.
- **Cost Management**: Identifies areas where costs can be reduced or managed better.

B. Balance Sheet

Overview:

- The balance sheet provides a snapshot of a company's financial position at a specific point in time, detailing assets, liabilities, and equity.

Key Components:

- **Assets**: Resources owned by the company, including cash, accounts receivable, inventory, and fixed assets (e.g., equipment, patents).
- **Liabilities**: Obligations the company owes, such as accounts payable, loans, and accrued expenses.
- **Equity**: The residual interest in the assets after deducting liabilities. It includes common stock, retained earnings, and additional paid-in capital.

Importance:

- **Financial Health**: Provides insights into the company's solvency and liquidity.
- **Asset Management**: Helps evaluate how effectively the company is using its assets.

C. Cash Flow Statement

Overview:

- The cash flow statement tracks the flow of cash in and out of a company, categorizing it into operating, investing, and financing activities.

Key Components:

- **Operating Activities**: Cash flows from core business operations, including sales receipts and expense payments.
- **Investing Activities**: Cash flows related to investments in assets, such as purchasing equipment or acquiring other companies.
- **Financing Activities**: Cash flows from transactions with shareholders and creditors, including issuing stock or repaying loans.

Importance:

- **Liquidity Management**: Helps assess the company's ability to generate cash to meet obligations.
- **Investment Decisions**: Provides insights into how cash is used for growth and expansion.

2. Key Financial Metrics

A. Gross Margin

Formula:

- Gross Margin = Revenue — COGS / Revenue x 100

Overview:

- Indicates the percentage of revenue remaining after deducting the cost of goods sold.

Importance:

- **Profitability Indicator**: Reflects the efficiency of production and cost control.

B. Operating Margin

Formula:

Operating Margin = Operating Earnings / Revenue x 100

Overview:

- Shows the percentage of revenue remaining after covering operating expenses.

Importance:

- **Operational Efficiency**: Highlights the profitability of core business operations.

C. Net Profit Margin

Formula:

Net Profit Margin = Net Income / Revenue x 100

Overview:

- Represents the percentage of revenue that remains as profit after all expenses.

Importance:

- **Overall Profitability**: Indicates the company's overall ability to generate profit.

D. Return on Assets (ROA)

Formula:

Return on Assets = Net Income / Total Assets x 100

Overview:

- Measures how efficiently a company uses its assets to generate profit.

Importance:

- **Asset Utilization**: Reflects how well the company uses its assets to generate profit.

E. Return on Equity (ROE)

Formula:

Return on Equity - Net Income / Shareholders Equity x 100

Overview:

- Shows the return on shareholders' equity, indicating how effectively thecompany is using equity to generate profit.

Importance:

- **Shareholder Value**: Demonstrates how well the company is generating returns to its shareholders.

F. Current Ratio

Formula:

Current Ratio = Current Assets / Current Liabilities

Overview:

- Assesses the company's ability to pay short-term obligations with short-term assets.

Importance:

Liquidity: Indicates the company's short-term financial health.

G. Quick Ratio

Formula:

- Quick Ratio = Quick Assets (Current Assets —Inventory) / Current Liabilities

Overview:

- A stricter measure of liquidity. Excludes inventory from current assets.

Importance:

- **Immediate Liquidity**: Provides a more conservative view of short-term financial health.

H. Debt-to-Equity Ratio

Formula:

Debt-to-Equity Ratio = Total Liabilities / Shareholder Equity

Overview:

- Shows the proportion of debt used to finance the company's assets relative to equity.

Importance:

- **Financial Leverage**: Indicates the company's reliance on debt for growth.

Conclusion

Understanding financial statements and key metrics is essential for pharmaceutical product managers to evaluate financial performance, make informed decisions, and drive business success. Mastery of these financial tools allows product managers to manage budgets effectively, assess profitability, and optimize resource allocation, ultimately supporting the successful development and commercialization of pharmaceutical products.

5.3 Budgeting Forecasting, and Financial Planning

Budgeting, forecasting, and Financial planning are crucial components of financial management for pharmaceutical product managers. These processes help ensure that resources are allocated efficiently, strategic goals are met, and financial targets are achieved. Here's a detailed overview of each component:

1. Budgeting

Overview:

- Budgeting is the process of creating a detailed financial plan for a specific fiscal year period. It involves setting financial targets, allocating resources, and monitoring spending to ensure the company remains within its financial limits.

Key Components:

- **Revenue Projections**: Estimating future sales based on historical data, market trends, and strategic goals.
- **Expense Forecasting**: Identifying and estimating costs associated with operations, including R&D, marketing, production, and administrative expenses.
- **Capital Expenditure (CapEx)**: Budgeting for investments in long-term assets like equipment, facilities, or technology.
- **Operational Expenditure (OpEx)**: Budgeting for day-to-day expenses necessary for business operations.

Importance:

- **Resource Allocation**: Ensures that funds are allocated effectively to various projects and departments.
- **Cost Control**: Helps monitor and control spending to stay within budget limits.
- **Performance Measurement**: Provides benchmarks for evaluating financial performance and identifying variances from the plan.

Example:

A pharmaceutical company preparing a budget for a new drug launch would estimate the costs of clinical trials, marketing campaigns, regulatory approvals, and manufacturing. The budget would allocate funds accordingly and set targets for revenue and profitability.

2. Forecasting

Overview:

- Forecasting involves predicting future financial outcomes based on historical data, market analysis, and statistical models. It helps anticipate future performance and plan accordingly.

Types of Forecasting:

- **Sales Forecasting**: Predicts future sales volumes based on market research, trends, and historical data.
- **Revenue Forecasting**: Estimates future revenue based on sales forecasts and pricing strategies.
- **Expense Forecasting**: Project future expenses, including cost of goods sold, operating expenses, and capital expenditures.
- **Cash Flow Forecasting**: Predicts cash inflows and outflows to ensure sufficient liquidity.

Importance:

- **Strategic Planning**: Provides insights for long-term planning and strategic decision-making.
- **Risk Management** helps identify potential financial risks and develop mitigation strategies.
- **Performance Tracking** allows comparison of actual performance against forecasts to evaluate accuracy and make adjustments.

Example:

A pharmaceutical company forecasting the sales of new diabetes medication would analyze historical sales data of similar products, market demand, and competitive landscape to predict future sales volumes and revenue.

3. Financial Planning

Overview:

- Financial planning is developing a comprehensive strategy to achieve long-term financial goals. It integrates budgeting, forecasting, and strategic planning to ensure financial stability and growth.

Key Components

- **Strategic Goals**: Setting long-term financial objectives, such as profitability, market share, and return on investment.

- **Financial Models**: Creating financial models to simulate different scenarios and assess their impact on financial performance.

- **Investment Planning**: Identifying and evaluating investment opportunities, including R&D projects, acquisitions, or new market entries.

- **Risk Management**: Developing strategies to manage financial risks such as currency fluctuations, interest rate changes, or market volatility.

Importance:

- **Long-Term Strategy**: Aligns financial resources with the company's strategic goals and vision.

- **Decision-Making**: Provides a framework for making informed financial decisions and prioritizing investments.

- **Sustainability**: Ensure financial stability and growth by balancing short-term needs with long-term objectives.

Example:

A pharmaceutical company developing a financial plan for a portfolio of drugs might evaluate different investment opportunities, such as funding for new clinical trials or expanding into emerging markets, to achieve its long-term growth objectives.

Integration and Application

Integrating Budgeting, Forecasting, and Financial Planning:

- **Consistency**: Ensure that budgeting, forecasting, and financial planning processes are aligned with the company's overall strategy and objectives.
- **Monitoring and Adjusting**: Regularly review and adjust budgets and forecasts based on actual performance and changing market conditions.
- **Collaboration**: Coordinate with different departments (e.g., R&D, marketing, finance) to gather input and ensure comprehensive financial planning.

Example of Integration:

A pharmaceutical product manager might use budgeting to allocate funds for a new drug launch, forecasting future sales and expenses, and financial planning to align the launch with long-term strategic goals. This integrated approach helps ensure the drug launch is successful and financially sustainable.

Conclusion

Mastering budgeting, forecasting, and financial planning is essential for pharmaceutical product managers to effectively manage resources, drive product success, and achieve strategic goals. By integrating these processes, product managers can make informed decisions, optimize financial performance, and navigate the complexities of the pharmaceutical industry.

5.4 Case Studies and Real-World Applications of Business Acumen and Financial Management

Some hypothetical case studies and real-world examples illustrate the critical role of business acumen and financial management in pharmaceutical product management.

Case Study #1: Budgeting and Forecasting at Accu Health Pharmaceuticals

Background

Accuhealth Pharmaceuticals successfully navigates the complexities of budgeting and forecasting to sustain its market position and drive innovation. This case study highlights AccuHealth's strategic financial management practices.

Strategic Budgeting:

AccuHealth implements a zero-based budgeting approach, which requires each department to justify all expenses for each new period. This approach ensures that resources are allocated efficiently and aligned with strategic priorities.

Forecasting Techniques:

Accuhealth employs rolling forecasts to adapt to changing market conditions. These forecasts are updated quarterly, allowing the company to respond to new data and adjust financial plans accordingly. Advanced analytics and machine learning models enhance the accuracy of these forecasts.

Outcome:

By adopting a dynamic budgeting and forecasting process, AccuHealth maintains financial flexibility and allocates resources to high-priority projects, such as R&D for new drug development. This proactive financial management has contributed to the successful launch of several blockbuster drugs.

Case Study #2: Profit and Loss Management at Pioneer Pharma

Background:

Pioneer Pharma illustrates robust profit and loss (P&L) management across its diverse pharmaceutical, medical device, and consumer health products portfolio.

P&L Management Strategies:

Pioneer Pharma's decentralized management structure allows each business unit to operate autonomously, managing its P&L statements. This structure fosters accountability and encourages business units to focus on profitability.

Cost Control and Margin Improvement:

Pioneer Pharma emphasizes cost control through initiatives like Lean Six Sigma, which streamline operations and reduce waste. By improving operational efficiency, Pioneer Pharmaceutical enhances its profit margins and reinvests savings into R&D and market expansion.

Outcome:

Effective P&L management enables Pioneer Pharma to maintain strong financial health and invest in innovation. The company's consistent profitability supports its reputation as a stable and reliable industry leader.

Case Study #3: Strategic Investment Decisions at InnoHealth

Background:

InnoHealth, known for its innovative therapies, strategically manages its portfolio's investment decisions to balance risk and reward.

Investment in R&D:

InnoHealth allocates a significant portion of its budget to R&D, focusing on high-potential therapeutic areas like oncology, neuroscience, and gene therapy. The company uses rigorous portfolio management techniques to evaluate and prioritize R&D projects based on potential ROI and strategic fit.

Capital Allocation:

InnoHealth employs a disciplined capital allocation framework to invest in growth opportunities. This includes M&A activities to acquire promising biotech firms and expand its pipelines. Financial modeling and scenario analysis guide these investment decisions.

Outcome:

Strategic investment decisions have positioned InoHealth as a leader in innovative therapies, with a robust pipeline and successful product launches, driving long-term shareholder value.

Case Study #4: CureWell Pharmaceuticals' Financial Management Practices

Background:

CureWell Pharmaceuticals, a biotechnology firm, leverages financial management practices to support its strategic goals and maintain a competitive edge.

Financial Planning and Analysis (FP&A):

CureWell's FP&A team conducts detailed financial analysis to support decision-making. This includes evaluating the financial impact of strategic initiatives, monitoring key performance indicators (KPIs), and providing actionable insights to senior management.

Risk Management:

CureWell uses financial risk management techniques to mitigate potential risks. This includes hedging foreign exchange exposures, managing interest rate risks, and ensuring compliance with regulatory requirements.

Outcome:

By integrating financial planning analysis and risk management into its financial practices, CureWell enhances its financial stability and resilience, supporting sustainable growth and innovation.

Real-World Applications of Business Acumen

A. Precision Medicine: Pharmaceutical companies increasingly incorporate genomic data and personalized medicine approaches. For example, oncology treatments often involve genetic profiling of tumors to identify mutations and select targeted therapies, improving treatment outcomes.

B. AI in Drug Discovery: Companies like Insilico Medicine and BenevolentAI utilize artificial intelligence to accelerate drug discovery. AI algorithms analyze vast datasets to identify potential drug candidates, predict efficacy, and optimize drug design, significantly reducing the time and cost of drug development.

C. Collaborative Research Networks: Pharmaceutical companies often collaborate with academic institutions, biotech firms, and research organizations. For instance, the Accelerating COVID-19 Therapeutic Inventions and Vaccines (ACTIV) partnership brought together various stakeholders to expedite the development of COVID-19 treatments and vaccines.

Conclusion

These case studies and real-world applications illustrate the critical role of business acumen and financial management in pharmaceutical product management. Effective financial practices, including strategic budgeting, forecasting, P&L management, and strategic investment decisions, enable pharmaceutical companies to sustain innovation, navigate market dynamics, and drive long-term success.

Project Management Skills

In the dynamic world of pharmaceutical product management, effective project management is crucial for product development, launch, and lifecycle management. Given the complexity of pharmaceutical projects— from drug discovery to regulatory approvals and market launches, project management skills are indispensable for navigating the challenges that arise.

Project management in the pharmaceutical industry involves more than just overseeing timelines and budgets; it requires a strategic approach to coordinating cross-functional teams, managing risks, and ensuring compliance with regulatory requirements. As pharmaceutical products progress through various development and marketing stages, product managers must employ various skills to ensure that projects are executed efficiently and effectively.

This chapter will explore the essential project management skills that pharmaceutical product managers must master. We will explore:

- **Core Principles of Project Management**: Understanding the foundational principles that underpin successful project management, including project planning, execution, monitoring, and closure.

- **Tools and Techniques**: The various tools and techniques available to manage projects, ranging from Gantt charts and Kanban boards to project management software and Agile methodologies.

- **Managing Cross-Functional Teams**: Strategies for leading and coordinating teams comprised of diverse professionals, such as scientists, marketers, and regulatory experts, to ensure alignment and collaboration.

- **Risk Management**: Identifying, assessing, and mitigating risks that impact project timelines and outcomes, focusing on developing contingency plans and proactive problem-solving.

- **Communication Strategies**: Effective communication practices for keeping stakeholders informed, managing expectations, and facilitating transparent and productive interactions among team members.

Effective project management is the backbone of successful pharmaceutical product development and marketing. By honing these skills, product managers can drive their projects to successful outcomes, meet market demands, and ultimately contribute to advancing healthcare solutions.

6.1 Essential Project Management Principles

Project management principles provide the foundational framework for effectively managing and executing projects. These principles ensure that projects are completed on time, within budget, and to the desired quality. Here's a detailed description of essential project management principles:

1. Clear Objectives and Scope

Principle:

- Define clear, specific, and achievable project objectives and scope.

Description:

- **Objective Definition**: Establish the project's goals, including specific deliverables and outcomes.
- **Scope Management**: Outline what is included and excluded from the project to prevent scope creep.

Importance:

- **Focus**: Provides a clear direction for the project and aligns with stakeholders' expectations.
- **Control**: Helps manage and control project activities, preventing unnecessary work or changes.

Example:

The objective of a new drug launch is successfully launching the drug in a specific market within a set timeframe, including regulatory approval, marketing, and distribution.

2. Comprehensive Planning

Principle:

- Develop a detailed project plan that outlines tasks, timelines, resources, and budgets.

Description:

- **Work Breakdown Structure (WBS)**: Decompose the project into manageable tasks and activities.
- **Timeline**: Create a detailed schedule with milestones and deadlines.

- **Resource Allocation**: Identify and allocate the necessary resources, including personnel and materials.
- **Budgeting**: Develop a budget that includes all costs associated with the project.

Importance:

- **Organization**: Provides a structured approach to managing the project and ensures all aspects are considered.
- **Coordination**: Helps coordinate activities and resources effectively.

Example:

A product manager would create a project plan for a clinical trial, detailing the phases of recruitment, testing, and data analysis, along with associated timelines and budgets.

3. Effective Communication

Principle:

- Ensure clear and effective communication among all project stakeholders.

Description:

- **Stakeholder Communication**: Regularly update stakeholders on project progress, changes, and issues.
- **Team Collaboration**: Facilitate communication within the project team to ensure alignment and resolve conflicts.
- **Reporting**: Provide consistent and accurate reports on project status, risks, and performance.

Importance:

- **Transparency**: Promotes straightforwardness and trust among stakeholders.
- **Alignment**: Ensures everyone is on the same page regarding project goals and progress.

Example:

During a new drug development project, a product manager might hold weekly team meetings to discuss progress, address issues, and update stakeholders.

4. Risk Management

Principle:

- Identify, assess, and mitigate risks that could impact the project.

Description:

- **Risk Identification**: Recognize potential risks that could affect the project.
- **Risk Assessment**: Evaluate the likelihood and impact of identified risks.
- **Mitigation Strategies**: Develop and implement strategies to minimize or manage risks.
- **Contingency Planning**: Prepare plans to address risks if they occur.

Importance:

- **Prevention**: Helps prevent issues from escalating and ensures smooth project execution.
- **Preparedness**: Enables proactive management of potential problems.

Example:

A product manager might identify risks, such as regulatory delays, during product launches and develop strategies to expedite approvals.

5. Time Management

Principle:

- Manage project timelines effectively to ensure timely completion.

Description:

- **Scheduling**: Develop a detialed project schedule with deadlines and dependencies.
- **Prioritization**: Prioritize tasks based on their importance and urgency.
- **Time Tracking**: Monitor progress and adjust schedules to stay on track.

Importance:

- **Efficiency**: Ensures that tasks are completed on time and resources are used effectively.
- **Goal Achievement**: Helps meet project deadlines and achieve objectives within the specified timeframe.

Example:

A product manager may use time management techniques to coordinate the phases of a drug launch, ensuring that marketing, regulatory, and distribution activities are completed as scheduled.

6. Quality Management

Principle:

- Ensure that project deliverables meet the required quality standards.

Description:

- **Quality Planning**: Define quality standards and criteria for the project deliverables.
- **Quality Assurance**: Implement processes and procedures to meet quality standards.
- **Quality Control**: Monitor and evaluate deliverables to ensure they meet quality requirements.

Importance:

- **Customer Satisfaction**: Ensures that the final product meets or exceeds stakeholder expectations.
- **Compliance**: Helps maintain compliance with regulatory and industry standards.

Example:

Quality management would involve rigorous testing and validation to ensure a new drug meets safety and efficacy standards.

7. Budget Management

Principle:

- Monitor and control project costs to stay within the allocated budget.

Description:

- **Budget Planning**: Develop a detailed budget that includes all project costs.
- **Cost Tracking**: Monitor actual expenditures against the budget and identify variances.
- **Financial Reporting**: Provide regular financial reports to stakeholders.

Importance:

- **Cost Control**: Helps prevent overspending and ensures efficient use of financial resources.
- **Financial Accountability**: Maintains accountability and transparency in financial management.

Example:

A product manager might track costs associated with clinical trials and marketing to ensure the project stays within budget and make adjustments as necessary.

8. Performance Monitoring

Principle:

- Continuously monitor and assess project performance to ensure it meets objectives.

Description:

- **Performance Metrics**: Define and track key performance indicators (KPIs) to measure project success.
- **Progress Reviews**: Conduct regular reviews to assess progress and identify areas for improvement.
- **Issue Resolution**: Address any issues or deviations from the plan promptly.

Importance:

- **Control**: Provides insight into project performance and helps ensure objectives are met.
- **Improvement**: Enables timely adjustments and improvements to enhance project outcomes.

Example:

During a drug development project, a product manager might use performance metrics to assess the progress of clinical trials and adjust the project plan as needed.

9. Stakeholder Management

Principle:

- Effectively manage relationships with all project stakeholders.

Description:

- **Stakeholder Identification**: Identify all individuals or groups affected by the project.
- **Engagement**: Engage stakeholders throughout the project to ensure their needs and expectations are addressed.
- **Feedback**: Collect and incorporate feedback from stakeholders to improve project outcomes.

Importance:

- **Support**: Ensures stakeholders are supportive and invested in the project's success.
- **Alignment**: Helps align project activities with stakeholder expectations and requirements.

Example:

A product manager might engage with healthcare professionals, regulatory bodies, and patients to gather feedback and ensure that a new drug meets their needs and expectations.

10. Adaptability and Flexibility

Principle:

- Be adaptable and flexible in response to changing circumstances and requirements.

Description:

- **Change Management**: Implement processes to manage and adapt to changes in project scope, timeline, or resources.

- **Problem-Solving**: Develop solutions to address unforeseen challenges and issues.
- **Continuous Improvement**: Embrace opportunities for improvement and innovation.

Importance:

- **Resilience**: Enhances the ability to handle changes and challenges effectively.
- **Optimization**: Promotes ongoing improvements and adjustments to optimize project outcomes.

Example:

If a regulatory requirement changes during a drug development project, the product manager must adapt the project plan and ensure compliance with the new regulations.

Conclusion

The essential principles of project management provide a structured approach to planning, executing, and controlling projects. By applying these principles, pharmaceutical product managers can ensure successful project outcomes, meet objectives, and drive business success. Effective project management is crucial for navigating the complexities of drug development, launches, and lifecycle management in the pharmaceutical industry.

6.2 Tools and Techniques for Effective Project Management

Effective project management relies on a combination of tools and techniques to plan, execute, and monitor projects. Here's an overview of some essential tools and techniques that can help product managers streamline their processes and ensure project success:

1. Project Management Software

Description:

- Project management software provides a centralized platform for planning, executing, and tracking project activities. These tools help manage tasks, timelines, resources, and communication.

Examples:

- **Microsoft Project**: Offers comprehensive project scheduling, task management, and resource allocation features.
- **Asana**: Facilitates task tracking, project planning, and team collaboration.
- **Trello**: Uses boards, lists, and cards to organize tasks and track project progress.
- **Monday.com**: Provides customizable workflows and visual project tracking.

Benefits:

- **Organization**: Centralizes project information and tasks.
- **Tracking**: Enables real-time tracking of progress and deadlines.
- **Collaboration**: Enhances team communication and collaboration.

2. Gantt Charts

Description:

- Gantt charts are visual tools that display project tasks, their durations, and dependencies in a timeline format. They help in planning and scheduling tasks.

Features:

- **Task Scheduling**: Shows start and end dates for each task.
- **Dependencies**: Illustrates the relationship between tasks and their dependencies.
- **Milestones**: Highlights key project milestones and deadlines.

Benefits:

- **Visualization**: Provides a clear overview of the project timeline.
- **Planning**: Helps in scheduling tasks and identifying potential delays.
- **Coordination**: Facilitates better coordination among team members.

Example:

A Gantt chart could show the timeline for a drug launch project's regulatory approval, marketing activities, and distribution phases.

3. Work Breakdown Structure (WBS)

Description:

- A Work Breakdown Structure (WBS) is a hierarchical decomposition of the project scope into smaller, manageable components or tasks.

Features:

- **Hierarchical Structure**: Break down the project into major deliverables and sub-tasks.
- **Task Definition**: Defines the scope and objectives for each component.
- **Responsibility Assignment**: Assign responsibilities for each task.

Benefits:

- **Clarity**: Provides a detailed view of project deliverables and tasks.

- **Organization**: Helps in organizing and managing project scope.

- **Control**: Enables better control and monitoring of project progress.

Example:

The WBS might include phases such as study design, participant recruitment, data collection, and analysis for a clinical trial.

4. Risk Management Tools

Description:

- Risk management tools help identify, assess, and mitigate risks that could impact the project.

Examples:

- **Risk Register**: A document that lists identified risks, their impact, likelihood, and mitigation strategies.

- **Risk Matrix**: A visual tool to assess and prioritize risks based on their impact and probability.

- **SWOT Analysis**: Identifies strengths, weaknesses, opportunities, and threats related to the project.

Benefits:

- **Proactive Management**: Helps in anticipating and addressing potential issues.

- **Mitigation**: Develop strategies to minimize or manage risks.

- **Preparation**: Ensures preparedness for unforeseen challenges.

Example:

A risk register might identify potential regulatory delays in a product launch and outline contingency plans.

5. Earned Value Management (EVM)

Description:

Earned Value Management (EVM) is a technique for assessing project performance by comparing planned progress with actual progress.

Key Metrics:

- **Planned Value (PV)**: The budgeted cost of work scheduled.
- **Earned Value (EV)**: The budgeted cost of work performed.
- **Actual Cost (AC)**: The actual cost incurred for work performed.

Benefits:

- **Performance Measurement**: Provides insight into cost and schedule performance.
- **Forecasting**: Helps forecast project outcomes based on current performance.
- **Control**: Enables better control of project costs and schedules.

Example:

EVM can evaluate whether a clinical trial is on budget and schedule, helping adjust resources or timelines if needed.

6. Communication Tools

Description:

- Communication tools facilitate effective information sharing and collaboration among project team members and stakeholders.

Examples:

- **Slack**: A messaging platform for real-time communication and collaboration.
- **Microsoft Teams**: Integrates chat, video conferencing, and document sharing.
- **Zoom**: Provides video conferencing and online meeting capabilities.

Benefits:

- **Collaboration**: Enhances team communication and information sharing.
- **Coordination**: Ensures alignment and resolves issues promptly.
- **Accessibility**: Allows remote teams to stay connected and engaged.

Example:

Using Slack or Microsoft Teams for regular updates and discussions during a drug development project can improve team coordination and response time.

7. Resource Management Tools

Description:

- Resource management tools help allocate and manage project resources, including personnel, equipment, and materials.

Examples:

- **Resource Leveling**: Adjusts project schedules to balance resource usage.
- **Resource Allocation Matrix**: Tracks the allocation of resources across different tasks and projects.
- **Microsoft Project**: Includes resource management features for tracking and optimizing resource use.

Benefits:

- **Optimization**: Ensures optimal use of resources and prevents overallocation.
- **Tracking**: Provides visibility into resource availability and utilization.
- **Planning**: Helps in planning and adjusting resource requirements.

Example:

Resource management tools can track the allocation of marketing personnel, budget, and materials in a marketing campaign to ensure efficient execution.

8. Agile Methodology

Description:

- Agile methodology is a flexible, iterative approach to project management that emphasizes collaboration, customer feedback, and incremental progress.

Key Practices:

- **Scrum:** A framework for managing and completing complex projects through iterative sprints.
- **Kanban:** A visual system for managing workflow and limiting work in progress.
- **Lean:** Focuses on delivering value by eliminating waste and improving processes.

Benefits:

- **Flexibility:** Adapts to changes and evolving project requirements.
- **Collaboration:** Encourages continuous feedback and collaboration with stakeholders.
- **Incremental Progress:** Delivers value incrementally through iterative cycles.

Example:

Using Scrum in a drug development project can help manage changing requirements and improve collaboration between research, development, and marketing teams.

9. Document Management Systems

Description:

- Document management systems store, organize, and manage project documents and records.

Examples:

- **SharePoint:** Provides document storage, sharing, and collaboration features.
- **Google Drive:** Offers cloud storage and collaborative document editing.
- **Dropbox:** Provides file storage and sharing capabilities.

Benefits:

- **Accessibility**: Ensures easy access to project documents.
- **Organization**: Helps in organizing and categorizing documents.
- **Collaboration**: Facilitates collaborative work on documents.

Example:

A document management system can store and share regulatory documents, clinical trial data, and marketing materials during a drug development project.

10. Project Management Frameworks

Description:

- Project management frameworks provide structured methodologies and best practices for managing projects.

Examples:

- **PMBOK (Project Management Body of Knowledge)**: A comprehensive project management standards and practices guide.
- **PRINCE2 (Projects in Controlled Environments)**: A structured project management method focusing on organization and control.
- **PMI (Project Management Institute)**: Provides standards and certifications for project management professionals.

Benefits:

- **Guidance**: Offers structured approaches and best practices for managing projects.
- **Standardization**: Ensures consistency and quality in project management.
- **Certification**: Provides credentials and recognition for project management skills.

Example:

Applying PMBOK guidelines can help standardize project management practices and improve project outcomes.

Conclusion

These tools and technologies enhance project management effectiveness by providing structure, visibility, and control. By leveraging these resources, pharmaceutical product managers can improve project planning, execution, and monitoring, leading to successful project outcomes and business success.

6.3 Managing Cross-Functional Teams and Timelines

In pharmaceutical product management, handling cross-functional teams and adhering to project timelines are crucial for successfully developing and launching pharmaceutical products. Here's a detailed description of how to effectively manage these aspects:

Managing Cross-Functional Teams

1. **Understanding Cross-Functional Teams**

Definition:

- Cross-functional teams consist of members from various departments or specialties who work together to achieve common project goals. In pharmaceutical product management, these teams typically include members from research and development, clinical trials, regulatory affairs, marketing, sales, and supply chain management.

Benefits:

- **Diverse Expertise**: Combines different skills and knowledge areas for comprehensive problem-solving.
- **Innovation**: Encourages creative solutions through diverse perspectives.
- **Efficiency**: Streamlines decision-making and coordination across functions.

2. **Building an Effective Team**

Steps:

1. **Define Roles and Responsibilities**: Clearly outline each team member's role and responsibilities to avoid overlap and ensure accountability.
2. **Foster Collaboration**: Create an environment encouraging open communication and collaboration among team members.

3. **Set Clear Objectives**: Establish clear, achievable objectives and goals for the team to ensure alignment and focus.

4. **Provide Resources and Support**: Ensure the team can access the necessary resources, tools, and training to perform their roles effectively.

Example:

For a new drug launch, a cross-functional team might include R&D scientists, regulatory experts, marketing specialists, and sales representatives collaborating to develop and execute the strategy.

3. Communication and Coordiantion

Strategies:

- **Regular Meetings**: Schedule regular team meetings to discuss progress, address issues, and align on objectives.

- **Centralized Communication Tools**: Use project management and communication tools (e.g., Slack, Microsoft Teams) to facilitate information sharing and collaboration.

- **Conflict Resolution**: Implement strategies to address and resolve conflicts promptly to maintain team harmony and productivity.

Example:

Implementing weekly status meetings and using a shared project management platform helps ensure all team members are updated on project developments and can coordinate their efforts effectively.

4. Performance Monitoring and Feedback

Approach:

- **Track Progress**: Use performance metrics and project milestones to monitor the team's progress and identify deviations from the plan.

- **Provide Feedback**: Offer constructive feedback and recognize achievements to motivate team members and improve performance.
- **Adjust Strategies**: Make necessary adjustments to team strategies or processes based on performance evaluations and feedback.

Example:

A project manager might use performance dashboards to track key performance indicators (KPIs) and provide regular feedback to the team on their progress toward project milestones.

Managing Timelines

1. Planning and Scheduling

Steps:

- **Define Project Phases**: Break down the project into distinct phases (e.g., research, development, testing, launch) and define the timelines for each phase.
- **Develop a Detailed Schedule**: Use tools like Gantt charts to create a detailed project schedule with specific deadlines, milestones, and dependencies.
- **Allocate Resources**: Assign resources and set deadlines for each task to ensure timely completion.

Example:

The timeline for a clinical trial includes phases such as protocol development, patient recruitment, data collection, and analysis, each with specific deadlines and dependencies.

2. Monitoring and Adjusting

Approach:

- **Track Progress**: Regularly monitor progress against the project schedule using project management tools.
- **Identify Delays**: Detect potential delays early and assess their impact on the timeline.

- **Implement Adjustments**: Make necessary adjustments to the schedule or resource allocation to address delays and keep the project on track.

Example:

If a delay occurs in patient recruitment for a clinical trial, the project manager might adjust the recruitment strategy or reallocate resources to mitigate the delay.

3. Risk Management

Strategies:

- **Identify Risks**: Potential risks impacting project timelines, such as regulatory delays or supply chain disruptions.

- **Develop Mitigation Plans**: Create risk mitigation plans to address identified risks and minimize their impact on the project schedule.

- **Monitor Risks**: Continuously monitor risks throughout the project and adjust mitigation strategies as needed.

Example:

Developing a contingency plan for potential regulatory delays by establishing alternative approval pathways or additional regulatory support can help manage timeline risks effectively.

4. Communication and Stakeholder Management

Approach:

- **Regular Updates**: Provide regular updates to stakeholders on project progress, timeline changes, and any issues encountered.

- **Manage Expectations**: Set realistic expectations with stakeholders regarding timelines and potential delays.

- **Seek Feedback**: Obtain feedback from stakeholders to ensure their concerns are addressed and to keep them engaged on the project.

Example:

Regularly updating senior management and regulatory bodies on the progress of a drug development project helps manage expectations and ensures that any necessary approvals or support are obtained on time.

Conclusion

Effectively managing cross-functional teams and timelines is essential for the success of pharmaceutical product management. By building cohesive teams, fostering collaboration, monitoring performance, and adhering to well-planned schedules, product managers can ensure that projects are completed successfully, within scope, and on time. Leveraging these strategies enhances the ability to navigate complex projects, mitigate risks, and achieve desired outcomes in the dynamic pharmaceutical industry.

6.4 Effective Project Management: Cases and Real-World Examples

Here are some case studies and real-world examples that illustrate effective project management in the pharmaceutical industry:

Case Study #5: Project Management in Drug Development: Pfizer's COVID-19 Vaccine

Background:

Developing the Pfizer-BioNTech COVID-19 vaccine (BNT 162b2) is a prime example of effective project management under extraordinary circumstances. Faced with the urgent global need for a COVID-19 vaccine, Pfizer and BioNTech embarked on an ambitious project to develop, test, and distribute a vaccine in record time.

Project Planning and Execution

To accelerate the timeline, Pfizer implemented a meticulous project management plan involving overlapping clinical trial phases (Phase I, II, and III). This approach, known as "parallel processing," significantly shortened the traditional sequential vaccine development timeline.

Risk Management

Given the unprecedented speed of the project, risk management was critical. Pfizer's project managers identified potential risks early, such as regulatory hurdles and supply chain disruptions, and developed mitigation strategies. For example, they proactively worked with regulatory bodies worldwide to ensure alignment and compliance and secured agreements with multiple suppliers to avoid bottlenecks.

Resource Allocation

The project required substantial human and financial resources. Pfizer reallocated resources from other projects and secured funding from various sources, including government contracts and private investments. This strategic allocation ensured the project was supported without compromising other ongoing projects.

Stakeholder Communication

A cornerstone of the project was regular and transparent communication with stakeholders, including regulatory agencies, government officials, and the public. Pfizer established dedicated communication channels and held frequent updates to maintain trust and collaboration.

Outcome

The Pfizer-BioNTech vaccine was developed, tested, and authorized for emergency use in less than a year—a process that typically takes several years. Effective project management strategies ensured a critical vaccine's rapid and safe delivery, demonstrating the power of well-executed project management in the pharmaceutical industry.

Case Study #6: Resource Allocation in a Pharmaceutical Product Launch — Gilead Sciences' Launch of Harvoni!

Background:

Gilead Sciences faced significant challenges in launching Harvoni, a breakthrough treatment for Hepatitis C. The launch required careful coordination of clinical trial data, regulatory approval, and market introduction.

Project Planning and Execution:

Gilead employed a comprehensive project plan with detailed timelines for each product launch phase, from clinical trials to marketing campaigns. The project managers ensured all departments were aligned with the launch objectives and timelines.

Risk Management:

The primary risks involved regulatory delays and market acceptance. Gilead's project managers worked closely with regulatory bodies to expedite approval processes and conducted extensive market research to understand patient and healthcare provider needs.

Resource Allocation:

Significant resources were allocated to clinical trials to gather robust data supporting Harvoni's efficacy and safety. Additionally, Gilead invested heavily in marketing and educational campaigns to raise awareness among healthcare providers and patients.

Stakeholder Communication:

Effective communication strategies were essential for the successful launch of Harvoni. Gilead's project managers ensured that all stakeholders, including internal teams, regulatory bodies, and external partners, were informed of progress and any potential issues.

Outcome:

Harvoni was successfully launched and quickly became a market leader in Hepatitis C treatment. Strategic project management practices, including resource allocation and stakeholder communication, were crucial in its successful introduction.

Case Study #7: Risk Management in Clinical Trials — Johnson & Johnson's Development of SGLT2 Inhibitors.

Background:

Johnson & Johnson's development of SGLT2 inhibitors for diabetes management required careful risk management, particularly concerning patient safety and regulatory compliance.

Project Planning and Execution:

The project involved extensive clinical trials to demonstrate the efficacy and safety of SGLT2 inhibitors. Johnson & Johnson's project managers meticulously planned each phase of the trials, ensuring strict adherence to protocols and timelines.

Risk Management:

Patient safety was the top priority. The Project managers implemented rigorous monitoring and risk mitigation strategies, such as regular safety assessments and contingency plans for adverse events. They also engaged with regulatory authorities early in the process to address potential concerns and expedite approvals.

Resource Allocation:

Substantial resources were dedicated to the clinical trial process, including staffing specialized roles for patient monitoring and data analysis. Johnson & Johnson also allocated funds for technological tools to enhance data collection and analysis.

Stakeholder Communication:

Transparent communication with regulatory bodies, healthcare professionals, and patient advocacy groups was maintained throughout the trials. Regular updates and open dialogues helped build trust and facilitated smoother approval processes.

Outcome:

The SGLT2 inhibitors developed by Johnson & Johnson were successfully brought to market, offering a new and effective treatment option for diabetes. The thorough risk management and strategic resource allocation were pivotal in achieving this outcome.

Conclusion

These case studies highlight the critical role of effective project management in pharmaceutical product management. Whether it is accelerating vaccine development, launching a new drug, or managing clinical trials, robust project management practices ensure that projects are completed successfully, on time, and within budget. By learning from these real-world examples, pharmaceutical product managers can better navigate the complexities of their roles and contribute to the success of their organizations.

Understanding Regulatory Affairs

Navigating the complex web of regulations and compliance standards is crucial for successfully developing, approving, and marketing medical products in the pharmaceutical industry. Regulatory affairs, a specialized field within the industry, serves as the bridge between pharmaceutical companies and regulatory authorities, ensuring that products meet the necessary legal and scientific requirements before they reach the market.

This chapter provides an in-depth exploration of the regulatory landscape that governs the pharmaceutical industry. It begins with an overview of key regulatory bodies and the frameworks they enforce, such as the Food and Drug Administration (FDA) in the United States, the European Medicines Agency (EMA) in the European Union, and other global counterparts. Understanding these regulatory frameworks is essential for product managers, as they are pivotal in guiding the products through the approval process.

The chapter then delves into the regulatory requirements at various stages of drug development, from pre-clinical research to post-marketing surveillance. It highlights the critical milestones that pharmaceutical companies must achieve, including Investigational New Drug (IND) applications, New Drug Applications (NDA), and Marketing Authorization Applications (MAA). The chapter also emphasizes the importance of adhering to Good Manufacturing

Practices (GMP), Good Clinical Practices (GCP), and other regulatory standards to ensure product safety, efficacy, and quality.

In addition to understanding the regulatory requirements, this chapter explores the role of the product manager in ensuring compliance throughout the product lifecycle. This includes managing regulatory submissions, coordinating with cross-functional teams to gather necessary data, and staying informed about evolving regulations and guidelines. Effective communication with regulatory authorities and internal stakeholders is also discussed as a key aspect of the regulatory affairs function.

By the end of this chapter, readers will have a comprehensive understanding of the regulatory environment in which pharmaceutical products are developed and marketed. They will also gain insights into the strategic importance of regulatory affairs in bringing safe, effective, and compliant products to market and the product manager's critical role in this process.

7.1 Overview of Regulatory Requirements in Pharma

Regulatory requirements in the pharmaceutical industry are critical for ensuring drug product safety, efficacy, and quality. Regulatory agencies establish these requirements to safeguard public health and ensure that pharmaceuticals meet high standards before they are allowed to be marketed. Here's an overview of the key aspects of regulatory requirements.

1. Overview of Regulatory Affairs

Definition:

- Regulatory affairs involve the processes and activities required to ensure that pharmaceutical products meet the legal and regulatory standards set by government agencies before they can be marketed. This includes compliance with guidelines for drug development, clinical trials, manufacturing, labeling, and marketing.

Importance:

- **Compliance**: Ensures that products meet safety, efficacy, and quality standards.
- **Approval**: Facilitates the approval process with regulatory agencies.
- **Market Access**: Enables successful entry into global markets by adhering to regional regulations.
- **Risk Management**: Helps manage risks related to regulatory non-compliance.

2. Key Regulatory Agencies

A. U.S. Food and Drug Administration (FDA):

- **Role**: Regulates drug safety, efficacy, and quality in the United States.
- **Key Activities**: Approves new drug applications (NDAs), monitors post-market safety, and enforces labeling requirements.

B. European Medicines Agency (EMA)

- **Role**: Regulates drugs in the European Union (EU).
- **Key Activities**: Reviews marketing authorization applications, provides scientific advice, and monitors post-market safety.

C. Pharmaceutical and Medical Devices Agency (PMDA) - Japan

- **Role**: Oversees drug and medical device approval in Japan.
- **Key Activities**: Evaluates clinical trial data, provides regulatory guidance, and ensures compliance with Japanese regulations.

D. National Medical Products Administration (NMPA) - China

- **Role**: Regulates pharmaceuticals and medical devices in China.
- **Key Activities**: Approves drug applications, inspects manufacturing facilities, and monitors post-market safety.

E. Central Drugs Standard Control Organization (CDSCO) - India

- **Role**: Regulates pharmaceuticals and medical devices in India.
- **Key Activities**: Approves drug applications, ensures compliance with Indian regulations and monitors post-market safety.

3. Drug Development and Approval Process

A. Preclinical Research:

- **Regulatory Oversight**: Ensures preclinical studies adhere to Good Laboratory Practice (GLP) standards and obtain necessary approvals before proceeding to clinical trials.

B. Clinical Trials:

- **Regulatory Oversight**: This requires the approval of Investigational New Drug (IND) applications in the US or Clinical Trial Authorization (CTA) in Europe. It also ensures adherence to Good Clinical Practice (GCP) standards and ethical considerations.

C. New Drug Application (NDA) / Marketing Authorization Application (MAA):

- **Regulatory Oversight**: Involves submitting detailed data on the drug's safety, efficacy, and manufacturing processes. The regulatory agency reviews this data to determine whether to approve marketing.

D. Post-Market Surveillance:

- **Regulatory Oversight**: Monitors the drug's performance in the market, including adverse events, and ensures compliance with post-market surveillance requirements such as Risk Evaluation and Mitigation Strategies (REMS) in the US.

4. **Key Regulatory Considerations**

A. Labeling and Advertising:

- **Regulatory Requirements**: Ensure that product labeling and advertising meet regulatory standards for accuracy, clarity, and compliance with approved indications.

B. Intellectual Property:

- **Regulatory Considerations**: Address patent rights, exclusivity periods, and protection of proprietary information during regulatory submissions and market entry.

C. International Regulations:

- **Regulatory Considerations**: Navigate different regulatory requirements across countries for global market access, including understanding regional variations and obtaining necessary approvals.

5. Role of the Product Manager in Regulatory Affairs

A. Regulatory Strategy Development:

- **Responsibilities**: Develop strategies to meet regulatory requirements and facilitate product approval. Engage with regulatory agencies and stay updated on regulatory changes.

B. Documentation and Submission:

- **Responsibilities**: Oversee the preparation and submission of regulatory documents, including NDAs, MAAs, and clinical trial applications. Ensure accuracy and compliance with regulatory guidelines.

C. Cross-Functional Coordination:

- **Responsibilities**: Collaborate with regulatory affairs teams, clinical development, and manufacturing to ensure that all aspects of the drug development process comply with regulatory standards.

D. Risk Management:

- **Responsibilities**: Identify and mitigate regulatory risks, address compliance issues, and effectively implement strategies to handle regulatory challenges.

E. Regulatory Intelligence:

- **Responsibilities**: Monitor and analyze regulatory trends, guidelines, and policies to anticipate and adapt to product development and marketing changes.

Conclusion

Understanding regulatory requirements is crucial for pharmaceutical product managers to ensure successful product development, approval, and market entry. By navigating the complex regulatory landscape, product managers can help ensure that drugs meet safety and efficacy standards, comply with legal requirements, and achieve successful market outcomes.

7.2 Navigating the Approval Process

The approval process for pharmaceutical products is a multi-stage journey designed to ensure that the new drugs are safe, effective, and manufactured to high standards before they are allowed on the market. Navigating this process effectively is crucial for pharmaceutical product managers. Here is a detailed overview of the key stages involved in the approval process:

1. Preclinical Research and Development

1.1　Preclinical Research

- **Objective**: Assess the drug's safety and efficacy in non-human models.

- **Requirement**: Conduct studies under Good Laboratory Practice (GLP). Data from these studies are used to support the drug's safety in human trials.

- **Key Documents**: Preclinical study reports, Safety Assessment Reports.

1.2　Investigational New Drug (IND) Application

- **Objective**: Obtain permission from regulatory agencies to begin clinical trials.

- **Requirements**: Submit detailed data on preclinical studies, manufacturing processes, and proposed clinical trial protocols.

- **Key Documents:** IND Application, Investigator's Brochures, Clinical Trial Protocol.

2. Clinical Trials

2.1　Clinical Trial Phases

- **Phase I**: Test safety and dosage in a small group of healthy volunteers or patients.

- **Key Activities**: Assess safety, tolerability, pharmacokinetics, and pharmacodynamics.

- **Phase II**: Evaluate efficacy and side effects in a larger group of patients with the targeted disease.

- **Key Activities**: Determine the drug's effectiveness, optimal dose, and safety profile.
- **Phase III**: Confirm efficacy, monitor side effects, and compare with standard treatments in a larger population.
- **Key Activities**: Provide comprehensive evidence of efficacy and safety for regulatory approval.
- **Phase IV**: Post-marketing studies to gather additional information on long-term safety and effectiveness.
- **Key Activities**: Surveillance of drug's performance in the general population, monitoring adverse effects.

2.2 Clinical Trial Documentation

- **Objective**: Ensure compliance with regulatory requirements and facilitate review.
- **Requirements**: Maintain records of trial protocols, informed consent forms, case report forms (CRFs), and data analysis results.

3. Regulatory Submission and Review

3.1 New Drug Application (NDA) / Marketing Authorization Application (MAA)

- **Objective**: Obtain formal approval to market the drug.
- **Requirement**: Submit comprehensive data from clinical trials, preclinical studies, and manufacturing processes, including proposed labeling and risk management plans.
- **Key Documents**: NDA or MAA, Clinical Study Reports, Chemistry, Manufacturing, and Controls (CMS) data, proposed labeling.

3.2 Regulatory Review Process

- **Objective**: Evaluate the drug's safety, efficacy, and quality.
- **Requirements**: Regulatory agencies review the submission, request additional information if necessary, and may inspect manufacturing facilities.
- **Key Activities**: Review meetings, advisory committee consultations, and potential approval conditions.

3.3 Approval and Post-Approval Activities

- **Objective**: Finalize the drug's market authorization and ensure ongoing compliance.

- **Requirements**: Address post-approval commitments such as Risk Evaluation and Mitigation Strategies (REMS), additional clinical studies, or labeling updates.

- **Key Activities**: Implement approved labeling, launch post-marketing surveillance, and comply with reporting requirements for adverse events.

4. Strategies for Successful Navigation

4.1 Early Engagement with Regulatory Agencies

- **Objective**: Gain guidance on regulatory requirements and expectations.

- **Strategies**: Conduct pre-IND meetings, request scientific advice, and participate in regulatory consultations.

4.2 Comprehensive Documentation and Submission

- **Objective**: Ensure accuracy and completeness of submission materials.

- **Strategies**: Maintain rigorous documentation practices, engage in thorough quality control, and seek expert input for complex sections.

4.3 Cross-Functional Collaboration

- **Objective**: Coordinate efforts across clinical, regulatory, and manufacturing teams.

- **Strategies**: Establish clear communication channels, align objectives, and manage timelines effectively.

4.4 Proactive Risk Management

- **Objective**: Anticipate and address potential challenges and regulatory hurdles.

- **Strategies**: Monitor regulatory changes, implement risk mitigation plans, and prepare for potential issues during the review process.

4.5 Post-Market Suveillance

- **Objective**: Ensure ongoing safety and effectiveness of the drug.
- **Strategies**: Conduct post-marketing studies, monitor adverse events, and maintain compliance with regulatory commitments.

Conclusion

Navigating the approval process in the pharmaceutical industry requires careful planning through documentation and strategic collaboration. By understanding each stage and adhering to regulatory requirements, pharmaceutical product managers can effectively guide their products through the approval process and ensure a successful market entry. This involves a continuous focus on safety, efficacy, and compliance to meet the standards set by regulatory agencies and to deliver high-quality therapeutic solutions to patients.

7.3 The Product Manager's Role in Regulatory Compliance

Regulatory compliance is a crucial aspect of pharmaceutical product management, ensuring that products meet the required safety, efficacy, and quality standards before they reach the market. Product managers are pivotal in navigating regulatory requirements, facilitating successful product approvals, and maintaining ongoing compliance. Here's a detailed look at the product manager's responsibilities in this area:

1. Understanding Regulatory Requirements

1.1 Staying Informed

- **Objective**: Ensure awareness of current and evolving regulations that impact the product.

Responsibilities:

- Review guidelines from regulatory bodies such as the FDA (US Food and Drug Administration), EMA (European Medicines Agency), and other relevant authorities worldwide regularly.

- Attend industry conferences, webinars, and training sessions focused on regulatory updates.

1.2. Translating Regulations

- **Objective**: Interpret regulatory requirements and apply them to product development and marketing strategies.

Responsibilities:

- Work closely with regulatory affairs teams to understand specific drug development, labeling, and post-market surveillance regulations.

- Ensure that product documentation, including labeling and promotional materials, adheres to regulatory standards.

2. Managing the Regulatory Submission Process

2.1. Preparing Submission Materials

- **Objective**: Compile and organize the necessary documentation for regulatory submissions.

Responsibilities:

- Collaborate with clinical, manufacturing, and quality assurance teams to gather data required for Investigational New Drug (IND) applications, New Drug applications (NDAs), or Marketing Authorization applications (MAAs).

- Oversee the preparation of submission documents, ensuring they are complete, accurate, and aligned with regulatory requirements.

2.2. Coordinating with Regulatory Authorities

- **Objective**: Facilitate effective communication with regulatory agencies.

Responsibilities:

- Manage interactions with regulatory authorities, including submissions, responses to queries, and addressing feedback.

- Coordinate meetings and discussions with regulatory agencies, including pre-submission meetings and advisory committee reviews.

3. Ensuring Compliance Throughout the Product Lifecycle

3.1 Post-Market Surveillance

- **Objective**: Monitor the product's performance and ensure ongoing compliance after market entry.

Responsibilities:

- Oversee post-marketing studies and pharmacovigilance activities to track adverse events and long-term safety.

- Ensure that all adverse events are reported to regulatory authorities promptly.

3.2. Labeling and Advertising Compliance

- **Objective**: Ensure that product labeling and promotional materials meet regulatory standards.

Responsibilities:

- Review and approve labeling and marketing materials to ensure they are truthful, non-misleading, and compliant with regulations.
- Monitor advertising and promotional activities to ensure they adhere to regulatory guidelines.

4. Facilitating Cross-Functional Collaboration

4.1. Aligning Teams

- **Objective**: Coordinate efforts across various departments to ensure compliance.

Responsibilities:

- Work with regulatory affairs, clinical development, manufacturing, and quality assurance teams to align on regulatory requirements and timelines.
- Facilitate communication and collaboration among cross-functional teams to address regulatory issues and ensure compliance.

4.2. Risk Management

- **Objective**: Identify and mitigate regulatory risks associated with the product.

Responsibilities:

- Develop and implement risk management plans to address potential regulatory challenges.
- Monitor regulatory changes and assess their impact on the product and its compliance status.

5. Ensuring Training and Awareness

5.1. Training Teams

- **Objective**: Educate internal teams on regulatory requirements and compliance practices.

Responsibilities:

- Develop and deliver training programs on regulatory compliance, product development, and marketing teams.
- Ensure that all relevant personnel are aware of regulatory obligations and best practices.

5.2. Promoting a Compliance Culture

- **Objective**: Foster a culture of regulatory compliance within the organization.

Responsibilities:

- Advocate for the importance of regulatory compliance and promote adherence to regulatory standards across the organization.
- Encourage a proactive approach to compliance, including early identification and resolution of potential issues.

Conclusion

The product manager's role in regulatory compliance is integral to the successful development, approval, and ongoing management of pharmaceutical products. By staying informed about regulatory requirements, managing the submission process, ensuring compliance throughout the product lifecycle, facilitating cross-functional collaboration, and promoting regulatory awareness, product managers play a crucial role in bringing safe and effective products to market while maintaining adherence to regulatory standards. This multifaceted responsibility helps ensure that pharmaceutical products meet the highest quality and safety standards for patients and healthcare professionals.

Marketing and Sales Expertise

In the dynamic and highly regulated pharmaceutical industry, marketing and sales expertise is crucial to the success of pharmaceutical product management. As the landscape evolves with technological advancements and shifting market demands, pharmaceutical product managers must robustly understand marketing and sales strategies to navigate this complex environment effectively.

Pharmaceutical product management extends beyond developing and launching new drugs; it encompasses a product's comprehensive lifecycle, from market research and strategic planning to execution and post-launch evaluation. Integrating marketing and sales expertise is essential in this process, as it ensures that a product meets patients' clinical needs and achieves commercial success.

Marketing and Sales Expertise in Context

In the pharmaceutical sector, marketing involves understanding and addressing the needs of diverse stakeholders, including healthcare professionals (HCPs), patients, payers, and regulators. Effective pharmaceutical marketing requires a deep comprehension of market trends, competitive landscapes, and the regulatory environment. It involves creating compelling value propositions, designing targeted campaigns, and employing various promotional strategies to communicate a product's benefits and differentiate it from competitors.

On the other hand, sales expertise focuses on driving product adoption and maximizing revenue. This involves building and maintaining relationships with HCPs, understanding their prescribing behaviors, and providing them with the necessary tools and information to make informed decisions. Sales strategies are closely aligned with marketing efforts but are executed through direct interactions and engagement with the healthcare community.

The Importance of Integration

Successful pharmaceutical product management hinges on seamlessly integrating marketing and sales strategies. Product managers must coordinate these efforts to ensure marketing initiatives translate into tangible sales outcomes. This integration involves aligning marketing messages with sales tactics, leveraging data and insights to refine strategies, and continuously adapting to market changes and feedback.

Key Areas of Focus

1. **Market Research and Analysis**: Understanding market needs, competitor activities, and customer preferences to inform strategic decisions.

2. **Strategic Planning**: Developing comprehensive marketing and sales plans that align with overall business objectives and regulatory requirements.

3. **Execution and Implementation**: Implementing marketing campaigns and sales strategies effectively while ensuring compliance with industry regulations.

4. **Performance Measurement**: Monitoring and evaluating the success of marketing and sales activities, making data-driven adjustments to optimize performance.

In summary, marketing and sales expertise are integral to pharmaceutical product management and pivotal in successfully commercializing pharmaceutical products. As the industry evolves, product managers must stay abreast of emerging trends and best practices in marketing and sales to drive growth and deliver value to all stakeholders involved.

8.1 Fundamentals of Pharmaceutical Marketing

Pharmaceutical marketing involves promoting and selling pharmaceutical products while navigating complex regulations and diverse stakeholder needs. Understanding the fundamentals of pharmaceutical marketing is crucial for product managers to position their products effectively, engage with healthcare professionals (HCPs), and improve patient outcomes. Here's a detailed overview of the core elements of pharmaceutical marketing:

1. Market Analysis and Research

1.1. Market Research

- **Objective**: Gather information on market dynamics, including patient needs, HCP preferences, and competitive landscape.
- **Methods**: Surveys, focus groups, interviews, secondary data analysis.
- **Importance**: Provides insights into market opportunities, challenges, and trends that inform strategic decision-making.

1.2. Competitor Analysis

- **Objective**: Understand the strengths and weaknesses of competing products and companies.
- **Methods**: SWOT analysis (Strengths, Weaknesses, Opportunities, Threats), benchmarking.
- **Importance**: Helps identify gaps in the market and opportunities for differentiation.

2. Product Positioning and Branding

2.1. Positioning Strategy

- **Objective**: Define how the product will be perceived relative to competitors.
- **Components**: Unique selling propositions (USPs), key benefits, and differentiators.

- **Importance**: Creates a clear and compelling narrative that resonates with target audiences.

2.2. Branding

- **Objective**: Build a strong brand identity that reflects the product's value and differentiates it from competitors.
- **Components**: Brand name, logo, tagline, visual identity.
- **Importance**: Enhances brand recognition and loyalty among HCPs and patients.

3. Regulatory Compliance

3.1. Understanding Regulations

- **Objective**: Ensure marketing activities adhere to legal and ethical standards.
- **Key Regulations**: FDA regulations, EMA guidelines, and local regulatory requirements.
- **Importance**: Prevents legal issues and ensures that promotional materials are accurate and truthful.

3.2. Developing Compliant Materials

- **Objective**: Create promotional content that meets regulatory standards.
- **Components**: Claims substantiation, proper labeling, and disclosure of risks.
- **Importance**: Maintains credibility and trust with HCPs and patients.

4. Promotional Strategies

4.1. Direct-to-Consumer (DTC) Advertising

- **Objective**: Engage patients directly through various media channels.
- **Methods**: Television, print ads, and online advertising.
- **Importance**: Increases awareness and generates demand among patients.

4.2. Professional Promotion

- **Objective**: Educate and influence HCPs about the product's benefits and uses.
- **Methods**: Sales representative visits, medical conferences, continuing medical education (CME).
- **Importance**: Drives adoption and recommendation of the product in clinical practice.

4.3. Digital Marketing

- **Objective**: Utilize digital channels to reach and engage target audiences.
- **Methods**: Search Engine Marketing (SEM), social media, email marketing, and content marketing.
- **Importance**: Provides measurable and targeted outreach and enhances engagement with HCPs and patients.

5. Sales Strategies and Management

5.1. Sales Planning

- **Objective**: Develop strategies to achieve sales targets and drive revenue growth.
- **Components**: Sales forecasts, territory management, target setting.
- **Importance**: Aligns sales efforts with business goals and optimizes resource allocation.

5.2. Sales Force Effectiveness

- **Objective**: Maximize the impact of the sales team through training and support.
- **Components**: Sales training programs, performance metrics, incentive schemes.
- **Importance**: Enhances the ability of sales representatives to engage with HCPs and close sales.

6. Customer Relationship Management (CRM)

6.1. Building Relationships

- **Objective**: Foster strong connections with HCPs and other Stakeholders.

- **Methods**: Personalized communication, regular follow-ups, loyalty programs.

- **Importance**: Strengthens customer loyalty and encourages repeat business.

6.2. Managing Customer Data

- **Objective**: Utilize CRM systems to track interactions and manage relationships.

- **Components**: Data collection, analysis, segmentation.

- **Importance**: Provides insights into customer needs and behaviors, facilitating targeted marketing efforts.

7. Market Access and Pricing

7.1. Pricing Strategies

- **Objective**: Determine optimal pricing to balance profitability and accessibility.

- **Methods**: Cost-based pricing, value-based pricing, competitive pricing.

- **Importance**: Affects market penetration and product adaptation.

7.2. Market Access

- **Objective**: Ensure product availability and reimbursement in target markets.

- **Methods**: Negotiating with payers, health economics evaluations, and formulary submissions.

- **Importance**: Facilitates access to the product for patients and supports commercial success.

8. Performance Measurement and Optimization

8.1. Monitoring and Evaluation

- **Objective**: Assess the effectiveness of marketing and sales activities.
- **Metrics**: Sales performance, market share, ROI, customer feedback.
- **Importance**: Enables data-driven decision-making and continuous improvement.

8.2. Adjusting Strategies

- **Objective**: Refine marketing and sales strategies based on performance data.
- **Methods**: Analyzing performance reports, conducting market reviews, and adapting tactics.
- **Importance**: Ensures that strategies remain effective and aligned with market conditions.

Conclusion

Pharmaceutical marketing is a multifaceted discipline that requires a deep understanding of market dynamics, regulatory requirements, promotional strategies, and sales management. By mastering these fundamentals, pharmaceutical product managers can effectively position their products, engage with stakeholders, and drive commercial success.

8.2 Sales Strategies and HCP Engagement

Pharmaceutical sales strategies and HCP (Healthcare Professional) engagement are critical components of a successful marketing approach. They help drive product adoption, ensure effective communication, and improve patient outcomes. Here's a detailed overview of these elements:

1. Sales Strategies

1.1. Developing a Sales Plan

- **Objective**: Create a comprehensive plan to achieve sales targets and drive revenue growth.

- **Components**:

- **Market Analysis**: Understand market dynamics, including competitor activities and market trends.

- **Sales Goals**: Set specific, measurable, achievable, relevant, and time-bound (SMART) goals.

- **Territory Management**: Define sales territories based on potential market size and current coverage.

- **Resource Allocation**: Distribute resources effectively to maximize impact.

1.2. Sales Tactics

- **Direct Selling**: Engage HCPs through one-on-one meetings, detailing sessions, and product demonstrations.

- **Key Account Management**: Focus on high-value accounts to build long-term relationships and drive higher sales volume.

- **Sales Promotions**: Implement promotional offers, discounts, or incentives to boost sales.

1.3. Sales Force Effectiveness

- **Training and Development**: Provide ongoing training to sales representatives on product knowledge, sales techniques, and compliance.

- **Performance Metrics**: To evaluate effectiveness, use key performance indicators (KPIs) such as sales volume, call frequency, and conversion rates.
- **Incentive Programs**: Design incentive schemes to motivate and reward top-performing sales representatives.

1.4. Sales Forecasting

- **Objective**: Predict future sales based on historical data, market trends, and business insights.
- **Components**:
- **Quantitative Forecasting**: Use statistical models and historical sales data.
- **Qualitative Forecasting**: Incorporate expert opinions and market research insights.
- **Importance**: Helps in planning production, inventory management, and budget allocation.

2. HCP Engagement

2.1. Understanding HCP Needs and Preferences

- **Objective**: Tailor engagement strategies based on HCPs' needs, preferences, and behaviors.
- **Methods**:
- **Segmentation**: Categorize HCPs based on specialty, practice size, and prescribing behavior.
- **Surveys and Feedback**: Collect insights from HCPs to understand their needs and preferences.

2.2. Effective Communication Channels

- **Objective**: Utilize various communication channels to reach and engage HCPs effectively.
- **Channels**:
- **Face-to-face Meetings**: Personalized interactions to discuss product benefits and answer questions.

- **Digital Channels**: Email, webinars, and online platforms provide information and updates.
- **Medical Conferences**: Engage with HCPs during industry events and conferences.

2.3. Providing Value-Added Content

- **Objective**: Offer content that is valuable and relevant to HCPs.
- **Types of Content**:
- **Educational Materials**: Provide the latest research, treatment guidelines, and clinical data information.
- **Clinical Case Studies**: Share real-world examples of how the product is used in clinical practice.
- **Patient Resources**: Offer tools and resources to assist HCPs in patient management.

2.4. Building Strong Relationships

- **Objective**: Develop long-term, trust-based relationships with HCPs.
- **Strategies**:
- **Personalized Interactions**: Tailor interactions based on the HCP's interests and needs.
- **Regular Follow-ups**: Maintain consistent communication to stay top-of-mind and address any concerns.
- **Recognition and Appreciation**: Acknowledge and reward HCPs for their contributions and engagement.

2.5. Feedback and Adaptation

- **Objective**: Continuously improve engagement strategies based on feedback from HCPs.
- **Methods**:
- **Surveys and Interviews**: Collect feedback on engagement experiences and areas for improvement.

- **Data Analysis**: Analyze engagement data to identify trends and refine strategies.
- **Importance**: Ensures that engagement strategies remain effective and aligned with HCP needs.

Conclusion

Effective sales strategies and HCP engagement are essential for the success of pharmaceutical products. By developing comprehensive sales plans, utilizing various communication channels, and building strong relationships with HCPs, pharmaceutical companies can drive product adoption, enhance market presence, and ultimately improve patient outcomes. Product managers play a pivotal role in executing these strategies and ensuring alignment with overall business goals.

8.3. Integrating Marketing and Sales for Product Success

Integrating marketing and sales is crucial for maximizing product success in the pharmaceutical industry. This synergy ensures that both functions work toward common objectives, leveraging complementary strengths and optimizing resources. Here's a comprehensive look at how to effectively integrate marketing and sales:

1. Aligning Goals and Objectives

1.1. Shared Vision

- **Objective**: Establish a unified vision and strategic goals that align marketing and sales efforts.

- **Action Steps**:

- **Collaborative Planning**: Conduct joint planning sessions to set common objectives and strategies.

- **Goal Setting**: Define specific, measurable goals that both marketing and sales teams aim to achieve.

- **Importance**: Ensures both teams work towards the same targets and reduces conflicts.

1.2. Clear Communication Channels

- **Objective**: Facilitate open and transparent communication between marketing and sales.

- **Action Steps**:

- **Regular Meetings**: Schedule meetings to discuss progress, challenges, and updates.

- **Shared Platforms**: Use collaborative tools and platforms to share information and track performance.

- **Importance**: Enhances coordination and ensures both teams are informed and aligned.

2. Collaborative Strategy Development

2.1. Joint Market Analysis

- **Objective**: Conduct joint analysis to understand market dynamics and customer needs.

- **Action Steps**:

- **Data Sharing**: Share market research data, customer feedback, and competitive intelligence.

- **Insights Integration**: Combine insights from both marketing and sales to develop a comprehensive strategy.

- **Importance**: Provides a holistic view of the market and customer needs, leading to more effective strategies.

2.2. Coordinated Campaigns

- **Objective**: Develop integrated marketing and sales campaigns to drive product success.

- **Action Steps**:

- **Campaign Planning**: Collaborate on campaign objectives, target audiences, and messaging.

- **Execution**: Ensure marketing and sales teams work together to implement and promote campaigns.

- **Importance**: Creates a unified message and approach, enhancing campaign effectiveness.

3. Aligning Messaging and Positioning

3.1. Consistent Messaging

- **Objective**: Ensure consistent messaging across all marketing and sales materials.

- **Action Steps**:

- **Develop Messaging Guidelines**: Create guidelines that outline key messages, value propositions, and branding.

- **Review Materials**: Regularly review marketing and sales materials to ensure consistency.

- **Importance**: Provides a cohesive brand experience and reinforces key product benefits.

3.2. Target Audience Understanding

- **Objective**: Align marketing and sales efforts based on a deep understanding of target audiences.
- **Action Steps**:
- **Customer Segmentation**: Collaborate on customer segmentation and profiling.
- **Tailored Approaches**: Develop tailored marketing and sales approaches based on audience insights.
- **Importance**: Enhances the relevance and effectiveness of marketing and sales efforts.

4. Leveraging Data and Analytics

4.1. Performance Tracking

- **Objective**: Monitor and analyze the performance of marketing and sales alternatives.
- **Action Steps**:
- **United Metrics**: Establish key performance indicators (KPIs) that reflect marketing and sales goals.
- **Data Integration**: Use integrated data systems to track and analyze performance.
- **Importance**: Provides insights into the effectiveness of strategies and allows for data-driven decision-making.

4.2. Feedback Loop

- **Objective**: Create a feedback loop to improve marketing and sales strategies continuously.
- **Action Steps**:
- **Regular Reviews**: Regularly review performance data and feedback from both teams.

- **Adjust Strategies**: Adjust marketing and sales strategies based on the feedback and performance analysis.
- **Importance**: Ensures continuous improvement and alignment with market needs.

5. Collaborative Training and Development

5.1. Joint Training Programs

- **Objective**: Provide training that addresses the needs of both marketing and sales teams.
- **Action Steps**:
- **Develop Training Content**: Create training programs covering key marketing and sales aspects.
- **Conduct Workshops**: Organize workshops and seminars to enhance skills and knowledge.
- **Importance**: Builds a shared understanding of strategies, tools, and practices.

5.2. Knowledge Sharing

- **Objective**: Facilitate the exchange of knowledge and expertise between marketing and sales.
- **Action Steps**:
- **Knowledge Repositories**: Create shared knowledge repositories, best practices, case studies, and insights.
- **Cross-Training**: Encourage cross-training to enhance understanding of each team's roles and responsibilities.
- **Importance**: Enhances collaboration and ensures both teams are well-informed.

6. Optimizing Resource Allocation

6.1. Budget Coordination

- **Objective**: Coordinate budget allocation to optimize resources for marketing and sales activities.

- **Action Steps:**
- **Joint Budget Planning**: Collaborate on budget planning and allocation decisions.
- **Resource Tracking**: Monitor and track resource usage to ensure alignment with goals.
- **Importance**: Ensures efficient use of resources and maximizes the impact of marketing and sales efforts.

6.2. Resource Sharing

- **Objective**: Share resources and tools between marketing and sales teams.
- **Action Steps:**
- **Shared Tools**: Use tools for CRM, analytics, and campaign management.
- **Collaborative Resources**: Develop and share resources such as sales enablement materials and marketing collateral.
- **Importance**: Enhances efficiency and ensures consistency across teams.

Conclusion

Integrating marketing and sales is essential for driving pharmaceutical product success. By aligning goals, collaborating on strategies, and optimizing resources, pharmaceutical companies can enhance their product offerings, improve market presence, and achieve better patient outcomes. Product managers are key in facilitating this integration, ensuring that both functions work together towards common objectives.

Part III. Decoding Social Sciences for Pharma

Decoding social sciences for pharmaceutical marketing involves understanding how various branches of social science contribute to better decision-making and strategy development in the pharmaceutical industry. Social sciences offer valuable insights into human behavior, cultural influences, and societal trends that can be pivotal for effective marketing and communication strategies.

Here is a detailed description of how different social sciences can be applied to pharmaceutical marketing:

1. Sociology

1.1. Understanding Social Structures and Institutions

- **Objective**: Gain insights into how social structures and institutions (e.g., family, healthcare systems) influence health behaviors and treatment choices.

- **Application**: Tailor marketing messages and product offerings to align with the norms and values of different social groups.

1.2. Social Behavior and Influence

- **Objectives**: Explore how social behavior and peer influence affect healthcare decisions.

- **Application**: Develop campaigns that leverage social proof and testimonials to encourage the adoption of new treatments.

1.3. Social Trends and Demographics

- **Objective**: Analyze social trends and demographic changes to identify emerging market opportunities.

- **Application**: Customize marketing strategies to target specific demographics and address their unique needs.

2. Anthropology

2.1. Cultural Understanding

- **Objective**: Understand cultural beliefs, practices, and values related to health and illness.

- **Application**: Design culturally sensitive marketing campaigns that resonate with diverse populations and avoid cultural insensitivity.

2.2. Ethnographic Research

- **Objective**: Conduct ethnographic research to observe and understand patient and healthcare behaviors in natural environments.

- **Application**: Use insights from ethnographic studies to develop more effective patient education materials and engagement strategies.

2.3. Health Practices and Rituals

- **Objective**: Explore how health practices and rituals vary across different cultures.

- **Application**: Adapt product messaging and educational content to align with cultural health practices and beliefs.

3. Behavioral Economics

3.1. Decision-Making Processes

- **Objective**: Analyze how people make decisions, especially under conditions of uncertainty and risk.

- **Application**: Develop marketing strategies that address cognitive biases and behavioral tendencies to influence treatment choices.

3.2. Incentives and Motivations

- **Objective**: Understand what motivates patients and healthcare providers to make certain choices.

- **Application**: Design incentive-based programs that encourage adherence to treatment regimens and participation in clinical trials.

3.3. Choice Architecture:

- **Objective**: Explore how the presentation of options affects decision-making.

- **Application**: Use choice architecture principles to present treatment options in a way that guides patients toward preferred choices.

4. Psychology

4.1. Patient Perceptions and Attitudes

- **Objective**: Study how patients perceive diseases and treatments and their attitudes toward healthcare providers.
- **Application**: Tailor marketing messages to address common misconceptions and build positive perceptions of the product.

4.2. Behavioral Change Theories

- **Objective**: Apply behavioral change theories to develop strategies to encourage adherence to treatment and lifestyle modifications.
- **Application**: Implement behavioral change interventions in marketing campaigns to improve patient outcomes.

4.3. Emotional and Cognitive Factors

- **Objective**: Explore how emotional and cognitive factors influence healthcare decisions.
- **Application**: Create emotionally compelling marketing materials that resonate with patients and motivate them to engage with the treatment.

5. Communication Studies

5.1. Effective Messaging

- **Objective**: Learn about effective communication strategies and how to craft messages that resonate with target audiences.
- **Application**: Develop clear, concise, and persuasive marketing messages communicating the product's benefits and value.

5.2. Media and Channel Selection

- **Objective**: Understand which media channels and communication methods are most effective for reaching different audiences.

- **Application**: Choose the appropriate channels for disseminating marketing messages based on audience preferences and media consumption habits.

5.3. Crisis Communication:

- **Objective**: Prepare for managing communication during a crisis or a product recall.

- **Application**: Develop crisis communication plans and strategies to handle potential issues and maintain stakeholder trust.

6. Application in Pharmaceutical Marketing

6.1. Market Segmentation and Targeting

- **Objective**: Use insights from social sciences to segment the market and target specific patient and healthcare provider groups effectively.

- **Application**: Develop tailored marketing strategies and interventions for different segments based on social and cultural characteristics.

6.2. Patient Engagement and Education

- **Objective**: Enhance patient engagement and education through culturally and behaviorally informed approaches.

- **Application**: Create educational materials and programs that address patients' unique needs, preferences, and barriers to adherence.

6.3. Healthcare Provider Communication

- **Objective**: Improve communication strategies with healthcare providers by understanding their motivations, behaviors, and decision-making processes.

- **Application**: Develop targeted communication and engagement strategies that align with healthcare providers' preferences and practices.

Conclusion

Decoding social sciences for pharmaceutical marketing provides valuable insights into human behavior, cultural influences, and decision-making processes. By leveraging these insights, pharmaceutical product managers can develop more effective marketing strategies, improve patient engagement, and enhance overall product success. Understanding and applying social sciences is essential for creating marketing campaigns that resonate with diverse audiences and drive positive outcomes in the pharmaceutical industry.

Understanding Sociology

Understanding sociology is crucial for pharmaceutical product management. It provides insights into how social structures, relationships, and cultural norms influence health behaviors, treatment adherence, and market dynamics. Sociology helps design and implement effective strategies catering to diverse populations and healthcare systems.

Here is a detailed description of how understanding sociology impacts pharmaceutical product management:

1. Social Structures and Institutions

1.1. Social Structures

- **Objective**: Understand how social structures, such as family, community, and social networks, impact health behaviors and treatment decisions.
- **Application**: Tailor marketing strategies to align with the social structures of different target populations, ensuring that promotional materials and campaigns resonate with their specific social contexts.

1.2. Healthcare Systems

- **Objective**: Explore how various healthcare systems (e.g., public vs. private) influence access to medications and patient care.

- **Application**: Develop strategies that consider the healthcare system's structure in different regions, address access issues, and adapt product offerings accordingly.

2. Social Behavior and Influence

2.1. Social Behavior

- **Objective**: Analyze how social behavior and interactions affect healthcare decisions and treatment adherence.

- **Application**: Use insights into social behavior to design marketing campaigns that leverage peer influence, social proof, and community endorsements to encourage product adoption.

2.2. Social Inlfuence

- **Objective**: Study how social networks and influencers impact health-related behaviors and choices.

- **Application**: Collaborate with key opinion leaders (KOLs) and influencers to endorse products and create positive social proof, enhancing credibility and acceptance among target audiences.

3. Social Trends and Demographics

3.1. Demographic Analysis

- **Objective**: Examine demographic factors such as age, gender, socioeconomic status, and ethnicity to understand their impact on health behaviors and needs.

- **Application**: Segment the market based on demographic data and create targeted marketing strategies that address different demographic groups' specific needs and preferences.

3.2. Social Trends

- **Objective**: Identify and analyze emerging social trends and their implications for healthcare and pharmaceutical marketing.

- **Application**: Adjust marketing strategies to align with current social trends, such as increased focus on wellness or digital health, to stay relevant and appeal to contemporary consumer preferences.

4. Cultural Influences

4.1. Cultural Norms and Values

- **Objective**: Understand how cultural norms and values shape health beliefs, practices, and treatment preferences.

- **Application**: Develop culturally sensitive marketing messages and materials that respect and align with the cultural values of different patient populations.

4.2. Health Practices and Beliefs

- **Objective**: Explore how cultural health practices and beliefs influence patients' attitudes toward treatments and healthcare providers.

- **Application**: Adapt product messaging and educational content to address cultural health practices and beliefs, improving engagement and acceptance.

5. Patient and Provider Perspectives

5.1. Patient Perspectives

- **Objective**: Gain insights into patients' experiences, expectations, and challenges related to healthcare and treatment.

- **Application**: Design patient-centric marketing strategies and support programs that address patients' needs, improve their experiences, and enhance adherence.

5.2. Healthcare Provider Perspectives

- **Objective**: Understand healthcare providers' roles, patient interactions, and decision-making processes.

- **Application**: Develop strategies that support healthcare providers in their practice, such as providing educational resources and facilitating efficient communication channels.

6. Social Determinants of Health

6.1. Socioeconomic Factors

- **Objective**: Analyze how socioeconomic factors such as income, education, and employment impact health outcomes and access to care.

- **Application**: Address socioeconomic disparities in marketing strategies by offering affordable options, financial assistance programs, and tailored educational content.

6.2. Environmental and Social Conditions

- **Objective**: Examine how environmental and social conditions, such as living conditions and social support systems, affect health and its utilization.

- **Application**: Design interventions and support programs considering these conditions, improving access and adherence among populations facing environmental and social challenges.

Conclusion

Understanding sociology equips pharmaceutical product managers with the knowledge to design and implement effective marketing and engagement strategies. By appreciating how social structures, behaviors, trends, and cultural influence shape health decisions, product managers can develop more targeted, culturally sensitive, and impactful campaigns. This sociological insight is essential for addressing diverse patient needs, improving healthcare outcomes, and ensuring the success of pharmaceutical products in a complex and varied market landscape.

9.1 The Role of Sociology in Pharma Marketing

Sociology is vital in pharmaceutical marketing by providing insights into how social structures, relationships, and cultural contexts influence health behaviors, treatment adherence, and consumer preferences. Understanding these sociological factors allows pharma marketers to design more effective, targeted, and culturally sensitive marketing strategies. Here's how sociology impacts pharma marketing:

1. Social Structures and Health Behavior

1.1. Social Structures

- **Influence on Behavior**: Social structures, such as family, community, and social networks, influence individuals' health behaviors and decisions. For example, family dynamics can impact a patient's willingness to seek treatment and adhere to a prescribed regimen.

- **Marketing Application**: Marketers can leverage this understanding to tailor campaigns that resonate with different social structures, such as family-oriented messaging or community-based health initiatives.

1.2. Healthcare Systems

- **Influence on Access**: Different healthcare systems (public vs. private) and structures affect how patients access medications and healthcare services.

- **Marketing Applications**: Develop strategies that account for variations in healthcare access, such as focusing on affordability and accessibility in regions with less comprehensive healthcare systems.

2. Social Influence and Networks

2.1. Social Influence

- **Impact on Choices**: Social networks and influencers play a significant role in shaping health-related choices and perceptions. Positive endorsements from peers or influencers can drive product adoption.

- **Marketing Application**: Utilize social proof by engaging key opinion leaders (KOLs), healthcare influencers, and patient advocates to endorse products and share positive experiences.

2.2. Community Engagement

- **Local Insights**: Communities have unique health concerns and cultural practices influencing treatment preferences and adherence.

- **Marketing Application**: Create community-specific campaigns that address local health issues and align with cultural norms, enhancing the relevance and effectiveness of marketing efforts.

3. Cultural Norms and Values

3.1. Cultural Influences

- **Health Beliefs**: Cultural norms and values shape how people perceive health, illness, and treatment. For instance, cultural attitudes towards pharmaceuticals vs. traditional remedies can affect product acceptance.

- **Marketing Application**: Design culturally sensitive messaging and educational content that aligns with the values and beliefs of different cultural groups, improving engagement and acceptance.

3.2. Health Practices

- **Adherence Factors**: Cultural practices and beliefs can impact adherence to prescribed treatments. Understanding these factors helps in addressing barriers to adherence.

- **Marketing Application**: Develop support programs and culturally relevant educational materials that address specific beliefs and practices that may influence treatment adherence.

4. Social Determinants of Health

4.1. Socioeconomic Status

- **Access and Outcomes**: Socioeconomic factors such as income, education, and employment significantly affect health outcomes and access to healthcare.

- **Marketing Application**: Tailor marketing strategies to address socioeconomic disparities by offering financial assistance programs, targeted educational content, and affordable product options.

4.2. Environmental Conditions

- **Impact on Health**: Environmental and social conditions, including living conditions and social support networks, influence health and healthcare utilization.

- **Marketing Application**: Consider these conditions when designing interventions and support programs, such as outreach programs in underserved areas or partnerships with local community organizations.

5. Patient and Provider Perspectives

5.1. Patient Perspectives

- **Experience and Expectations**: Understanding patients' experiences, expectations, and challenges helps create more effective marketing strategies.

- **Marketing Application**: Develop patient-centric campaigns that address specific needs and concerns, enhancing patient education and engagement.

5.2. Healthcare Provider Perspectives

- **Provider Interactions**: Healthcare providers' interactions with patients and their decision-making processes influence treatment choices and adherence.

- **Marketing Application**: Create Strategies that support healthcare providers, such as providing relevant educational resources and facilitating effective communication channels.

Conclusion

Incorporating sociological insights into pharmaceutical marketing helps craft strategies sensitive to social dynamics, cultural contexts, and patient needs. By addressing how social structures, cultural norms, and socioeconomic factors influence health behaviors and treatment decisions, pharma marketers can develop more targeted, impactful, and patient-centered marketing campaigns. This approach enhances product acceptance and adherence and improves health outcomes and market success.

9.2 Social Structures and Their Impact on Healthcare

Social structures refer to the organized patterns of relationships and institutions within a society that shape individuals' behaviors and interactions. These structures are crucial in influencing healthcare success, utilization, and outcomes. Understanding these structures can help pharmaceutical product managers and healthcare providers design more effective and targeted interventions. Here's how social structures impact healthcare.

1. Family Structure and Health Decisions

1.1. Influence on Health Behavior

- **Decision-Making**: Family members often play a significant role in healthcare decisions. For example, a spouse or parent may influence a person's choice to seek medical treatment or adhere to a medicaiton regimen.

- **Support Systems**: Families provide emotional and practical support, such as reminders for medication or transportation to medical appointments, which can impact treatment adherence and overall health outcomes.

1.2. Marketing Application

- **Family-Centric Application**: Tailor marketing strategies to address family dynamics and involve family members in health-related decisions. For instance, provide educational materials to share within the family or offer family-oriented support programs.

2. Community and Social Networks

2.1. Peer Influence

- **Health Behaviors**: Social networks and peer groups can significantly influence health behaviors and attitudes towards treatment. Positive experiences shared within a social network can increase acceptance and adoption of health interventions.

- **Health Information**: Social networks are a primary source of health information and can impact how individuals perceive and respond to health messages.

2.2. Marketing Application

- **Social Proof and Endorsements**: Utilize testimonials, endorsements, and recommendations from influential figures within communities or social networks to build trust and encourage product adoption.

- **Community-Based Programs**: Engage with local communities to provide health education and support tailored to their needs and concerns.

3. Socioeconomic Status

3.1. Access and Utilization

- **Healthcare Access**: Socioeconomic status, including income, education, and occupation, affects healthcare services. Individuals with higher socioeconomic status generally have better access to healthcare and resources than those with lower status.

- **Treatment Adherence**: Economic constraints can impact a person's ability to afford medications, leading to issues with adherence and health outcomes.

3.2. Marketing Application

- **Affordable Options**: Develop strategies that address socioeconomic disparities, such as offering financial assistance programs or tiered pricing to make medications more accessible to lower-income populations.

- **Educational Outreach**: Provide targeted educational resources and support for individuals in lower socioeconomic groups to improve health literacy and access to care.

4. Healthcare Systems and Institutions

4.1. System Structure

- **Healthcare Delivery**: Different healthcare systems (public vs. private) and institutional structures affect how healthcare is delivered and accessed. For instance, access may be more regulated in public healthcare systems, while private systems may offer more personalized care.

- **Regulatory Influence**: The structure of healthcare institutions and regulatory bodies influences the approval, availability, and promotion of pharmaceutical products.

4.2. Marketing Application

- **Tailored Strategies**: Design marketing strategies that align with healthcare systems' specific characteristics and regulations. For example, customize approaches based on whether the market is predominantly public or private.

- **Institutional Partnerships**: Collaborate with healthcare institutions to support initiatives that align with their goals and regulatory requirements.

5. Cultural and Social Norms

5.1. Health Beliefs

- **Cultural Influences**: Cultural beliefs and norms influence health practices and attitudes towards treatment. For example, cultural views on pharmaceuticals versus traditional remedies can impact product acceptance.

- **Health Practices**: Societal norms dictate health practices and behaviors, such as preventive measures and attitudes toward chronic disease management.

5.2. Marketing Application

- **Culturally Sensitive Messaging**: Develop marketing materials and campaigns that respect and align with cultural values and beliefs. Ensure that messaging is culturally appropriate and sensitive to diverse populations.

- **Educational Initiatives**: Implement educational initiatives addressing cultural barriers and promoting understanding of new treatments within specific contexts.

6. Social Support Systems

6.1. Support Networks

- **Impact on Health Outcomes**: Access to social support networks, including friends, family, and community resources, can positively impact health outcomes by providing emotional and practical support.

- **Influence on Treatment**: Support systems can influence an individual's motivation to adhere to treatment plans and engage in healthy behaviors.

6.2. Marketing Application

- **Support Programs**: Develop support programs that leverage social networks to assist and encourage patients.

- **Community Engagement**: Partner with community organizations to create support systems that address specific health needs and challenges.

Conclusion

Understanding social structures is essential for pharmaceutical product managers as it enables them to create more effective and targeted marketing strategies that address the complexities of health behavior and access. By considering factors such as family dynamics, community influence, socioeconomic status, and cultural norms, marketers can develop interventions that resonate with their target audiences, improve health outcomes, and enhance the overall effectiveness of their products.

9.3 Applying Sociological Insights to Marketing Strategies

Sociological insights offer a valuable understanding of how social structures, relationships, and cultural norms influence behavior. By integrating these insights into marketing strategies, pharmaceutical companies can create more effective and tailored campaigns that resonate with target audiences. Here's how sociological insights can be applied to marketing strategies in the pharmaceutical industry:

1. Understanding Social Norms and Cultural Values

1.1. Cultural Sensitivity

- **Tailoring Messages**: Recognize and respect cultural values and beliefs that impact health behaviors. Customize marketing messages to align with cultural norms and preferences, avoiding stereotypes and ensuring cultural appropriateness.

- **Example**: In markets where traditional medicine is preferred, campaigns might emphasize integrating new treatments with existing practices rather than positioning them as replacements.

1.2 Social Acceptance

- **Building Trust**: Address societal concerns and promote acceptance by demonstrating how a product aligns with social norms and expectations. Highlight endorsements from respected figures or institutions to build credibility.

- **Example**: An anti-smoking campaign might feature testimonials from community leaders and emphasize how quitting aligns with positive social values.

2. Leveraging Social Networks and Influences

2.1. Influencer Engagement

- **Leveraging Key Opinion Leaders**: Collaborate with healthcare professionals, patient advocates, and influencers who can impact social networks and provide credible product endorsements.

- **Example**: Partner with well-known health bloggers or medical experts to share information about a new treatment on social media platforms, leveraging their reach and influence.

2.2. Social Proof and Peer Recommendations

- **Utilizing Testimonials**: Use testimonials and success stories from patients or healthcare providers to showcase the effectiveness of products. Social proof can significantly impact decision-making.

- **Example**: Highlight patient success stories in marketing materials to encourage others to consider the treatment based on positive experiences shared by peers.

3. Addressing Social Determinants of Health

3.1. Economic and Social Barriers

- **Targeted Solutions**: Develop strategies that address economic and social barriers to healthcare access. Offer solutions such as financial assistance programs, tiered pricing, or community-based support.

- **Example**: Implement patient assistance programs that provide subsidies or discounts to lower-income individuals to make medications more affordable.

3.2. Tailored Educational Initiatives

- **Educational Outreach**: Provide education and resources tailored to specific social groups or communities, addressing their unique health challenges and needs.

- **Example**: Create localized health education materials that address prevalent health issues in specific regions or communities.

4. Engaging with Family and Social Structures

4.1. Family-Centric Campaigns

- **Involving Families**: Develop campaigns that involve family members in health decisions and support. Provide resources that can be shared with family members to facilitate collective decision-making.

- **Example**: Offer family-focused educational resources on managing chronic conditions, emphasizing the role of family support in improving health outcomes.

4.2. Community Engagement

- **Building Community Programs**: Engage with communities to create programs that address collective health needs and provide support. Leverage local organizations and community leaders to reach target audiences.

- **Example**: Partner with community health organizations to host educational workshops or health fairs that address specific health concerns relevant to the community.

5. Enhancing Product Adoption Through Social Influence

5.1. Peer Group Influence

- **Influencing Adoption**: Utilize insights into peer group dynamics to design campaigns encouraging product adoption through positive peer influence.

- **Example**: Develop referral programs where existing patients can refer friends or family members to a new treatment, leveraging peer recommendations to drive adoption.

5.2. Social Engagement Strategies

- **Interactive Campaigns**: Create interactive and engaging campaigns that foster social interaction and community-building around health topics. Use social media platforms to facilitate discussions and share information.

- **Example**: Launch social media campaigns that encourage users to share their health journeys and engage with branded content related to specific treatments.

6. Understanding Healthcare Access and Utilization

6.1. Access Barriers

- **Addressing Gaps**: Identify and address barriers to healthcare access within specific social groups, such as transportation, language, or financial constraints.

- **Example**: Provide multilingual educational materials and support services to non-English speaking patients to access and understand healthcare information.

6.2. Utilization Patterns

- **Behavioral Insights**: Analyze healthcare utilization patterns within different social groups to tailor marketing strategies and interventions.

- **Example**: Target marketing efforts towards populations that underutilize preventive care services, highlighting the benefits and importance of early intervention.

Conclusion

Applying sociological insights to marketing strategies allows pharmaceutical companies to create more effective, personalized campaigns that resonate with diverse audiences. By understanding and addressing cultural values, social networks, economic barriers, and family dynamics, marketers can develop strategies that enhance product adoption, improve health outcomes, and build stronger relationships with target audiences. Integrating these insights into marketing efforts helps ensure that campaigns are relevant, respectful, and impactful, contributing to the success of pharmaceutical products.

The Role of Anthropology

Anthropology, a study of human societies, cultures, and their development, provides invaluable insights for pharmaceutical product management. By understanding cultural practices, social norms, and human behavior, pharmaceutical product managers can create more effective strategies that resonate with diverse populations. Here's how anthropology plays a crucial role in pharmaceutical product management:

1. Understanding Cultural Contexts

1.1. Cultural Sensitivity

- **Tailoring Strategies**: Anthropology helps product managers understand cultural differences and nuances influencing health behaviors and perceptions. This understanding allows for the development of culturally sensitive marketing strategies and educational materials.

- **Example**: Product managers might emphasize how the drug complements rather than replaces traditional treatments when launching a new drug in a region with strong traditional medicine practices.

1.2. Avoiding Missteps

- **Preventing Cultural Insensitivity**: Insight into cultural practices and taboos helps avoid marketing messages or product claims that could be offensive or misunderstood in different cultural contexts.

- **Example**: In some cultures, discussing certain medical conditions openly might be considered a taboo. Anthropological insights can guide how to approach such topics delicately.

2. Enhancing Patient Engagement

2.1. Patient-Centered Design

- **Designing for Local Needs**: Anthropology helps us understand patients' daily lives, health practices, and needs. This information can be used to design more relevant and effective products, services, and support systems.

- **Example**: In regions with limited access to healthcare facilities, product managers might develop solutions like telemedicine options or mobile health units to reach underserved populations.

2.2. Behavior and Attitudes

- **Influencing Behavior**: Understanding how cultural and social factors influence health-related behaviors and attitudes can help craft messages that resonate with patients and encourage adherence to treatment regimens.

- **Example**: If a culture values family involvement in health decisions, marketing campaigns might focus on how a product supports the whole family's well-being.

3. Facilitating Market Entry and Expansion

3.1. Market Assessment

- **Cultural Adaptation**: Anthropological research provides insights into local health needs, market gaps, and the competitive landscape. This information is crucial for developing strategies that align with local market conditions.

- **Example**: Before entering a new geographical market, a pharmaceutical company might use anthropological studies to understand local health priorities and adjust its product offerings accordingly.

3.2. Local Partnerships

- **Building Relationships**: Understanding local cultural and social dynamics helps forge partnerships with healthcare providers, community leaders, and other stakeholders. This can enhance market acceptance and facilitate smoother product introductions.

- **Example**: Collaborating with local health organizations to promote a drug can be more effective when understanding local customs and practices informs the partnership.

4. Informing Product Development

4.1. User Experience

- **Cultural Preferences**: Insights into cultural preferences and practices can guide product design and development to ensure that it meets the needs and expectations of different user groups.

- **Example**: If certain delivery methods (e.g., pills, injections) are less acceptable in some cultures, product management might focus on alternative forms like patches or oral solutions.

4.2. Safety and Compliance

- **Cultural Safety Considerations**: Understanding cultural attitudes towards medicine and health can help anticipate potential safety concerns and ensure that products comply with local regulations and expectations.

- **Example**: In cultures with specific dietary restrictions, it is important to ensure that a drug does not contain culturally prohibited ingredients for safety and acceptance.

5. Enhancing Communication Strategies

5.1. Effective Messaging

- **Cultural Relevance**: Anthropology helps craft culturally relevant messages and resonates with target audiences, increasing the effectiveness of communication efforts.

- **Example**: Advertisements and educational materials that reflect cultural values and address specific health concerns can be more persuasive and engaging.

5.2. Engaging Stakeholders

- **Local Engagement**: Understanding the role of various stakeholders (e.g., family and community leaders) in health decisions can improve how products are marketed and promoted.

- **Example**: In some cultures, community leaders play a significant role in health decisions. Engaging with these leaders can help gain community support for a new treatment.

6. Supporting Ethical Practices

6.1. Ethical Considerations

- **Respecting Traditions**: Anthropology ensures that marketing and product development practices respect local traditions and ethical norms, avoiding exploitation or misrepresentation.

- **Example**: Conducting research respecting local customs and involving community members in the development process helps maintain ethical standards and build trust.

6.2. Informed Consent

- **Cultural Sensitivity in Research**: Anthropology helps design research protocols sensitive to cultural differences and ensures that informed consent is obtained respectfully and understandably.

- **Example**: Tailoring consent forms and procedures to fit cultural expectations ensures that participants fully understand their involvement and give consent appropriately.

Conclusion

Anthropology provides valuable insights into cultural, social, and behavioral factors that impact pharmaceutical product management.

By applying anthropological principles, product managers can develop more culturally sensitive strategies, enhance patient engagement, support market entry and expansion, and ensure ethical practices. Integrating anthropological insights into pharmaceutical product management helps create relevant, respectful, and effective campaigns, leading to better outcomes and success in diverse markets.

10.1. Cultural Influences on Healthcare and Treatment

Cultural influences significantly shape how individuals perceive, access, and engage with healthcare and treatment. Understanding these influences is crucial for pharmaceutical product managers and healthcare professionals to develop effective, culturally sensitive strategies. Here's a detailed look at various cultural factors and their impact on healthcare and treatment:

1. Beliefs and Attitudes Toward Health

1.1. Traditional vs. Modern Medicine

- **Traditional Medicine**: In many cultures, traditional medicine practices, such as herbal remedies or alternative therapies, play a significant role. These practices may coexist with or even supersede modern medical treatments.

- **Example**: In some Asian cultures, traditional Chinese medicine (TCM) is widely used alongside conventional treatments. Pharmaceutical companies must consider these practices when introducing new products.

1.2. Attitude Toward Illness

- **Perceptions of Disease**: Cultural beliefs influence how diseases are perceived and managed. For example, some cultures might view illness as a spiritual or moral issue rather than a biological one.

- **Example**: In certain Indigenous cultures, illness might be attributed to spiritual imbalances, leading to a preference for traditional healing practices over pharmaceutical interventions.

2. Health Behaviors and Practices

2.1. Preventive Health Measures

- **Health Practices**: Cultural practices affect how individuals approach preventive health measures such as vaccinations or screenings.

- **Example**: Skepticism toward vaccinations due to cultural beliefs or past experiences might exist in some communities, affecting vaccination rates and requiring targeted educational efforts.

2.2. Adherence to Treatment

- **Medication Adherence**: Cultural factors can influence patients' adherence to prescribed treatments. Beliefs about medicaiton efficacy, potential side effects, or trust in healthcare providers play a role.

- **Example**: If a culture values natural remedies over pharmaceuticals, patients might be less inclined to adhere to prescribed medicaiton regimens.

3. Communication and Interaction with Healthcare Providers

3.1. Communication Styles

- **Patient-Provider Interaction**: Cultural norms influence communication styles between patients and healthcare providers. Some cultures emphasize indirect communication or hierarchical relationships, which can impact how patients can express their needs or concerns.

- **Example**: In cultures where respect for authority is paramount, patients might be reluctant to question or challenge healthcare providers, potentially affecting treatment outcomes.

3.2. Health Literacy

- **Understanding Health Information**: Cultural factors can affect how individuals interpret and understand health information. Language barriers, literacy levels, and familiarity with medical terminology play a role in health literacy.

- **Example**: Multilingual communities might require health information in multiple languages or culturally adapted formats to ensure understanding and compliance.

4. Dietary Practices and Lifestyle

4.1. Dietary Restrictions

- **Cultural Diets**: Certain cultures have specific dietary practices or restrictions that can impact treatment options, particularly for conditions requiring dietary adjustments or specific medications.

- **Example**: In Hindu culture, many people are vegetarian, which can affect their ability to use certain medications or supplements derived from animal sources.

4.2. Lifestyle Factors

- **Cultural Practices**: Lifestyle factors such as physical activity, sleep patterns, and social practices influenced by culture can affect health and treatment efficacy.

- **Example**: Cultural celebrations involving large meals and sedentary activities might impact managing chronic conditions like diabetes.

5. Access to Healthcare Services

5.1. Healthcare Utilization

- **Access and Barriers**: Cultural factors influence how and when individuals seek medical care. Economic, geographic, and social barriers can affect access to healthcare services.

- **Example**: In some communities, stigma around certain health conditions might discourage individuals from seeking medical help, leading to delayed diagnoses and treatment.

5.2. Trust in Healthcare Systems

- **Trust and Skepticism**: Cultural beliefs can impact trust in the healthcare system and pharmaceutical products. Historical or systematic issues might lead to skepticism toward modern medicine.

- **Example**: Communities that have historically mistrusted medical institutions might require additional outreach and education to build confidence in new treatments or clinical trials.

6. Family and Community Influence

6.1. Family Dynamics

- **Decision-Making**: In some cultures, healthcare decisions are made collectively by family rather than individuals. This can influence treatment choices and adherence.

- **Example**: In cultures with strong family bonds, family members may be involved in discussions and decisions regarding a patient's treatment plan, impacting how products are marketed and communicated.

6.2. Community Health Practices

- **Collective Practices**: Community health practices and beliefs can shape individual health behaviors and attitudes toward treatment.

- **Example**: Community-based health initiatives and peer support networks can promote health and treatment adherence significantly.

Conclusion

Cultural influences profoundly affect healthcare and treatment, impacting beliefs, behaviors, communication, access, and adherence. Understanding these cultural factors is essential for pharmaceutical product managers and healthcare providers to develop effective, culturally sensitive strategies that improve patient outcomes and enhance the acceptance and success of pharmaceutical products. By integrating cultural insights into product development, marketing, and patient engagement, pharmaceutical companies can better address the diverse needs of global populations and foster positive health outcomes.

10.2 Ethnographic Research in Pharma Marketing

Ethnographic research is a qualitative method to understand people's behaviors, needs, and experiences in their natural environments. In pharmaceutical marketing, ethnographic research provides deep insights into how patients, healthcare providers (HCPs), and other stakeholders interact with and perceive pharmaceutical products and healthcare systems. Here is a detailed overview of ethnographic research in pharma marketing:

1. Overview of Ethnographic Research

1.1. Definition and Purpose

- **Definition**: Ethnographic research involves immersive, long-term observation and interaction with individuals or groups to gain a comprehensive understanding of their behaviors, attitudes, and cultural contexts.

- **Purpose**: The goal is to uncover insights inaccessible through traditional quantitative research methods, providing a richer, more nuanced understanding of how products and services fit into people's lives.

1.2. Methodology

- **Participant Observation**: Researchers engage with participants in their natural settings, observing their behaviors and interactions without intervening.

- **Interviews and Conversations**: In-depth, semi-structured, and informal conversations are conducted to gather personal experiences and perspectives.

- **Field Notes and Documentation**: Detailed notes and records are kept to capture observations, interactions, and contextual information.

2. Applications in Pharma Marketing

2.1. Understanding Patient Experiences

- **Patient Journeys**: Ethnographic research helps map the journey from diagnosis to treatment, revealing challenges, unmet needs, and decision-making processes.

- **Example**: A study observing patients with chronic conditions might uncover difficulties in managing their treatment regimens and the emotional impact of their conditions, informing the development of more supportive and effective interventions.

2.2. Insight into Healthcare Provider Practices

- **Clinical Practices**: Observing HCPs in their clinical settings provides insights into their prescribing behaviors, patient interactions, and decision-making processes.

- **Example**: Ethnographic research could reveal how HCPs integrate new treatments into their practice, including their concerns, preferences, and barriers to adoption.

2.3. Evaluating Product Use and Acceptance

- **Product Interaction**: Researchers can observe how patients and HCPs use pharmaceutical products, including adherence to dosage instructions and response to side effects.

- **Example**: By observing patients using a new medicaiton, researchers might identify challenges in the administration process or areas where additional patient education is needed.

2.4. Identifying Market Opportunities

- **Market Gaps**: Ethnographic research helps identify unmet needs and market gaps by understanding the daily experiences and challenges of patients and HCPs.

- **Example**: Insights gained from observing patient support groups might highlight areas where existing treatments fall short, leading to the development of new solutions.

3. Benefits of Ethnographic Research in Pharma Marketing

3.1. Deep Understanding of Context

- **Rich Data**: Ethnographic research provides a detailed, contextual understanding of how patients and HCPs experience healthcare, revealing nuances that quantitative methods might miss.

- **Example**: Understanding cultural practices and beliefs around health can help tailor marketing messages and product features to better align with the target audience's values and needs.

3.2. Enhanced Product Development

- **User-Centered Design**: Insights from ethnographic research inform the design and development of products that better meet the needs and preferences of end-users.

- **Example**: Observations of patients managing chronic conditions might lead to the development of more user-friendly delivery mechanisms or support tools.

3.3. Improved Communication Strategies

- **Tailored Messaging**: By understanding the language, concerns, and priorities of patients and HCPs, pharmaceutical companies can create more effective and resonant marketing messages.

- **Example**: Messaging that addresses specific pain points or challenges observed during ethnographic research can be more compelling and persuasive.

3.4. Building Stronger Relationships

- **Empathy and Trust**: Ethnographic research fosters a deeper empathy for patients and HCPs, helping to build trust and strengthen relationships.

- **Example**: Understanding patients' real-world experiences and challenges can enhance brand loyalty and patient engagement.

4. Challenges and Considerations

4.1. Time and Resource Intensive

- **Duration**: Ethnographic research requires significant time and resources, involving long-term engagement and observation.

- **Resource Allocation**: Companies must allocate adequate resources for conducitng and analyzing ethnographic research.

4.2. Ethical Considerations

- **Informed Consent**: Researchers must ensure that participants provide informed consent and understand the purpose and scope of the research.
- **Privacy**: Maintaining the privacy and confidentiality of participants is crucial.

4.3. Interpretation and Application

- **Complex Data**: Analyzing and interpreting qualitative data from ethnographic research can be complex and requires skilled researchers.
- **Integration**: Translating ethnographic insights into actionable strategies and product improvements can be challenging.

Conclusion

Ethnographic research offers valuable insights for pharmaceutical marketing by providing a deep, contextual understanding of patients' and HCPs' experiences. By immersing in the environments of these stakeholders, pharmaceutical companies can uncover unmet needs, enhance product development, and create more effective marketing strategies. Despite its challenges, ethnographic research remains a powerful tool for gaining a holistic view of the healthcare landscape and improving engagement and outcomes in the pharmaceutical industry.

10.3 Leveraging Anthropological Insights for Global Campaigns

Anthropological insights offer invaluable perspectives for designing and executing global pharmaceutical marketing campaigns. By understanding the diverse cultural, social, and behavioral contexts of different populations, pharmaceutical companies can create campaigns that resonate more deeply and effectively across various regions. Here's a detailed overview of how anthropological insights can be leveraged in global pharma campaigns:

1. Understanding Cultural Diversity

1.1. Cultural Norms and Values

- **Definition**: Cultural norms and values shape individuals' behaviors, beliefs, and attitudes toward health and medicine.
- **Application**: Understanding these norms helps tailor marketing messages to align with local values and expectations.
- **Example**: In some cultures, traditional medicine is highly valued, and integrating or acknowledging these practices in marketing strategies can enhance acceptance and trust.

1.2. Health Beliefs and Practices

- **Definition**: Different cultures have unique beliefs about health, illness, and treatment methods.
- **Application**: Marketers can design culturally sensitive and relevant messages to address these beliefs.
- **Example**: In cultures that strongly believe in holistic health, emphasizing a pharmaceutical product's natural or holistic aspects can be more effective.

2. Adapting Messaging and Communication

2.1. Language and Communication Styles

- **Definition**: Language differences and communication styles vary significantly across cultures.

- **Application**: Adapting language and tone to match local communication preferences ensures that messages are understood and well-received.

- **Example**: In some cultures, direct communication is preferred, while in others, a more indirect or nuanced approach is more effective.

2.2. Symbolism and Imagery

- **Definition**: Symbols and imagery have different meanings in different cultures.

- **Application**: Using culturally appropriate symbols and images helps avoid misunderstandings and ensures that marketing materials are respectful and engaging.

- **Example**: A symbol associated with health or wellness in one culture might have a completely different or even negative connotation in another.

3. Navigating Social Structures

3.1. Influencers and Decision-Makers

- **Definition**: Social structures determine who holds influence and makes decisions in healthcare settings.

- **Application**: Identifying and engaging with key influencers, such as community leaders or respectful healthcare professionals, can enhance campaign effectiveness.

- **Example**: In some regions, family and community elders play a significant role in healthcare decisions. Involving them in campaigns can build trust and facilitate acceptance.

3.2. Access to Healthcare and Treatment

- **Definition**: Access to healthcare and treatment can vary based on socioeconomic status, geographic location, and other factors.

- **Application**: Understanding these variations helps design campaigns that address barriers to access and promote equitable healthcare solutions.

- **Example**: In areas with limited access to healthcare facilities, campaigns might focus on mobile health solutions or partnerships with local clinics to improve reach.

4. Conducting Ethnographic and Field Research

4.1. Immersive Research

- **Definition**: Ethnographic research involves observing and interacting with people in their natural environments to gain insights into their behaviors and experiences.
- **Application**: Conducting ethnographic research in target markets provides a deep understanding of local health practices and preferences.
- **Example**: Observing how patients manage chronic conditions in different cultural settings can reveal insights into their needs and preferences, informing more effective product development and marketing strategies.

4.2. Participatory Approaches

- **Definition**: Participatory research involves engaging with local communities to co-create solutions and gather feedback.
- **Application**: Involving community members in developing marketing materials and strategies ensures they are culturally relevant and responsive to local needs.
- **Example**: Collaborating with local health advocates to co-create educational materials can enhance credibility and engagement.

Implications for Pharma Product Managers

1. Strategic Planning

- **Implication**: Incorporating anthropological insights into strategic planning ensures that global campaigns are culturally sensitive and aligned with local needs and preferences.

- **Action**: Product managers should collaborate with anthropologists and local experts to gather insights and adapt strategies accordingly.

2. Marketing Execution

- **Implication**: Effective execution of global campaigns requires understanding cultural nuances and local practices.
- **Action**: Product managers should oversee the localization of marketing materials, ensuring they are culturally appropriate and resonate with target audiences.

3. Building Local Partnerships

- **Implication**: Developing partnerships with local organizations and influencers can enhance the effectiveness of global campaigns.
- **Action**: Product managers should seek out and collaborate with local stakeholders to build trust and facilitate campaign success.

Case Studies and Real-World Examples

The following case studies illustrate how anthropological insights can be harnessed to design and execute effective global marketing campaigns in the pharmaceutical industry. By understanding and integrating cultural and societal contexts, pharmaceutical companies can enhance the relevance and impact of their campaigns and improve patient outcomes and engagement.

Case Study #9: A Pharma Major's Global Campaign for Diabetes Medication

Overview:

A global pharmaceutical company launched a global campaign for diabetes medication, leveraging anthropological insights to adapt its messaging and outreach strategies to various cultural contexts.

Key Elements:

1. Cultural Adaptation:

- **Challenge**: Diabetes management practices vary widely across cultures, influenced by dietary habits, lifestyle, and perceptions of the disease.

- **Insight**: In many cultures, traditional dietary practices impact diabetes management. For instance, in some Asian cultures, rice is a staple food, which can affect blood sugar levels.

- **Solution**: The company adapted its campaign by incorporating culturally relevant dietary advice and highlighting how its medication could fit into local diets.

2. Local Partnerships

- **Challenge**: Building trust and credibility in diverse markets.

- **Insight**: Local health influencers and community leaders are crucial in healthcare decisions in many cultures.

- **Solution**: The company collaborated with local healthcare professionals and community leaders to endorse their medication and provide culturally tailored educational materials.

Outcome

- **Impact**: The culturally adapted campaign resulted in higher engagement and acceptance of the medicaiton. Feedback indicated improved patient adherence and satisfaction due to the relevance of the campaign's messaging and educational content.

Case Study #10: BreatheWell Pharma Uses Anthropological Research to Design a Global Campaign!

Overview

BreatheWell Pharma utilized anthropological research to design a respiratory health campaign that addressed environmental and lifestyle factors affecting respiratory conditions in various regions.

Key Elements:

1. **Environmental Factors**
   - **Challenge**: Respiratory issues are influenced by local environmental factors such as air quality and climate.
   - **Insight**: Respiratory issues are more prevalent in regions with high levels of air pollution. Cultural practices, like indoor smoking or the use of traditional heating methods, can also exacerbate respiratory problems.
   - **Solution**: BreatheWell Pharma's campaign included specific advice on managing respiratory health in polluted environments and educational content about reducing exposure to environmental triggers.

2. **Lifestyle Integration**
   - **Challenge**: Different cultures have distinct lifestyle practices that impact respiratory health.
   - **Insight**: Cultural practices around physical activity, indoor vs. outdoor living, and traditional remedies can affect respiratory health.
   - **Solution**: The campaign integrated advice on incorporating respiratory health practices into daily routines and offered solutions compatible with the local lifestyle.

Outcome:

- **Impact**: The tailored campaign increased awareness of respiratory health issues and increased engagement with BreatheWell Pharma's products. The campaign effectively addressed local environmental and lifestyle factors, improving the relevance and impact of the messaging.

Case Study #11: A Leading Pharma Company Focusing on Cardiology Launches Heart Health Initiative

Overview:

A leading pharmaceutical company focusing on cardiology used anthropological insights to tailor their heart health initiative to different cultural contexts, focusing on dietary habits, social norms, and health beliefs.

Key Elements

1. Dietary Habits

- **Challenge**: Heart health is significantly influenced by diet, which varies across cultures.
- **Insight**: In Mediterranean cultures, diets high in olive oil are linked to better heart health, whereas in other regions, high consumption of processed foods can increase heart disease risk.
- **Solution**: The company adapted its campaign to emphasize dietary recommendations relevant to each culture and included local recipes and tips for heart-healthy eating.

2. Social Norms and Health Beliefs

- **Challenge**: Different cultures have varying attitudes towards heart disease and preventative measures.
- **Insights**: Some cultures may prioritize traditional remedies over conventional medical advice or have differing views on preventive health.
- **Solution**: The campaign incorporated local health beliefs and respected traditional practices, prompting evidence-based heart health strategies.

Outcome: The culturally nuanced campaign led to increased adoption of heart health practices and higher engagement with the company's products. The integration of local dietary and health beliefs enhanced the campaign's effectiveness.

Case Study #12: Immuno Biotech's Global Vaccination Campaign

Overview:

Immuno Biotech, a leading biopharmaceutical company, employed anthropological research to develop a global vaccination campaign that addressed cultural attitudes and access issues related to vaccination.

Key Elements

1. **Cultural Attitudes Towards Vaccination**

   - **Challenge**: Attitudes towards vaccination can vary significantly, with some cultures having skepticism about vaccines.

   - **Insight**: In some regions, mistrust of pharmaceutical companies or government health initiatives can impact vaccine uptake.

   - **Solution**: Immuno Biotech's campaign included educational efforts to build trust, featuring endorsements from local healthcare leaders and providing transparent information about vaccine safety and efficacy.

2. **Access and Barriers**

   - **Challenge**: Access to vaccination services can be limited in certain areas due to logistical, economic, or infrastructural barriers.

   - **Insight**: Understanding local challenges, such as transportation issues and economic constraints, was crucial for effective campaign design.

   - **Solution**: The campaign included strategies to improve vaccine access, such as mobile vaccination units and partnerships with local organizations to reach underserved communities.

Outcome:

- **Impact**: The campaign improved vaccine coverage rates and increased public trust in vaccination. Its culturally sensitive approach and focus on overcoming access barriers were key factors in its success.

Conclusion

Anthropological insights are crucial for designing and executing successful global pharmaceutical marketing campaigns. Pharmaceutical companies can create culturally sensitive, relevant, and impactful campaigns by understanding and integrating cultural, social, and behavioral contexts. Leveraging these insights helps ensure that marketing strategies resonate with diverse populations, leading to more effective engagement and improved health outcomes.

Insights from Behavioral Economics

In an increasingly competitive and complex pharmaceutical landscape, understanding the factors that drive consumer and healthcare professional behavior is critical for effective product management. Behavioral economics offers valuable tools and concepts to decipher these behaviors. This chapter delves into how principles of behavioral economics can be applied to pharmaceutical product management, providing product managers with a strategic advantage in designing marketing campaigns, optimizing patient engagement, and enhancing overall product strategy.

Behavioral Economics: Bridging Psychology and Economics

Behavioral economics challenges the traditional economic assumption that individuals always act rationally. Instead, it acknowledges that people often make decisions based on cognitive biases, emotional responses, and social influences. By incorporating these insights, behavioral economics provides a more nuanced understanding of decision-making processes.

Relevance to Pharma Product Management

Pharma product managers face the unique challenge of navigating a highly regulated industry while striving to meet the needs of diverse stakeholders, including healthcare professionals (HCPs), patients, and payers.

Behavioral economics can illuminate how these stakeholders make decisions and what factors influence their choices. This understanding can drive more effective marketing strategies, improve product adoption, and foster patient adherence.

Key Behavioral Economics Concepts

1. **Heuristics and Biases**:
   - Explore how mental shortcuts and biases, such as overconfidence or loss aversion, impact decision-making in healthcare settings. Understanding these biases can help product managers design strategies that align with how stakeholders naturally think and behave.

2. **Nudging and Choice Architecture**:
   - Learn about nudging, which subtly guides stakeholders toward desired behaviors without restricting their choices. In pharma, this can enhance patient adherence or encourage HCPs to consider new treatment options.

3. **Framing Effects**:
   - Investigate how the presentation of information can affect decision-making. Effective framing can highlight the benefits of a new drug or treatment, influencing HCPs' prescribing habits and patients' treatment choices.

4. **Social Proof and Norms**:
   - Understand how social proof and perceived norms influence behavior. Leveraging testimonials, case studies, and peer influence can drive product adoption and enhance credibility.

Applying Behavioral Economics in Pharma

Product managers can apply these behavioral economics principles in various ways:

- **Marketing Strategies**: Tailoring marketing messages to align with cognitive biases and decision-making processes can increase engagement and effectiveness.

- **Patient Adherence**: Designing interventions that account for common behavioral biases can improve adherence to treatment regimens.
- **HCP Engagement**: Crafting messages and materials considering HCP's decision-making heuristics can enhance product uptake and advocacy.

Conclusion

Integrating behavioral economics into pharmaceutical product management offers a powerful lens for understanding and influencing stakeholder behavior. By leveraging these insights, product managers can create more effective strategies, improve stakeholder interactions, and ultimatley drive better patient and industry outcomes. As we explore the principles and applications of behavioral economics in the following sections, you will gain practical knowledge on harnessing these insights for strategic advantage.

11.1 Behavioral Economics and Decision-Making in Healthcare

Behavioral economics integrates psychological insights into economic theory, offering a deeper understanding of how people make decisions and behave in the real world. These insights can be crucial for pharmaceutical product managers in shaping strategies that effectively influence healthcare professionals (HCPs), patients, and other stakeholders. Here is a detailed look at how behavioral economics can impact pharma product management:

1. Understanding Decision-Making Processes

1.1. Prospect Theory

- **Concept**: People value gains and losses differently, with losses typically having a greater psychological impact than equivalent gains.
- **Implication**: In pharma marketing, framing a drug as preventing a loss (e.g., avoiding serious health complications) can be more compelling than emphasizing potential gains.

1.2. Loss Aversion

- **Concept**: Individuals are more motivated to avoid losses than to achieve gains.
- **Implication**: Messaging that highlights the avoidance of negative outcomes (e.g., the consequences of untreated conditions) can be more effective in persuading HCPs and patients to adopt a new medication.

2. Influencing Patient Behavior

2.1. Nudging

- **Concept**: Small changes in how choices are presented can significantly influence decision-making without restricting options.
- **Implication**: Pharma companies can use nudges to encourage adherence to treatment plans, such as

reminders or easy-to-use apps that integrate with daily routines.

2.2. Default Options

- **Concept**: People are likelier to choose pre-set options due to inertia or the perceived norm.
- **Implication**: Setting default choices for drug regimens or treatment protocols in electronic health records can improve adherence rates.

3. Enhancing HCP Engagement

3.1. Incentives and Rewards

- **Concept**: The type and timing of rewards can influence behavior. Immediate rewards are often more effective than delayed ones.
- **Implication**: Providing timely and relevant incentives for HCPs, such as CME credits or access to cutting-edge research, can drive engagement with new products.

3.2. Social Proof

- **Concept**: People tend to follow the behavior of others, especially peers or influencers.
- **Implication**: Leveraging endorsements from respected HCPs or showcasing peer usage of a drug can increase the credibility and acceptance of new treatments.

4. Addressing Cognitive Biases

4.1. Anchoring

- **Concept**: People rely heavily on the first piece of information they receive (the anchor) when making decisions.
- **Implication**: Highlighting positive clinical trial results early in the marketing materials can set a favorable context for the product.

4.2. Overconfidence

- **Concept**: Individuals often overestimate their knowledge or ability to make correct decisions.

- **Implication**: Providing clear, evidence-based information and addressing common misconceptions can help counteract overconfidence and support informed decision-making.

5. Designing Effective Marketing Strategies

5.1. Behavioral Segmentation

- **Concept**: Segmenting the target audience based on behavioral traits rather than demographic factors.
- **Implication**: Tailoring marketing strategies to different behavioral segments, such as early adopters vs. late adopters, can enhance the effectiveness of campaigns.

5.2. Simplification

- **Concept**: Reducing complexity can improve decision-making and compliance.
- **Implication**: Simplifying the communication of drug benefits, risks, and usage instructions can help HCPs and patients make informed decisions.

11.2 Nudging Behavior: Practical Applications in Pharma

Nudging is subtly guiding individuals toward desired behaviors without restricting their freedom of choice. From behavioral economics, nudging leverages insights into human psychology to improve decision-making and behavior. In the pharmaceutical industry, nudging can enhance patient adherence, optimize healthcare practices, and improve overall outcomes by positively influencing behavior. Here is a detailed look at practical applications of nudging behavior in pharma:

1. Enhancing Medication Adherence

1.1. Automated Reminders

- **Concept**: Use automated systems to remind patients about their medicaiton schedules.

- **Application**: Text messages or app notifications that remind patients to take their medications at specific times can help maintain adherence.

- **Example**: Programs that send personalized reminders with specific instructions or motivational messages based on the patient's treatment plan.

1.2. Simplified Regimens

- **Concept**: Design medicaiton regimens that are easy to follow.

- **Application**: Combination pills or once-daily dosing can reduce the complexity of medicaiton schedules.

- **Example**: Single-pill combinations for chronic conditions like hypertension or diabetes reduce the number of pills patients need to manage.

1.3. Default Options

- **Concept**: Set beneficial options as defaults to encourage adherence.

- **Application**: Automatically enroll patients in refill programs unless they opt-out.
- **Example**: Pharmacies that automatically refill prescriptions and provide a "one-click" option for patients to continue with medications.

2. Improving Patient Engagement

2.1. Personalized Health Communications

- **Concept**: Tailor health information and communication strategies to individual preferences and needs.
- **Application**: Use patient data to send personalized messages that resonate with their specific health conditions and treatment goals.
- **Example**: Personalized emails or app notifications that provide tailored health tips, reminders, and motivational content based on patient behavior and preferences.

2.2. Visual and Emotional Appeals

- **Concept**: Use visuals and emotional appeals to motivate behavioral change.
- **Application**: Design educational materials that include impactful images and stories.
- **Example**: Infographics showing the positive effects of medication adherence or testimonials from other patients who have benefited from sticking to their treatment.

2.3. Social Proof

- **Concept**: Leverage the influence of peers and community.
- **Application**: Highlight testimonials or endorsements from other patients or healthcare providers.
- **Example**: Displaying reviews or success stories from patients who have benefited from a treatment on patient portals or educational materials.

3. Optimizing Healthcare Professional Practices

3.1. Decision Aids

- **Concept**: Provide tools that guide healthcare professionals toward best practices.

- **Application**: Integrate decision support systems that use nudging principles to recommend evidence-based treatment options.

- **Example**: Clinical decision support systems highlighting recommended treatment options based on current guidelines and patient data.

3.2. Behavioral Feedback

- **Concept**: Provide feedback on professional practices and outcomes.

- **Application**: Use dashboards or reports to show healthcare professionals how their practices compare to peers or benchmarks.

- **Example**: A system that tracks and displays the frequency of adherence to clinical guidelines, encouraging HCPs to align with best practices.

3.3. Incentivization

- **Concept**: Use incentives to promote desired behaviors.

- **Application**: Implement reward systems for healthcare professionals who consistently follow best practices or achieve certain outcomes.

- **Example**: Recognition programs or financial incentives for HCPs who meet adherence or patient satisfaction targets. However, such incentives must comply with the regulatory guidelines and industry practices.

4. Enhancing Marketing Strategies

4.1. Framing Effect

- **Concept**: Present information in a way that emphasizes positive outcomes or minimizes negative outcomes.

- **Application**: Frame marketing messages to highlight the benefits of treatment or the risks of not using it.

- **Example**: Marketing campaigns that frame medication adherence as a way to achieve better health and avoid complications, rather than just focusing on the medication itself.

4.2. Choice Architecture

- **Concept**: Design choices in a way that guides individuals toward better decisions.

- **Application**: Structure options to make beneficial choices more prominent or easier to select.

- **Example**: Creating educational materials or decision aids that help patients understand the benefits of different treatment options.

4.3. Behavioral Nudges in Advertising

- **Concept**: Use behavioral insights to design effective advertising strategies.

- **Application**: Apply principles like scarcity or urgency in advertisements to motivate action.

- **Example**: Ads that emphasize limited-time offers or urgent health messages to encourage quick responses from target audiences.

Practical Examples

1. Medication Adherence Programs

- **Overview**: Pharmaceutical companies have implemented adherence programs that use nudges and reminders to encourage patients to follow prescribed treatments.

- **Outcome**: Improved adherence rates and better health outcomes due to more consistent medication use.

2. HCP Engagement Campaigns

- **Overview**: Utilizing social proof by featuring testimonials and endorsements from leading experts in marketing campaigns.
- **Outcome**: Increased acceptance and uptake of new therapies among HCPs, driven by the influence of respected peers.

3. Simplified Treatment Protocols

- **Overview**: Simplifying dosing regimens and providing clear, easy-to-understand instructions for patients.
- **Outcome**: Enhanced patient compliance and improved treatment outcomes due to reduced complexity in following the prescribed regimens.

Conclusion

Nudging behavior offers a powerful approach to influencing decisions and improving outcomes in the pharmaceutical industry. By understanding and applying nudging principles, pharmaceutical product managers can design interventions that enhance patient adherence, optimize healthcare practices, and create effective marketing strategies. This approach aligns with behavioral insights to subtly guide individuals and professionals toward better choices, leading to improved health outcomes and more efficient use of healthcare resources.

11.3 Designing Marketing Strategies Based on Behavioral Insights

Behavioral insights are derived from understanding how people make decisions and how various psychological, social, and environmental factors influence their choices. By leveraging these insights, pharmaceutical companies can design marketing strategies that more effectively engage target audiences and drive desired behaviors. Here is a detailed guide on how to design marketing strategies based on behavioral insights:

1. Understanding Behavioral Insights

1.1. Key Concepts

- **Behavioral Biases**: Cognitive biases that affect decision-making, such as overconfidence, loss aversion, and anchoring.
- **Nudging**: Subtle changes in how choices are presented to encourage better decision-making.
- **Social Proof**: The tendency of individuals to look to others for guidance on how to behave.

1.2. Importance in Pharma

- **Patient Adherence**: Improve medication adherence by understanding patient behavior.
- **HCP Engagement**: Enhance communication strategies with healthcare professionals.
- **Market Positioning**: Enhance communication strategies with healthcare professionals.

2. Applying Behavioral Insights to Marketing Strategies

2.1. Segmenting and Targeting

2.1.1. Behavioral Segmentation

- **Concept**: Segment audiences based on behavioral patterns and decision-making processes.

- **Application**: Identify groups based on adherence patterns, health behaviors, or responses to previous campaigns.
- **Example**: Creating tailored messaging for high-risk patients who struggle with adherence versus those who are more compliant.

2.1.2. Personalization

- **Concept**: Customize marketing messages based on individual preferences and behaviors.
- **Application**: Use data to deliver personalized content that resonates with specific needs and interests.
- **Example**: Sending targeted emails about new research or treatments based on a healthcare professional's specialty.

2.2. Crafting Messages

2.2.1. Framing and Anchoring

- **Concept**: Present information to emphasize positive outcomes or minimize negative consequences.
- **Application**: Frame marketing messages to highlight benefits rather than risks.
- **Example**: Framing a new treatment as a way to promote quality of life rather than just reducing symptoms.

2.2.2. Use of Social Proof

- **Concept**: Incorporate testimonials, endorsements, or user reviews to influence behavior.
- **Application**: Highlight positive experiences from other patients or healthcare professionals.
- **Example**: Showcasing success stories from patients who have benefited from a new medication.

2.2.3. Emotional Appeals

- **Concept**: Tap into emotions to create a stronger connection with the audience.
- **Application**: Use stories, visuals, or narratives that evoke empathy or inspiration.

- **Example**: Creating campaigns featuring patient journeys and treatment's positive impact on their lives.

2.3. Designing Interventions

2.3.1. Choice Architecture

- **Concept**: Structure choices to make the desired option more attractive or accessible.

- **Application**: Simplify decision-making processes for patients and HCPs.

- **Example**: Designing easy-to-use apps for medication management or simplifying the process for requesting sample products.

2.3.2. Default Options

- **Concept**: Set beneficial options as defaults to increase their uptake.

- **Application**: Automatically enroll patients in refill programs or default settings for automatic renewals.

- **Example**: Providing automatic refills for medications unless the patient opts out.

2.3.3. Incentive and Rewards

- **Concept**: Use incentives to encourage desired behaviors.

- **Application**: Implement reward systems for actions like signing up for programs or adhering to treatment plans.

- **Example**: Offering discounts or additional services for patients with consistent medication adherence.

3. Implementing and Evaluating Strategies

3.1. A/B Testing

- **Concept**: Test different versions of marketing materials to see which performs better.

- **Application**: Use A/B testing to refine messages, visuals, or calls-to-action based on audience response.

- **Example**: Testing different email subject lines to determine which generates higher open rates.

3.2. Monitoring and Analytics

- **Concept**: Track performance metrics to evaluate the effectiveness of strategies.
- **Application**: Use analytics tools to measure engagement, conversion rates, and overall impact.
- **Example**: Analyzing data from social media campaigns to assess engagement levels and adjust strategies accordingly.

3.3. Feedback Loops

- **Concept**: Continuously gather and use feedback to improve marketing strategies.
- **Application**: Collect feedback from patients and HCPs to refine and enhance marketing efforts.
- **Example**: Surveys or focus groups to understand marketing messages' impact and identify improvement areas.

4. Examples

4.1. Medication Adherence Programs

- **Overview**: Programs that use behavioral insights to improve adherence rates.
- **Application**: Automated reminders, simplified regimens, and default options to encourage medication adherence.
- **Outcome**: Increased adherence rates and improved patient outcomes.

4.2. Personalized Health Campaigns

- **Overview**: Campaigns that use behavioral insights for targeted messages.
- **Application**: Personalized emails, tailored content, and emotional appeals based on patient behavior and preferences.

- **Outcome**: Higher engagement and better response rates.

4.3. Decision Support Tools for HCPs

- **Overview**: Tools that integrate behavioral insights to assist HCPs in decision-making.

- **Application**: Clinical decision support systems with nudges and recommendations based on best practices.

- **Outcome**: Enhanced decision-making and improved patient care.

Conclusion

Designing marketing strategies based on behavioral insights enables pharmaceutical companies to create more effective, engaging, and impactful campaigns. By understanding and applying principles from behavioral economics, companies can better meet the needs of patients and healthcare professionals, ultimately leading to improved outcomes and more successful marketing initiatives. Embracing these insights ensures that marketing efforts are aligned with how individuals make decisions and interact with healthcare information, resulting in more effective and efficient communication strategies.

Part IV. Scientific Foundation

In the dynamic field of pharmaceutical product management, a profound understanding of scientific principles is crucial for steering the successful development, marketing, and lifecycle management of pharmaceutical products. Integrating scientific knowledge into product management enhances decision-making and ensures products are developed and marketed effectively, aligning with scientific rigor and marketing needs.

The Role of Science in Pharma Product Management

Pharmaceutical product management requires a comprehensive grasp of various scientific disciplines, including pharmacology, clinical research, epidemiology, biostatistics, pharmacoeconomics, and more. This scientific foundation empowers product managers to make informed decisions throughout the product lifecycle, from initial development to market launch and beyond.

1. **Pharmacology**: Understanding drug mechanisms of action, therapeutic classes, and interactions is essential for product managers to position their products within the market effectively. Knowledge of pharmacokinetics and pharmacodynamics helps develop compelling product claims and address potential safety concerns.

2. **Clinical Research**: Insights into clinical trial design, methodology, and results are vital for assessing a product's efficacy and safety. Product managers use this information to support regulatory submissions, develop marketing strategies, and communicate with stakeholders.

3. **Epidemiology**: Analyzing disease patterns, prevalence, and incidence rates aids in identifying market opportunities and targeting interventions. Epidemiological data helps product managers understand the context in which their products will be used and tailor strategies accordingly.

4. **Biostatistics**: Proficiency in data analysis and interpretation is crucial for evaluating clinical trial outcomes and other

research data. Biostatistical knowledge enables product managers to make evidence-based decisions and present data effectively to stakeholders.

5. **Pharmacoeconomics**: Pharmacoeconomics plays a crucial role in the scientific foundation of pharmaceutical marketing by evaluating the cost-effectiveness and overall value of medications and healthcare interventions. It systematically assesses the costs and outcomes associated with pharmaceutical products, helping determine their economic impact on healthcare systems and patients. For pharma product managers, understanding pharmacoeconomics is essential as it provides insights into the economic benefits of their products compared to alternatives. This knowledge aids in making informed decisions about pricing strategies, market positioning, and value propositions. By incorporating pharmacoeconomic evaluations into their strategies, product managers better advocate for their products in a competitive market, justify investment in new treatments, and support payer negotiations with robust data on cost-effectiveness and patient outcomes.

Importance of Scientific Knowledge

Scientific knowledge provides the backbone for strategic decision-making and effective communication in pharmaceutical product management. It allows product managers to:

Develop Robust Product Strategies: By understanding the underlying science, product managers can create strategies that effectively leverage their products' unique strengths and address market needs.

Ensure Regulatory Compliance: Knowledge of scientific principles supports compliance and regulatory requirements and facilitates successful interactions with regulatory bodies.

Enhance Communication: A strong scientific foundation enables product managers to articulate complex concepts clearly and convincingly to healthcare professionals, patients, and other stakeholders.

Drive Innovation: Understanding scientific advancements and trends helps product managers identify opportunities for innovation and stay ahead of competitors in the market.

Conclusion

The scientific foundation in pharmaceutical product management is integral to the success of pharmaceutical products. By integrating scientific knowledge into every aspect of product management, from development to marketing, product managers can ensure that their products meet the best standards of efficacy, safety, and market relevance. As the pharmaceutical industry continues to evolve, a solid grasp of scientific principles will remain a key asset for navigating the complexities of the field and achieving long-term success.

Importance of Pharmacology

Pharmacology is the branch of medicine that studies drugs and their effects on the human body. Understanding pharmacology is crucial for pharmaceutical product managers because it directly impacts various drug development, marketing, and management aspects. Here is why pharmacology is so important:

1. Drug Mechanism and Efficacy

1.1. Mechanism of Action

- **Definition**: Pharmacology helps understand how a drug works at the molecular and cellular levels. This includes the interaction between the drug and its target receptors or enzymes.

- **Implication**: Knowledge of the mechanism of action aids in determining how effectively the drug treats a particular condition and predicting potential side effects. For product managers, this understanding is vital for accurately positioning and communicating the drug's benefits and risks.

1.2. Therapeutic Indications

- **Definition**: Pharmacology provides insights into the specific diseases and conditions a drug is intended to treat based on its mechanism of action.

- **Implication**: Understanding therapeutic indications helps define the target patient population and develop appropriate marketing strategies.

2. Drug Development and Testing

2.1. Preclinical Research:

- **Definition**: Pharmacology is involved in preclinical research to assess the safety and efficacy of drug candidates in animal models before human trials.

- **Implication**: This stage is crucial for identifying potential issues early and guiding modifications to improve drug safety and efficacy. Product managers use these findings to prepare for clinical trials and regulatory submissions.

2.2. Clinical Trials

- **Definition**: During clinical trials, pharmacology informs dose-ranging studies, safety monitoring, and efficacy evaluations.

- **Implication**: A solid understanding of pharmacology helps design clinical trials, analyze results, and interpret data to ensure that drugs are safe and effective for human use.

3. Pharmacokinetics and Pharmacodynamics

3.1. Pharmacokinetics (ADME)

- **Definition**: Pharmacokinetics studies how a drug is absorbed, distributed, metabolized, and excreted in the body.

- **Implication**: Knowledge of pharmacokinetics influences the dosing regimen and helps predict drug interactions. This information is essential for product managers to develop dosing guidelines and ensure safe and effective use.

3.2. Pharmacodynamics

- **Definition**: Pharmacodynamics examines a drug's effects on the body and how these effects relate to its concentration.

- **Implication**: Understanding pharmacodynamics is crucial for assessing the drug's efficacy and optimizing therapeutic dosing. This information helps communicate the drug's benefits to healthcare professionals and patients.

4. Safety and Side Effects

4.1. Adverse Effects

- **Definition**: Pharmacology helps identify potential adverse effects and toxicity associated with a drug.

- **Implication**: Knowing the possible side effects allows product managers to develop risk management strategies and provide comprehensive information to healthcare providers and patients. It also helps in preparing for regulatory discussions and labeling requirements.

4.2. Drug Interactions

- **Definition**: Pharmacology studies how drugs, food, and other substances interact.

- **Implication**: Understanding drug interactions is crucial for developing safe prescribing guidelines and educating healthcare providers about potential interactions that could affect patient safety.

5. Marketing and Communication

5.1. Evidence-Based Marketing

- **Definition**: Pharmacology provides the scientific basis for marketing claims and promotional materials.

- **Implication**: Product managers use pharmacological knowledge to create accurate and compelling marketing messages, ensuring they are grounded in scientific evidence.

5.2. Healthcare Provider Education

- **Definition**: Pharmacology informs the development of healthcare providers' educational materials and training programs.

- **Implication**: Effective education helps providers understand the drug's benefits, mechanism of action, and proper use, leading to better patient outcomes and enhanced product adoption.

6. Strategic Planning and Decision-Making

6.1. Market Positioning

- **Definition**: Pharmacology aids in positioning a drug within the market, highlighting its unique benefits, and comparing it with existing treatments.

- **Implication**: Understanding pharmacological profiles helps develop strategies to differentiate drugs from competitors and address unmet medical needs.

6.2. Risk Management

- **Definition**: Pharmacology helps assess and manage risks associated with the drug's use.

- **Implication**: Effective risk management strategies are essential for ensuring the drug's safety profile and gaining regulatory approval.

Conclusion

Pharmacology is fundamental to pharmaceutical product management, providing critical insights into drug action, safety, efficacy, and interactions. This knowledge supports various aspects of drug development, regulatory compliance, marketing, and strategic planning. For product managers, a thorough understanding of pharmacology ensures informed decision-making, effective communication, and successful product management throughout the drug's lifecycle.

12.1 Overview of Pharmacology and Its Relevance to Pharma

Pharmacology is a branch of medicine that studies drugs and their effects on biological systems. It encompasses the understanding of how drugs interact with the body and how the body affects drugs. This field is crucial for developing, assessing, and managing pharmaceuticals and their therapeutic use. Here's an overview of pharmacology and its relevance to the pharmaceutical industry:

1. Core Concepts in Pharmacology

1.1. Pharmacokinetics (ADME)

- **Absorption**: How a drug enters the bloodstream after administration.

- **Distribution**: How the drug spreads through the body's tissues.

- **Metabolism**: How the drug is broken down, typically in the liver.

- **Excretion**: How the drug is eliminated from the body, usually through urine or feces.

1.2. Pharmacodynamics

- **Definition**: The study of the effects of drugs on the body, including the mechanism of action at the molecular, cellular, and systemic levels.

- **Receptors and Enzymes**: Understanding how drugs bind to specific receptors or inhibit enzymes to produce therapeutic or side effects.

1.3. Drug Classification and Mechanism of Action

- **Drug Classes**: Categorizing drugs based on their chemical structure, mechanism of action, or therapeutic use.

- **Mechanisms**: How drugs interact with specific biological targets to alter physiological processes.

2. Relevance to Pharmaceutical Development

2.1. Drug Discovery and Development:

- **Target Identification**: Pharmacology helps identify biological targets (e.g., receptors, enzymes) that drugs can modulate to treat diseases.

- **Preclinical Testing**: Understanding pharmacology is essential for designing preclinical studies to evaluate drug safety and efficacy in animal models before human trials.

2.2. Clinical Trials

- **Dose Optimization**: Pharmacokinetic and pharmacodynamic data guide the selection of dose and regimen design for clinical trials.

- **Safety and Efficacy**: Monitoring adverse effects and therapeutic responses to ensure the drug's safety and effectiveness.

2.3. Regulatory Approval

- **Data Submission**: Comprehensive pharmacological data is required for regulatory submissions to agencies like the FDA or EMA.

- **Labeling and Guidelines**: Pharmacology informs the development of drug labeling, including indications, dosage recommendations, and safety warnings.

3. Relevance to Pharmaceutical Marketing

3.1. Product Positioning and Differentiation

- **Mechanism of Action**: Knowing how a drug works helps position it effectively against competitors and highlights its unique benefits.

- **Therapeutic Advantages**: Understanding pharmacological properties enables marketing teams to communicate the drug's advantages and address specific medical needs.

3.2. Healthcare Provider Education

- **Educational Materials**: Developing accurate and informative materials for healthcare providers based on pharmacological data.

- **Training Programs**: Ensuring providers understand the drug's pharmacology, including its proper use, potential interactions, and side effects.

4. Impact on Patient Outcomes

4.1. Personalized Medicine

- **Tailored Therapies**: Pharmacology supports the development of personalized treatment approaches based on individual patient characteristics and genetic profiles.

- **Optimized Dosing**: Understanding pharmacokinetics helps adjust doses to achieve the desired therapeutic effect while minimizing adverse effects.

4.2. Safety Monitoring

- **Adverse Drug Reactions**: Ongoing pharmacological research and surveillance are crucial for identifying and managing adverse drug reactions.

- **Risk Management**: Implementing strategies to mitigate risks associated with drug therapy, including monitoring for potential interactions and long-term effects.

5. Strategic Planning and Decision-Making

5.1. Market Access and Pricing

- **Value Proposition**: Pharmacological insights help define the drug's value proposition, support pricing strategies, and negotiate reimbursement.

- **Health Economics:** Understanding pharmacokinetics and cost-effectiveness based on pharmacological data aids in justifying the drug's value to payers and stakeholders.

5.2. Product Lifecycle Management

- **Lifecycle Strategies**: Pharmacology informs strategies for managing a product throughout its lifecycle, including label extensions, line extensions, and lifecycle optimization.

Conclusion

Pharmacology is a foundational discipline in the pharmaceutical industry, influencing every stage of drug development, from discovery to commercialization. Its relevance extends to optimizing drug safety, efficacy, and therapeutic benefits and supporting effective marketing, regulatory compliance, and patient outcomes. For pharmaceutical product managers, a deep understanding of pharmacology is essential for making informed decisions, guiding product strategies, and ensuring the successful management of pharmaceutical products in a competitive market.

12.2 Drug Mechanisms of Action and Therapeutic Categories

1. Drug Mechanisms of Action

Drug mechanisms of action refer to how drugs produce their effects in the body. Understanding these mechanisms is crucial for developing effective therapies and managing their use. Here are some common mechanisms of action:

1.1. Receptor Interactions

- **Agonists**: Drugs that activate receptors to elicit a biological response. For example, beta-agonists like albuterol stimulate beta-adrenergic receptors in the lungs to relax bronchial muscles and improve airflow.

- **Antagonists**: Drugs that block receptors and prevent the activation by endogenous substances or other drugs. For example, beta-blockers like atenolol inhibit beta-adrenergic receptors, reducing heart rate and blood pressure.

- **Partial Agonists**: Drugs that activate receptors but produce a less than maximal response than full agonists. For example, pindolol acts as a partial agonist at beta-adrenergic receptors.

1.2. Enzyme Inhibition

- **Competitive Inhibition**: Drugs compete with the natural substrate for the active site of an enzyme. For example, statins inhibit HMG-CoA reductase, an enzyme in cholesterol synthesis.

- **Non-Competitive Inhibition**: Drugs bind to an allosteric site on the enzyme, altering its activity without competing with the substrate. For example, aspirin inhibits cyclooxygenase (COX) enzymes, reducing the production of prostaglandins involved in inflammation and pain.

1.3. Ion Channel Modulation

- **Channel Blockers**: Drugs that block ion channels affect ions' flow across cell membranes. For example, calcium channel blockers like amlodipine prevent calcium from entering cells, reducing blood pressure.

- **Channel Openers**: Drugs that increase the activity of ion channels. For example, nitroglycerin dilates blood vessels by increasing nitric oxide levels, which opens potassium channels and reduces vascular tone.

1.4. Transporter Modulation

- **Reuptake Inhibitors** are drugs that block the reuptake of neurotransmitters, increasing their availability in the synaptic cleft. For example, selective serotonin reuptake inhibitors (SSRIs) like fluoxetine increase serotonin levels to treat depression.

- **Inhibitors of Transport Proteins**: Drugs that inhibit the transport of molecules across membranes. For example, proton pump inhibitors (PPIs) like omeprazole block the H+/K+ATPase in the stomach lining, reducing gastric acid secretion.

1.5. Gene Expression Modulation

- **Transcriptional Regulation**: Drugs that affect the expression of specific genes. For example, corticosteroids act on nuclear receptors to regulate gene expression in inflammation and immune responses.

2. Therapeutic Categories

Drugs are often categorized based on their therapeutic use, which helps in understanding their applications and effects. Here are some common therapeutic categories:

2.1. Analgesics and Antipyretics

- **Purpose**: Relieving pain (analgesics) and reducing fever (antipyretics).

- **Examples**: Acetaminophen, Ibuprofen, aspirin.

2.2. Antibiotics and Antimicrobials

- **Purpose**: Treat infections caused by bacteria, viruses, fungi, or parasites.
- **Examples**: Penicillin (antibiotic), acyclovir (antiviral), and fluconazole (antifungal).

2.3. Antihypertensives

- **Purpose**: Used to manage high blood pressure.
- **Examples**: ACE Inhibitors (e.g., lisinopril), beta-blockers (e.g., metoprolol), and calcium channel blockers (e.g., amlodipine).

2.4. Cardiovascular Agents

- **Purpose**: Used to manage heart conditions and improve cardiovascular health.
- **Examples**: Statins (e.g., atorvastatin), diuretics (e.g., furosemide), and anticoagulants (e.g., warfarin).

2.5. Antidiabetics

- **Purpose**: Used to manage diabetes and control blood glucose levels.
- **Examples**: Insulin, metformin, sulfonylureas (e.g., glipizide).

2.6. Respiratory Agents

- **Purpose**: Used to manage respiratory conditions such as asthma and chronic obstructive pulmonary disease (COPD).
- **Examples**: Bronchodilators (e.g., albuterol), corticosteroids (e.g., fluticasone), leukotriene receptor antagonists (e.g., montelukast).

2.7. Gastrointestinal Agents

- **Purpose**: Used to manage gastrointestinal conditions such as acid reflux, ulcers, and constipation.
- **Examples**: Proton pump inhibitors (e.g., omeprazole), antacids (e.g., magnesium hydroxide), laxatives (e.g., bisacodyl).

2.8. Psychiatric Medications

- **Purpose**: Used to treat mental health conditions such as depression, anxiety, and schizophrenia.
- **Examples**: Antidepressants (SSRIs like sertraline), antipsychotics (e.g., risperidone), and anxiolytics (e.g., diazepam).

2.9. Oncology Drugs

- **Purpose**: Used to treat various types of cancer.
- **Examples**: Chemotherapy agents (e.g., cisplatin), targeted therapies (e.g., Imatinib), and immunotherapies (e.g., pembrolizumab).

Conclusion

Understanding drug mechanisms of action and therapeutic categories is fundamental for pharmaceutical product management. It provides insights into how drugs work, their therapeutic uses, and how they can be effectively positioned and managed in the market. This knowledge is crucial for developing strategies, ensuring effective communication with healthcare providers, and optimizing patient outcomes.

12.3 The Product Manager's Role in Communicating Pharmacology

Pharmaceutical product managers are crucial in bridging the gap between complex pharmacological concepts and various stakeholders, including healthcare professionals (HCPs), regulatory authorities, patients, and internal teams. Effective communication of pharmacological concepts is essential for ensuring the successful development, marketing, and adoption of pharmaceutical products.

1. Understanding Pharmacological Concepts

1.1. Deep Knowledge:

- Product managers must thoroughly understand pharmacology to effectively communicate their products' mechanisms of action, therapeutic uses, side effects, and interactions.

1.2. Translation of Science:

- They must translate complex pharmacological data into understandable language that resonates with different audiences, from scientific experts to laypersons.

2. Communicating with Healthcare Professionals (HCPs)

2.1. Educational Materials:

- Product managers create and provide detailed educational materials such as product monographs, clinical trial data summaries, and scientific publications. These materials help HCPs understand the product's pharmacological profile and clinical relevance.

2.2. Training and Workshops:

- They organize and lead training sessions and workshops for HCPs, explaining the product's pharmacological basis, benefits, and place in therapy. These sessions often involve presenting data from clinical trials and real-world evidence.

2.3. Scientific Meetings and Conferences:

- Product managers represent their companies at scientific meetings and conferences, discussing the pharmacological aspects of their products and engaging with the scientific community to address questions and concerns.

3. Regulatory Communication

3.1. Regulatory Submissions:

- They prepare and review regulatory documents that include detailed pharmacological information required for drug approval. This includes explaining the rationale behind the drug's development and its expected impact on disease management.

3.2. Regulatory Interactions:

- Product managers must clearly articulate the pharmacological data and its implications for drug safety and efficacy during interactions with regulatory authorities.

4. Patient Communication

4.1. Patient Education:

- Product managers oversee the development of patient education materials that explain the drug's pharmacological aspects in layperson's terms. These may include brochures, websites, and patient-focused videos that simplify the drug's mechanism of action and benefits.

4.2. Support Programs:

- They might also develop patient support programs that provide additional information and resources to help patients understand how the drug works and manage their treatment.

5. Internal Team Communication

5.1. Cross-Functional Collaboration:

- Product managers communicate pharmacological concepts to internal teams such as marketing, sales, and R&D. They ensure all team members understand the product's scientific background to align their strategies and messaging.

5.2. Training Internal Teams:

- They conduct training sessions for sales and marketing teams to equip them with the knowledge needed to effectively promote the product and address questions from HCPs and patients.

6. Marketing and Promotional Materials

6.1. Developing Content:

- Product managers oversee creating marketing content that accurately represents the drug's pharmacological attributes. This includes brochures, advertisements, and digital content.

6.2. Ensuring Accuracy:

- They ensure that all promotional materials comply with regulatory requirements and accurately reflect the pharmacological data to avoid misleading information.

Implications

1. **Enhanced Product Understanding**: Clear communication of pharmacological concepts ensures that all stakeholders understand the product properly, which can enhance its adoption and effectiveness.

2. **Regulatory Compliance**: Accurate and transparent communication helps meet regulatory requirements and facilitates a smoother approval process.

3. **Informed Decision-Making**: Product managers provide comprehensive pharmacological information to enable HCPs to make informed treatment decisions, leading to better patient outcomes.

4. **Effective Marketing**: Well-communicated pharmacological concepts underpin effective marketing strategies, helping to differentiate the product in a competitive market.

5. **Patient Adherence**: Clear and accessible information about how a drug works can improve patient adherence and satisfaction with The treatment.

Conclusion

Pharmaceutical product managers are instrumental in communicating pharmacological concepts across various channels. Their role involves translating complex scientific data into actionable insights, ensuring all stakeholders have the necessary information to make informed decisions. By effectively managing this communication, product managers contribute to the successful development, approval, and market uptake of pharmaceutical products.

Relevance of Epidemiology, Biostatistics, and Pharmacoeconomics

In the dynamic landscape of pharmaceutical marketing and product management, understanding the interplay between scientific disciplines and their applications is essential for effective decision-making and strategy formulation. Among these disciplines, epidemiology, biostatistics, and pharmacoeconomics are critical areas that provide invaluable insights into drug development, market positioning, and patient outcomes.

Epidemiology, the study of disease patterns and determinants in populations, offers a comprehensive understanding of how diseases spread, the impact of various interventions, and the identification of health trends. For pharmaceutical product managers, epidemiology provides foundational knowledge that supports the development of targeted therapies and public health strategies. Product managers can make informed decisions about market opportunities, therapeutic needs, and potential impact by analyzing disease prevalence, incidence, and risk factors.

Biostatistics is pivotal in interpreting complex data from clinical trials, observational studies, and other research endeavors. It involves applying statistical methods to analyze data, draw conclusions, and make predictions. For pharmaceutical professionals, biostatistics is crucial for designing robust clinical trials, analyzing efficacy and safety data, and understanding the statistical significance of research findings. Mastery of biostatistical techniques ensures that product managers can accurately assess drug performance and support evidence-based marketing strategies.

Pharmacoeconomics evaluates the cost-effectiveness of pharmaceutical interventions, considering both the economic impact and the value of therapeutic outcomes. This discipline helps pharmaceutical companies assess the financial implications of new treatments and make strategic decisions regarding pricing, reimbursement, and market access. By understanding pharmacoeconomics, product managers can better position their products in the market, engage with payers and health authorities, and ensure that their strategies align with economic and clinical value.

In this chapter, we will explore the relevance of these three disciplines in the context of pharmaceutical product management. We will examine how epidemiological data shapes drug development and marketing strategy, how biostatistical methods underpin the analysis of clinical research, and how pharmacoeconomic evaluations influence product positioning and reimbursement. By integrating these scientific principles into their practice, pharmaceutical product managers can enhance their strategic decision-making, optimize product development processes, and ultimatley improve patient outcomes.

This chapter aims to thoroughly understand epidemiology, biostatistics, and pharmacoeconomics' contributions to pharmaceutical marketing and management. It offers practical insights and real-world examples to illustrate their application in the industry.

13.1 Epidemiology in Pharma: Understanding Disease Patterns

Epidemiology in Pharma: Understanding Disease Patterns

Epidemiology is the study of how diseases affect populations and the factors that influence their spread and impact. Understanding disease patterns through epidemiology is crucial for developing, marketing, and managing treatments effectively in the pharmaceutical industry. This knowledge helps pharmaceutical companies make informed decisions about drug development, market strategies, and public health interventions.

1. Definition and Scope

1.1. Definition:

- Epidemiology focuses on disease distribution, determinants, and frequency in populations. It seeks to identify patterns and causes of diseases and to apply this knowledge to control and prevent health problems.

1.2. Scope in Pharma:

- In the pharmaceutical sector, epidemiology helps identify unmet medical needs, design clinical trials, and assess the potential impact of new treatments on public health.

2. Understanding Disease Patterns

2.1. Disease Prevalence and Incidence:

- **Prevalence** refers to the total number of cases of a disease in a population at a given time. It helps understand the widespread condition and estimate the potential market size for new treatments.

- **Incidence** refers to the number of new cases of a disease in a population over a specific period. Tracking incidence helps understand trends and the emergence of new health issues.

2.2. Epidemiological Studies:

- **Descriptive Studies** describe the distribution of diseases by time, place, and person. These studies help identify patterns and trends in disease occurrence.

- **Analytical Studies** investigate the causes and risk factors associated with diseases. They include case-control and cross-sectional studies, providing insights into disease etiology and potential interventions.

- **Interventional Studies** assess the effectiveness of treatments or preventive measures. Randomized controlled trials (RCTs) are key for evaluating new drugs.

2.3. Risk Factors and Determinants:

- **Genetic Factors**: Genetic predisposition that increases susceptibility to certain diseases.

- **Environmental Factors**: Exposure to environmental elements such as pollution, toxins, or lifestyle factors (e.g., diet and exercise).

- **Socioeconomic Factors**: Influences of socioeconomic status, education, and access to healthcare on disease prevalence and outcomes.

3. Applications in Drug Development

3.1. Identifying Unmet Needs:

- Epidemiological data helps identify diseases with high prevalence or incidence and significant unmet medical needs. This information guides pharmaceutical companies in prioritizing drug development efforts.

3.2. Clinical Trial Design:

- **Target Population**: Understanding disease patterns helps define the target population for clinical trials, ensuring that the study participants are representative of the patient population that will use the drug.

- **Endpoints and Outcomes**: Epidemiological data informs the selection of relevant endpoints and outcomes for clinical trials, reflecting real-world disease impact.

3.3. Market Assessment:

- **Market Size Estimation**: Epidemiological data estimates potential market size by assessing the number of patients with a particular condition.
- **Geographic and Demographic Trends**: Identifies regions or populations with a higher disease burden, helping to plan targeted marketing strategically.

4. Public Health Impact

4.1. Disease Burden Assessment:

- Evaluating the burden of diseases on populations helps to understand the public health impact and the potential benefits of new treatments.

4.2. Policy and Guidelines:

- Epidemiological data informs healthcare policies and guidelines, influencing disease management and treatment protocol recommendations.

4.3. Preventive Measures:

- Identifying risk factors and patterns helps design preventive measures and public health campaigns to reduce disease incidence.

Role of Pharmaceutical Product Managers

1. **Strategic Planning**: Product managers use epidemiological data to inform strategic decisions, such as drug development priorities, target markets, and marketing strategies.
2. **Collaboration with Researchers**: They collaborate with epidemiologists and researchers to ensure that clinical trials are designed based on accurate and relevant disease data.

3. **Communication and Education**: Product managers communicate epidemiological findings to internal teams, stakeholders, and external partners, ensuring that all parties have a clear understanding of disease patterns and their implications.

Conclusion

Understanding disease patterns through epidemiology is fundamental to pharmaceutical product managers. It provides essential insights into the prevalence and incidence of disease, identifies unmet medical needs, informs clinical trial design, and guides strategic decisions. By leveraging epidemiological data, product managers can develop effective treatments, assess market opportunities, and contribute to improving public health outcomes.

13.2 Applying Epidemiological Insights to Marketing Strategies

Epidemiology, the study of how diseases affect populations, provides valuable insights for pharmaceutical marketing strategies. By applying epidemiological insights, product managers can better tailor their marketing efforts to address public health needs, target appropriate patient populations, and optimize resource allocation. Here's a detailed overview of how to integrate epidemiological insights into pharmaceutical marketing strategies:

1. Understanding Disease Patterns and Prevalence

1.1. Disease Epidemiology

- **Prevalence and Incidence Rates**: Use epidemiological data on disease prevalence (total cases) and incidence (new cases) to identify target markets and assess a drug's potential market size.

- **Disease Burden**: Evaluate the disease burden, including disability-adjusted life years (DALYs) and quality-adjusted life years (QALYs), to understand the impact of the disease on patients and the healthcare system.

1.2. Geographic and Demographic Trends:

- **Regional Variations**: Analyze regional differences in disease prevalence to target marketing efforts in high-burden areas. Tailor messaging to address specific regional health issues.

- **Population Segmentation**: Use demographic data (age, gender, socioeconomic status) to segment the market and design targeted campaigns that resonate with different patient populations.

2. Identifying Target Patient Populations

2.1. High-Risk Groups:

- **Risk Factors**: Identify high-risk populations based on epidemiological studies. Develop targeted marketing strategies for these groups to promote disease awareness and preventative measures.

- **Screening and Prevention:** Leverage data on screening practices and preventative measures to position the drug as part of a comprehensive disease management plan.

2.2. Patient Journey Mapping:

- **Disease Progression**: Map the patient's journey from diagnosis to treatment, considering how the disease progresses and the impact of different stages. Develop marketing messages that address patient needs at each stage.

- **Treatment Gaps**: Identify gaps in current treatment options and highlight how your product fills these gaps, improving patient outcomes.

3. Informing Product Development and Positioning

3.1. Unmet Medical Needs:

- **Gap Analysis**: Use epidemiological evidence to identify unmet medical needs and develop products that address these needs. Position your product as a solution to specific challenges faced by patients.

- **Evidence-Based Positioning**: Use epidemiological evidence to support claims about the drug's efficacy and safety, enhancing its credibility and appeal.

3.2. Market Access and Reimbursement

- **Cost-Effective Analysis**: Perform cost-effective analyses using epidemiological data to demonstrate the drug's economic value. Support reimbursement and access discussions with evidence of improved patient outcomes and reduced healthcare costs.

- **Health Economics**: Use data on disease burden and healthcare utilization to justify the drug's value proposition to payers and health authorities.

4. Designing Effective Awareness Campaigns

4.1. Educational Initiatives:

- **Public Health Campaigns**: Develop public health campaigns based on epidemiological insights to raise awareness about the disease and the availability of new treatments.

- **Healthcare Professional (HCP) Education**: Create educational materials for HCPs based on epidemiological data to enhance their understanding of disease prevalence, risk factors, and treatment options.

4.2. Patient Engagement:

- **Disease Education**: Provide patients with information about disease prevalence and risk factors to encourage early diagnosis and treatment. Use data to create relatable and compelling content.

- **Support Programs**: Develop programs that address patients' specific challenges, such as managing chronic conditions or accessing treatment.

5. Monitoring and Evaluating Marketing Effectiveness

5.1. Tracking Disease Trends:

- **Post-Marketing Surveillance**: Continuously monitor epidemiological data to assess changes in disease patterns and adjust marketing strategies accordingly.

- **Effectiveness Evaluation**: Evaluate the effectiveness of marketing campaigns by analyzing changes in disease awareness, diagnosis rates, and treatment uptake.

5.2. Feedback Loops:

- **Patient and HCP Feedback**: Collect feedback from patients and HCPs to understand the impact of marketing efforts and identify areas for improvement.

- **Data Integration**: Integrate epidemiological data with marketing analytics to refine strategies and enhance effectiveness.

Conclusion

Applying epidemiological insights to marketing strategies allows pharmaceutical companies to develop more targeted and effective campaigns. By understanding disease patterns, identifying target populations, informing product development, and designing impactful awareness programs, product managers can enhance their marketing efforts, improve patient outcomes, and achieve better market positioning. Integrating epidemiological data into marketing strategies ensures that resources are used efficiently and that the the drug addresses real-world health needs.

13.3 Biostatistics for Product Managers: Data Analysis and Interpretation

Biostatistics is a critical tool in the pharmaceutical industry. It is essential for analyzing and interpreting data from clinical trials, observational studies, and market research. Understanding biostatistics enables pharmaceutical product managers to make informed decisions based on data-driven insights, ensuring product development, marketing strategies, and regulatory submissions are grounded in solid evidence.

1. Overview of Biostatistics

1.1. Definition:

- Biostatistics applies statistical methods to biological, medical, and health-related research. It involves the design of studies, data collection and analysis, and results interpretation.

1.2. Relevance to Pharma Product Managers:

- Product managers must interpret statistical findings to understand drug efficacy, safety, market potential, and patient demographics, influencing product development and marketing strategies.

2. Core Statistical Concepts

2.1. Descriptive Statistics:

- **Measures of Central Tendency**: Mean, median, and mode summarize the central location of data.

- **Measures of Variability**: Standard deviation, variance, and range describe the dispersion or spread of data points.

- **Frequency Distributions**: Present how often each value occurs in the dataset.

2.2. Inferential Statistics:

- **Hypothesis Testing**: Determines if the observed effects or differences are statistically significant. Common tests include t-tests, chi-square tests, and ANOVA (Analysis of Variance).

- **Confidence Intervals (CIs)**: Provide a range of values within which the true effect or parameter is expected to lie, with a certain level of confidence (e.g., 95% CI).

2.3. Regression Analysis:

- **Linear Regression**: Examines the relationship between a dependent variable and one or more independent variables. Used to predict outcomes and assess the strength of associations.

- **Logistic Regression**: Used for binary outcomes (e.g., success/failure). Helps in understanding the odds of an event occurring based on predictor variables.

2.4. Survival Analysis:

- **Kaplan-Meier Estimator**: Estimates the survival function from lifetime data, commonly used in clinical trials to analyze time-to-event data.

- **Cox Proportional-Hazards Model**: Assesses the effect of several variables on the time a specified event takes to happen.

2.5. Bayesian Statistics:

- Bayesian Inference uses prior distributions and new data to update a hypothesis's probability. It provides a flexible framework for incorporating prior knowledge and adjusting estimates based on new information.

3. Applications in Drug Development

3.1. Clinical Trials:

- **Study Design**: Biostatistics helps design trials, including sample size calculations, randomization, and selection of appropriate endpoints.

- **Data Analysis**: Involves analyzing trial data to determine drug efficacy, safety, and potential side effects.

- **Interim Analysis**: Evaluates data at interim points to make decisions about continuing or modifying the trial.

3.2. Regulatory Submissions:

- **Statistical Evidence**: Ensures that statistical methods and results meet regulatory standards for drug approval. This includes presenting data in regulatory submissions and addressing statistical questions from agencies like the FDA or EMA.

3.3. Market Research:

- **Segmentation Analysis**: Identifies and analyzes different patient segments to tailor marketing strategies.

- **Predictive Modeling**: Forecasts market trends and drug uptake based on statistical models.

3.4. Post-Market Surveillance:

- **Safety Monitoring**: Analyzes adverse event data to ensure ongoing safety and effectiveness of the drug.

- **Long-Term Efficacy**: Assesses the long-term impact of the drug on patient outcomes.

4. Interpreting Biostatistical Data

4.1. Understanding Results:

- **Statistical Significance vs Clinical Relevance**: This distinction distinguishes between statistically significant and practical and important real-world results.

- **Effect Size**: Assesses the magnitude of the effect observed in the study, providing context beyond statistical significance.

4.2. Dealing with Data Limitations:

- **Bias and Confounding**: Identifies and mitigates potential sources of bias and confounding variables that could affect study results.

- **Missing Data**: Applies appropriate methods for handling missing data, such as imputation techniques or sensitivity analyses.

4.3. Communicating Findings:

- **Visualizations**: Uses charts, graphs, and tables to present data in an understandable and actionable format.

- **Stakeholder Communication**: Translates complex statistical findings into clear insights for stakeholders, including regulatory bodies, internal teams, and external partners.

Implications for Pharma Product Managers

1. **Decision Making**: Biostatistical knowledge enables product managers to make data-driven decisions about product development, marketing strategies, and market entry.

2. **Strategic Planning**: Helps in assessing the potential success of a product in the market by analyzing clinical and market data.

3. **Regulatory Compliance**: Ensures statistical analyses meet regulatory requirements and support successful drug approvals.

4. **Effective Communication**: Translating complex data into actionable insights for internal and external communication.

Conclusion

Biostatistics is a fundamental aspect of pharmaceutical product management, providing the tools and techniques necessary for analyzing and interpreting data across drug development and marketing processes. By understanding and applying biostatistical principles, product managers can ensure that their decisions are grounded in robust evidence, leading to more effective products and successful market outcomes.

13.4 Pharmacoeconomics: Economic Evaluations of Drug Therapies

Pharmacoeconomics is a branch of health economics that evaluates the cost-effectiveness, value, and economic impact of pharmaceutical products, treatments, and healthcare interventions. It provides decision-makers with evidence-based insights into allocating healthcare resources, pricing strategies, and reimbursement policies. Here is an overview of pharmacoeconomics and its key components:

Key Concepts in Pharmacoeconomics

1. Cost Analysis

- **Cost Identification**: Identifies and categorizes all relevant costs associated with pharmaceutical intervention, including direct medical costs (e.g., drug costs, healthcare utilization) and indirect costs (e.g., productivity loss, caregiver burden).

- **Cost Measurement**: Costs are quantified using standardized methods such as micro-costing (actual resource use) or gross-costing (aggregate costs), ensuring transparency and comparability across studies.

2. Outcome Measures

- **Clinical Outcomes**: Measures health outcomes resulting from pharmaceutical interventions, such as improvements in disease symptoms, quality-adjusted life years (QALYs), mortality rates, and hospitalizations avoided.

- **Patient-Reported Outcomes (PROs)**: This approach incorporates patient perceptions of treatment benefits and quality of life changes, enhancing the comprehensiveness of outcome assessments.

3. Types of Pharmacoeconomic Studies

- **Cost-Effectiveness Analysis (CEA)**: Compares the costs of alternative interventions with their clinical outcomes (e.g., cost per life saved, cost per QALY gained), helping prioritize healthcare investments based on efficiency criteria.

- **Cost-utility analysis (CUA)** evaluates interventions based on their impact on patient health-related quality of life. Outcomes are expressed using QALYs to facilitate comparisons across different diseases and interventions.

- **Cost-benefit analysis (CBA)** monetizes interventions' costs and benefits, determining whether benefits outweigh costs financially. It is often used in policy decision-making.

4. Decision-Analytical Modeling

- **Markov Models**: Simulates disease progression and treatment effects over time, integrating clinical data, costs, and outcomes to project long-term economic implications of healthcare interventions.

- **Budget Impact Analysis (BIA)**: Estimates financial consequences of adopting new therapies or technologies within healthcare budgets, forecasting expenditure changes and affordability.

Applications of Pharmacoeconomics

1. Healthcare Resource Allocation

- **Formulary Decisions** guide formulary committees and health insurers in selecting cost-effective therapies for coverage and reimbursement, balancing clinical benefits with budgetary constraints.

- **Priority Setting**: Supports healthcare policymakers and administrators in allocating resources to interventions that maximize population health outcomes and healthcare system efficiency.

2. Drug Pricing and Market Access

- **Pricing Strategy**: Informs pharmaceutical manufacturers on pricing strategies that reflect the value of therapies relative to their clinical benefits and economic impact.

- **Market Access**: Facilitates negotiations with payers and regulatory agencies by providing evidence of cost-effectiveness and value proposition, enhancing market acceptance and patient access.

3. **Health Technology Assessment (HTA):**

- **Economic Evaluations**: Contributes to HTA processes by generating evidence on the economic viability and societal value of new drugs and healthcare technologies, influencing coverage decisions and reimbursement policies.

- **Comparative Effectiveness Research (CER)**: Compares alternative treatments' relative benefits and costs to inform clinical practice guidelines and patient-centered care decisions.

Challenges and Future Directions

1. **Data Quality and Transparency**

- **Real-World Evidence (RWE)**: This approach incorporates diverse data sources (e.g., electronic health records, patient registries) and advanced analytics to enhance the robustness and generalizability of pharmacoeconomic evaluations.

- **Methodological Standards**: Advances in methodologies (e.g., value-based pricing, dynamic modeling) and guidelines (e.g., ISPOR Good Practices Task Force) promote consistency and reliability in pharmacoeconomic research.

2. **Globalization and Health Equity**

- **Access to Medicines**: Addresses disparities in healthcare access and affordability by evaluating interventions in diverse healthcare settings and patient populations, advocating for equitable distribution of healthcare resources.

- **Patient-Centered Outcomes Research (PCOR)**: Engage patients, caregivers, and communities in defining healthcare priorities and outcomes that matter most, fostering patient-centered decision-making and healthcare delivery.

Why Should Pharma Managers Understand Pharmacoeconomics Thoroughly

Pharmacoeconomics, the branch of health economics focusing on the cost and value of pharmaceuticals and drug therapies, is an essential knowledge area for pharmaceutical product managers with the tools to evaluate and communicate the economic impact of their products, which is critical in today's healthcare environment where cost-effectiveness is a significant consideration for stakeholders. Here are several reasons why this knowledge is relevant:

1. **Market Access and Reimbursement**
   - **Importance**: Pharmacoeconomic evaluations determine whether healthcare payers (insurance companies, government programs, etc.) will reimburse a drug. Understanding these evaluations helps product managers:
     - Develop strategies to demonstrate the economic value of their products.
     - Navigate the complex landscape of pricing and reimbursement negotiations.
     - Secure favorable pricing and reimbursement conditions, which are crucial for market access.
   - **Relevance**: Without a thorough understanding of pharmacoeconomics, product managers may struggle to effectively argue for their product's inclusion in formularies or justify its price point.

2. **Clinical and Economic Value Proposition**
   - **Importance**: Product managers must articulate their products' clinical and economic benefits. Pharmacoeconomics provides a framework for assessing:

- *Cost-effectiveness*: Comparing the costs and outcomes of a drug to alternatives.
- *Budget impact*: Estimating the financial implications of adopting a new therapy.
- **Relevance**: Demonstrating a product's economic value can differentiate it in a competitive market, making it more attractive to healthcare providers and payers.

3. Strategic Decision Making

- **Importance**: Understanding pharmacoeconomics adds to strategic decision-making throughout a product's lifecycle, from development to post-marketing. Product managers can:
  - Inform clinical trial design by incorporating economic endpoints.
  - Prioritize resource allocation based on cost-effective analysis.
  - Make data-driven decisions about product positioning and target markets.
- **Relevance**: Incorporating pharmacoeconomic insights ensures that strategic decisions align with market demands and economic realities, enhancing the product's success.

4. Health Technology Assessment (HTA) and Regulatory Approval

- **Importance**: Many countries require pharmacoeconomic data as part of Health Technology Assessments (HTAs) to approve new drugs. Understanding these requirements allows product managers to:
  - Prepare robust economic evidence for submission to HTA bodies.
  - Anticipate and address potential barriers to regulatory approval.

- **Relevance**: A lack of pharmacoeconomic understanding can result in delays or rejections during the regulatory process, hindering market entry and revenue generation.

5. Stakeholder Engagement and Communication

- **Importance**: Pharmacoeconomics provides a common language with diverse stakeholders, including:
  - *Payers*: Justifying the drug's cost and value.
  - *Healthcare Providers*: Explaining the economic benefits alongside clinical outcomes.
  - *Patients*: Highlighting cost savings and access benefits.
- **Relevance**: Effective communication with stakeholders using pharmacoeconomic evidence strengthens relationships and supports broader product adoption.

6. Competitive Advantage

- **Importance**: Pharmacoeconomics can be a differentiator in a crowded market. By demonstrating superior cost-effectiveness or budget impact compared to competitors, product managers can:
  - Enhance the product's competitive positioning.
  - Drive market uptake and loyalty among cost-conscious buyers.
- **Relevance**: Incorporating pharmacoeconomic insights into marketing and sales strategies can provide a significant competitive edge.

Conclusion

Pharmacoeconomics is an integral component of pharmaceutical product management. Product managers can make informed decisions, demonstrate value, secure reimbursement, and communicate effectively with stakeholders. In an industry where economic considerations are increasingly pivotal, a solid grasp of pharmacoeconomics is indispensable for ensuring the success and sustainability of pharmaceutical products.

Pharmaceutical Marketing Research

Pharmaceutical marketing research is a systematic process used to gather, analyze, and interpret data related to pharmaceutical products, market trends, and consumer behavior. It provides valuable insights that help pharmaceutical companies make informed decisions about product development, marketing strategies, and market positioning. Here is a detailed overview of pharmaceutical marketing research:

1. Purpose and Importance

1.1. Understanding Market Dynamics:

- **Market Assessment**: Research helps assess the current market landscape, including size, growth potential, and competitive environment. It provides a foundation for strategic planning and market entry.

- **Customer Insights**: Understanding patient needs, preferences, and behaviors allows companies to tailor their products and marketing strategies to meet market demands better.

1.2. Guiding Strategic Decisions:

- **Product Development**: Research insights inform product development decisions, including formulation, packaging, and features. This ensures that new products meet market needs and have a competitive edge.

- **Marketing Strategies**: Research helps design effective marketing campaigns by identifying key target audiences, preferred communication channels, and messaging strategies.

2. Types of Pharmaceutical Marketing Research

2.1. Exploratory Research:

- **Objective**: To gain a preliminary understanding of a market or issue. It is often used in the early stages of product development or market entry.
- **Methods**: Qualitative techniques include focus groups, expert interviews, and ethnographic studies.

2.2. Descriptive Research:

- **Objective**: To describe the characteristics of a market or target audience. It provides a snapshot of market conditions and consumer demographics.
- **Methods**: Quantitative techniques such as surveys, questionnaires, and observational studies.

2.3. Causal Research:

- **Objective**: It is used to determine cause-and-effect relationships between variables and understand marketing activities' impact on sales and market performance.
- **Methods**: Experimental designs and controlled trials.

2.4. Competitive Analysis:

- **Objective**: To assess the strengths and weaknesses of competitors and understand their strategies.
- **Methods**: SWOT (Strengths, Weaknesses, Opportunities, Threats) analysis, competitor profiling, and market share analysis.

3. Key Components of Pharmaceutical Marketing Research

3.1. Market Segmentation:

- **Objective**: To divide the market into distinct groups based on demographics, psychographics, behavior, or needs.

- **Benefits**: Allows targeted marketing efforts, personalized messaging, and efficient resource allocation.

3.2. Customer Insights:

- **Objective**: To understand patient and healthcare professional (HCP) needs, preferences, and behaviors.

- **Methods**: Surveys, interviews, and analysis of patient records, and feedback.

3.3. Product Positioning:

- **Objective**: To define how a product is perceived relative to competitors and how it should be positioned in the market.

- **Techniques**: Brand positioning studies, perceptual mapping, and value perception analysis.

3.4. Sales Forecasting:

- **Objective**: To predict future sales based on historical data, market trends, and planned marketing activities.

- **Methods**: Statistical modeling, trend analysis, and scenario planning.

3.5. Market Trends Analysis:

- **Objective**: To identify emerging trends and shifts in the pharmaceutical industry.

- **Sources**: Industry reports, scientific journals, and market research databases.

4. Methods and Tools

4.1. Quantitative Methods:

- **Surveys and Questionnaires**: Collect structured data from large samples to quantify market conditions and consumer preferences.

- **Statistical Analysis**: Use statistical techniques to analyze data and identify patterns and correlations.

4.2. Qualitative Methods:

- **Focus Groups**: Gather insights from small groups of participants to explore attitudes, perceptions, and motivations.

- **In-depth Interviews**: Conduct one-on-one interviews with key stakeholders to gain detailed insights.

4.3. Data Analytics:

- **Big Data Analysis**: This technique utilizes large datasets from various sources, such as electronic health records and social media, to identify trends and insights.

- **AI and Machine Learning**: Monitor competitor activities and market developments through secondary sources.

5. Applications in Pharmaceutical Marketing

5.1. Product Launch and Market Entry:

- **Strategy Development**: Use research to develop launch strategies, identify target markets, and position the product effectively.

- **Pre-Launch Testing**: Conduct pre-launch studies to gauge market readiness and refine marketing plans.

5.2. Ongoing Market Monitoring:

- **Performance Tracking**: Monitor sales performance, market share, and customer feedback to assess the effectiveness of marketing strategies.

- **Adjustments and Optimization**: Use research findings to adjust data-driven marketing tactics and strategy.

5.3. Regulatory Compliance:

- **Evidence Generation**: Provide evidence to support regulatory submissions and demonstrate the value and safety of the product.

6. Challenges and Considerations

6.1. Data Quality and Accuracy:

- **Challenge**: Ensuring the data collected is accurate, reliable, and representative of the target market.

- **Solution**: Implement rigorous data collection and analysis protocols.

6.2. Changing Market Dynamics:

- **Challenge**: Adapting to changing market conditions and emerging trends.
- **Solution**: Conduct regular research and stay updated with industry developments.

6.3. Ethical Considerations:

- **Challenge**: Ensuring that research practices comply with ethical standards and regulations.
- **Solutions**: Adhere to ethical guidelines and obtain necessary approvals for research activities.

Conclusion

Pharmaceutical marketing research is crucial for understanding market dynamics, guiding strategic decisions, and optimizing marketing efforts. By leveraging various research methods and tools, product managers can gain valuable insights into market conditions, customer preferences, and competitive landscapes. This knowledge enables them to make informed decisions, enhance product positioning, and ultimately achieve greater success in the pharmaceutical market.

14.1 The Role of Marketing Research in Pharma

Marketing research is a cornerstone of successful pharmaceutical marketing and strategy. It involves systematically gathering, analyzing, and interpreting data related to market dynamics, consumer behavior, and competitive landscapes. For pharmaceutical companies, marketing research is essential in guiding decision-making, optimizing strategies, and improving outcomes across various facets of the business. Here is a detailed description of its role.

1. Guiding Product Development and Innovation

1.1. Identifying Market Needs:

- **Objective**: Understand unmet medical needs, patient preferences, and healthcare professional (HCP) requirements.
- **Methods**: Surveys, focus groups, and interviews with patients and HCPs.
- **Impact**: It helps develop products that address specific market needs and gaps, leading to higher acceptance and success.

1.2. Concept Testing:

- **Objective**: Evaluate the potential of new product ideas and concepts before full-scale development.
- **Methods**: Prototype testing, concept testing surveys, and market simulations.
- **Impact**: Provides insights into the validity of new ideas, allowing for adjustments before investing in development.

2. Informing Marketing Strategies

2.1. Segmentation and Targeting:

- **Objective**: Divide the market into demographics, behaviors, and needs segments.
- **Methods**: Statistical Analysis and market segmentation studies.

- **Impact**: Enables targeted marketing efforts, personalized messaging, and efficient resource allocation.

2.2. Positioning and Branding:

- **Objective**: Determine how to position the product in the market and differentiate it from competitors.
- **Methods**: Brand positioning studies and competitive analysis.
- **Impact**: Develop a strong brand identity and clear positioning that resonates with target audiences.

2.3. Campaign Development:

- **Objective**: Create effective marketing campaigns based on consumer behavior and preferences insights.
- **Method**: A/B testing, message testing, and media planning.
- **Impact**: Ensures marketing campaigns are relevant, engaging, and effective in reaching and influencing the target audience.

3. Enhancing Sales and Distribution

3.1. Sales Forecasting:

- **Objective**: Predict future sales based on historical data, market trends, and planned marketing activities.
- **Methods**: Statistical modeling, trend analysis.
- **Impact**: Helps in planning production, managing inventory, and setting sales targets.

3.2. Channel Strategy:

- **Objective**: Determine the most effective distribution channels for reaching the target market.
- **Methods**: Channel analysis and distribution audits.
- **Impact**: Optimize distribution strategies to ensure product availability and maximize reach.

4. Monitoring and Adapting to Market Changes

4.1. Market Trends Analysis:

- **Objective**: Identify emerging trends and shifts in the pharmaceutical industry.

- **Methods**: Industry reports, trend analysis.
- **Impact**: Keeps the company informed about market dynamics and allows for timely adjustment to strategies.

4.2. Competitor Analysis:

- **Objective**: Assess competitors' strengths, weaknesses, and strategies.
- **Methods**: Competitive intelligence, SWOT analysis.
- **Impact**: Provides insights into competitive positioning and informs strategic decisions to gain a competitive advantage.

4.3. Customer Feedback:

- **Objective**: Gather feedback from patients and HCPs to assess satisfaction and identify areas for improvement.
- **Methods**: Surveys, interviews, and social media monitoring.
- **Impact**: Enhances product and service offerings based on real-world feedback, improving customer satisfaction and loyalty.

5. Supporting Regulatory and Compliance Efforts

5.1. Regulatory Submissions:

- **Objective**: Provide evidence to support regulatory submissions and approvals.
- **Methods**: Clinical trial data analysis and market research reports.
- **Impact**: Ensures that regulatory submissions are robust and well-supported by evidence.

5.2. Compliance Monitoring:

- **Objective**: Ensure that marketing practices comply with industry regulations and standards.
- **Methods**: Compliance audits and regulatory reviews.
- **Impact**: Minimizes the risk of non-compliance and legal issues.

6. Driving Strategic Decisions

6.1. Strategic Planning:

- **Objective**: Inform long-term strategic planning and decision-making.
- **Methods**: SWOT analysis, scenario planning.
- **Impact**: Provides a data-driven basis for strategic decisions, helping to align business goals with market realities.

6.2. Investment Decisions:

- **Objective**: Evaluate the potential return on investment for new initiatives and products.
- **Methods**: Financial analysis and market potential assessments.
- **Impact**: Supports informed investment decisions, ensuring that resources are allocated effectively.

Conclusion

Marketing research is pivotal in the pharmaceutical industry by providing insights that drive product development, marketing strategies, and business decisions. By systematically gathering and analyzing data, pharmaceutical companies can better understand market needs, optimize their marketing efforts, and improve patient outcomes. Integrating marketing research into strategic planning and operational activities enhances the ability to navigate the complexities of the pharmaceutical market and achieve sustainable success.

14.2 Key Methods and Techniques for Gathering Insights

Gathering insights in pharmaceutical marketing involves employing various methods and techniques to collect and analyze data. These methods help pharmaceutical companies understand market dynamics, patient needs, and the competitive landscape. Here is a detailed description of key methods and techniques:

1. Qualitative Research Methods

1.1. Focus Groups:

- **Description**: Involves moderated discussions with small groups of participants (patients, healthcare professionals, etc.) to gather in-depth insights.
- **Applications**: Exploring attitudes, perceptions, and opinions about new products, treatment options, or disease states.
- **Benefits**: Provides rich, detailed data on participants' experiences and motivations.

1.2. In-Depth Interviews:

- **Description**: One-on-one interviews with key stakeholders to gather detailed qualitative information.
- **Applications**: Understanding individual experiences, uncovering unmet needs, and obtaining feedback on product concepts.
- **Benefits**: Offers a deep understanding of individual perspectives and nuanced insights.

1.3. Ethnographic Research:

- **Description**: Observational research where researchers immerse themselves in the environment of the study subjects.
- **Applications**: Studying how patients use medications in real-world settings and understanding the context of their healthcare decisions.

- **Benefits**: Provides contextual insights into patient behavior and treatment experiences.

2. Quantitative Research Methods

2.1. Surveys and Questionnaires

- **Description**: Structured tools used to collect data from a large number of respondents.
- **Applications**: Measuring attitudes, preferences, and behaviors related to pharmaceutical products or health conditions.
- **Benefits**: Allows for statistical analysis and generalization of findings to larger populations.

2.2. Market Segmentation Analysis

- **Description**: Identifies and analyzes distinct groups within a market based on demographics, psychographics, and behavioral factors.
- **Applications**: Tailoring marketing strategies and product offerings to different market segments.
- **Benefits**: Enhances targeting and personalization of marketing efforts.

2.3. Conjoint Analysis

- **Description**: A technique used to determine how different attributes of a product influence consumer preferences.
- **Applications**: Evaluating the importance of various product features and pricing strategies.
- **Benefits**: Helps design products and pricing strategies that align with consumer preferences.

3. Secondary Research Methods

3.1. Literature Reviews

- **Description**: Systematic review of existing research, publications, and reports

- **Applications**: Understanding trends, identifying gaps in knowledge, and gathering background information on diseases and treatments.
- **Benefits**: Provides a comprehensive overview of existing knowledge and informs further research.

3.2. Competitive Analysis

- **Description**: Analyzing competitors' products, strategies, and market performance.
- **Applications**: Identifying competitive advantages, market positioning, and potential opportunities.
- **Benefits**: Helps in strategic planning and benchmarking against competitors.

3.3. Industry Reports

- **Description**: Reports published by market research firms, industry associations, and regulatory bodies.
- **Applications**: Gaining insights into market trends, regulatory changes, and industry benchmarks.
- **Benefits**: Provides up-to-date and authoritative information on industry dynamics.

4. Data Analytics Techniques

4.1. Big Data Analytics

- **Description**: Analyzing large and complex data sets to uncover patterns, trends, and insights.
- **Applications**: Predicting market trends, understanding patient behavior, and optimizing marketing strategies.
- **Benefits**: Provides comprehensive insights and supports data-driven decision-making.

4.2. Predictive Analytics

- **Description**: Using historical data and statistical algorithms to forecast future trends and behaviors.

- **Applications**: Forecasting sales, predicting patient outcomes, and assessing market opportunities.
- **Benefits**: Helps in proactive decision-making and strategic planning.

4.3. Sentiment Analysis

- **Description**: Analyzing text data from social media, reviews, and other sources to gauge public sentiment.
- **Applications**: Monitoring brand perception, understanding patient attitudes, and identifying potential issues.
- **Benefits**: Provides real-time insights into consumer opinions and feedback.

5. Emerging Research Techniques

5.1. AI and Machine Learning

- **Description**: Leveraging artificial intelligence and machine learning algorithms to analyze large data sets and uncover insights.
- **Applications**: Enhancing predictive models, personalizing marketing efforts, and optimizing clinical trials.
- **Benefits**: Offers advanced analytical capabilities and improves accuracy in data interpretation.

5.2. Blockchain for Data Integrity

- **Description**: Using blockchain technology to ensure the integrity and security of research data.
- **Applications**: Enhancing data transparency and preventing tampering in clinical trials and research studies.
- **Benefits**: Ensures data accuracy and builds trust in research findings.

Conclusion

Effective insight gathering in pharmaceutical marketing research involves a combination of qualitative and quantitative methods, secondary research, data analytics, and emerging technologies.

Each method and technique provides unique advantages and contributes to a comprehensive understanding of market dynamics, patient needs, and competitive positioning. By employing diverse research approaches, pharmaceutical companies can make informed decisions, develop targeted strategies, and achieve better outcomes in their marketing efforts.

14.3 Leveraging Research for Informed Decision-Making

Effective decision-making in pharmaceutical product management relies heavily on comprehensive and actionable research. Leveraging research helps product managers make data-driven decisions that can optimize strategies, enhance product success, and improve patient outcomes. Here is how pharmaceutical product managers can leverage research for informed decision-making:

1. Understanding Market Needs and Trends

1.1. Identifying Unmet Needs

- **Approach**: Conduct market research, including surveys and focus groups, to identify gaps in current treatment options and patient needs.
- **Benefit**: It helps develop products that address specific patient or healthcare provider (HCP) needs, improving the likelihood of market acceptance.

1.2. Monitoring Market Trends

- **Approach**: Utilize secondary research, industry reports, and competitive analysis to stay informed about emerging trends and shifts in the market.
- **Benefit**: Allows product managers to anticipate changes, adapt strategies, and stay ahead of competitors.

2. Strategic Product Development

2.1. Guiding Product Features and Design

- **Approach**: Apply insights from conjoint analysis and ethnographic research to design products with features that align with user preferences and needs.
- **Benefit**: Increases the chances of product success by ensuring it meets the expectations and requirements of the target audience.

2.2. Optimizing Pricing Strategies

- **Approach**: Use pricing research and elasticity studies to determine optimal pricing that balances profitability with market demand.

- **Benefit**: Ensures competitive pricing while maximizing revenue and market share.

3. Enhancing Marketing and Sales Strategies

3.1. Targeting and Segmentation

- **Approach**: Leverage market segmentation analysis to identify and target specific patient populations or healthcare professionals.

- **Benefits**: Allows more precise and effective marketing campaigns, improving engagement and conversion rates.

3.2. Developing Communication Strategies

- **Approach**: Use sentiment analysis and feedback from HCPs and patients to craft messaging that resonates with the audience.

- **Benefit**: Enhances the relevance and impact of marketing communications, leading to better brand perception and customer loyalty.

4. Optimizing Clinical Trials and Research

4.1. Designing Effective Clinical Trials

- **Approach**: Utilize epidemiological insights and biostatistics to design scientifically robust trials aligned with regulatory requirements.

- **Benefit**: Improves the likelihood of successful trials and regulatory approval, accelerating time-to-market for new products.

4.2. Monitoring Trial Progress

- **Approach**: Implement data analytics to monitor trial data in real-time and adjust as needed.

- **Benefit**: Ensures trials stay on track and addresses issues promptly, enhancing the reliability of results.

5. Improving Operational Efficiency

5.1. Streamlining Processes

- **Approach**: Apply insights from operational research and AI-driven analytics to optimize supply chain management and operational workflows.

- **Benefit**: Reduces costs, improves efficiency, and ensures timely delivery of products to market.

5.2. Enhancing Compliance

- **Approach**: Utilize research on regulatory requirements and best practices to ensure adherence to compliance standards.

- **Benefit**: Minimizes the risk of regulatory issues and ensures that marketing materials and product claims are accurate and compliant.

6. Guiding Strategic Decisions

6.1. Informing Strategic Planning

- **Approach**: Use comprehensive research to inform strategic decisions, such as market entry, product launches, and partnerships.

- **Benefit**: Provides a solid foundation for making strategic choices that align with market opportunities and organizational goals.

6.2. Evaluating Performance

- **Approach**: Leverage performance metrics and market feedback to assess the effectiveness of strategies and make necessary adjustments.

- **Benefit**: Helps refine strategies, improve performance, and achieve better results.

Conclusion

Leveraging research for informed decision-making in pharmaceutical product management involves utilizing various research methods and techniques to gather, analyze, and apply data. By integrating research insights into strategic planning, product development, marketing, and operational processes, product managers can make well-informed decisions that enhance product success, optimize performance, and improve patient outcomes. Translating research into actionable strategies is essential for navigating the complex and competitive pharmaceutical landscape.

Competitive Intelligence and Sentiment Analysis

In the rapidly evolving pharmaceutical industry, staying ahead of the competition requires more than innovative products and effective marketing strategies. It demands a deep understanding of the competitive landscape and the ability to gauge the sentiments of key stakeholders—healthcare professionals (HCPs), patients, payers, and regulators. Competitive intelligence (CI) and sentiment analysis have emerged as critical tools for pharmaceutical product managers to navigate these complexities.

Competitive Intelligence involves systematically gathering, analyzing, and applying information about competitors' activities, market dynamics, and industry trends. By understanding the competitive landscape, product managers can anticipate market shifts, identify potential threats and opportunities, and make informed strategic decisions. CI provides a framework for benchmarking products, predicting competitor moves, and developing strategies that capitalize on market gaps.

On the other hand, **Sentiment analysis** offers insights into how stakeholders perceive a brand, product, or therapeutic area. This technique uses advanced data analytics, including natural language processing (NLP) and machine learning, to analyze opinions, attitudes, and emotions expressed in various forms of communication, such as social media, patient forums, and medical literature. Understanding stakeholder sentiment allows product managers to tailor their messaging, improve patient and HCP engagement, and enhance brand loyalty.

This section explores the integration of competitive intelligence and sentiment analysis in pharmaceutical marketing. It delves into the methodologies, tools, and techniques that product managers managers can employ to gain a competitive edge.

15.1. Competitive Intelligence

Competitive intelligence (CI) in the pharmaceutical industry is a strategic process of gathering, analyzing, and interpreting data about competitors, market conditions, and trends to inform decision-making and maintain a competitive edge. Given the dynamic nature of the pharmaceutical industry, where the stakes are high due to the long drug development timelines, stringent regulatory requirements, and significant investment in research and development (R&D), competitive intelligence plays a crucial role in ensuring the success of pharmaceutical companies.

Importance of Competitive Intelligence in Pharma

The pharmaceutical industry is characterized by rapid scientific advancements, shifting regulatory landscapes, and fierce competition. In such an environment, competitive intelligence is crucial for several reasons:

1. **Informed Decision-Making**: Competitive intelligence gives pharma companies insights into competitors' strategies, pipeline developments, and market dynamics, enabling informed decision-making.

2. **Risk Mitigation**: By understanding competitors' moves and potential market disruptions, companies can anticipate challenges and mitigate risks effectively.

3. **Innovation and R&D Strategy**: Competitive intelligence helps companies identify market gaps, potential partnerships, and emerging technologies, guiding their R&D strategies.

4. **Regulatory Compliance**: Understanding regulatory changes and competitors' compliance strategies can help companies navigate the complex regulatory environment more effectively.

5. **Market Positioning**: Competitive intelligence allows companies to position themselves and their products more effectively by understanding the competitive landscape, customer needs, and market trends.

Conducting Competitive Intelligence in Pharma: A Step-by-Step Guide

1. **Define Objectives and Scope**

   - **Objective Setting**: The first step in conducting competitive intelligence is clearly defining the objectives. These may include understanding competitors' R&D pipelines, analyzing market trends, monitoring regulatory changes, or assessing potential threats and opportunities.

   - **Scope Definition**: Determine the scope of the competitive intelligence effort. This involves identifying the specific areas to focus on, such as therapeutic areas, geographic markets, or specific competitors.

2. **Identify Key Competitors**

   - **Competitor Mapping**: Create a list of key industry competitors. This may include direct competitors (those developing similar drugs or targeting the same markets) and indirect competitors (companies with alternative therapies or business models).

   - **Segmentation**: Segment competitors based on criteria such as size, market share, therapeutic focus, and geographic presence.

3. **Data Collection**

   - **Secondary Research**: Gather publicly available information from various sources such as scientific publications, patents, regulatory filings, company websites, press releases, industry reports, and financial statements.

   - **Primary Research**: Conduct interviews with industry experts, key opinion leaders (KOLs), and other stakeholders. Attend industry conferences, webinars, and networking events to gather firsthand information.

   - **Digital Tools and Databases**: Utilize digital tools and databases such as ClinicalTrials.gov, PubMed, FDA databases, and patent databases to gather detailed information about competitors' pipelines, clinical trials, and regulatory activities.

4. Data Analysis

- **Competitive Landscape Analysis**: Analyze the data collected to create a competitive landscape. This involves mapping competitors' products, R&D pipelines, market positioning, and strengths and weaknesses.

- **SWOT Analysis**: Conduct a SWOT (Strengths, Weaknesses, Opportunities, Threats) analysis for key competitors. This helps in understanding their strategic positioning and potential vulnerabilities.

- **Market Trend Analysis**: Analyze market trends, including changes in therapeutic areas, emerging technologies, regulatory developments, and shifts in patient demographics.

5. Strategic Insights and Scenario Planning

- **Strategic Insights**: Based on the analysis, generate strategic insights that can inform decision-making. These may include recommendations for product development, market entry, pricing strategies, and partnership opportunities.

- **Scenario Planning**: Develop potential scenarios based on competitors' likely actions and market dynamics. This helps in preparing for different outcomes and making contingency plans.

6. Reporting and Communication

- **Report Preparation**: Prepare detailed competitive intelligence reports summarizing the findings, analysis, and strategic insights. The reports should be clear, concise, and tailored to the needs of different stakeholders within the company.

- **Presentation to Stakeholders**: Present the competitive intelligence findings to key stakeholders, including senior management, R&D teams, marketing, and regulatory affairs. Use visual aids such as charts, graphs, and infographics to enhance understanding.

- **Regular Updates**: Competitive Intelligence (CI) is an ongoing process. Regularly update the CI reports to reflect new developments and ensure the company stays ahead of the competition.

7. Monitoring and Continuous Improvement

- **Ongoing Monitoring**: Continuously monitor competitors and the market for new information and changes. This involves tracking regulatory filings, clinical trial updates, patent publications, and marketing news.

- **Feedback and Improvement**: Gather feedback from stakeholders on the usefulness of the competitive intelligence efforts and identify areas for improvement. Adjust the CI process based on lessons learned and evolving business needs.

Conclusion

Competitive intelligence is critical in the pharmaceutical industry, enabling companies to navigate the complex competitive landscape effectively. Pharmaceutical companies can make informed decisions, mitigate risks, and capitalize on opportunities by systematically gathering, analyzing, and interpreting information about competitors and market conditions.

Competitive intelligence requires a structured approach, from defining objectives and identifying competitors to data collection, analysis, and reporting. With the right tools and techniques, pharmaceutical companies can leverage competitive intelligence to drive innovation, enhance their market position, and achieve long-erm success in an increasingly competitive industry.

15.2 Sentiment Analysis

Sentiment analysis, also known as opinion mining, is a technique used to analyze textual data to determine the sentiment or emotion expressed within it. In pharmaceutical marketing, sentiment analysis plays a crucial role in understanding the attitudes, perceptions, and opinions of stakeholders such as healthcare providers (HCPs), patients, caregivers, and the general public towards pharmaceutical products, brands, treatments, and healthcare services. Here's an overview of sentiment analysis and its application in pharmaceutical marketing:

Understanding Sentiment Analysis

1. **Objective**:
   - **Sentiment Classification**: This process classifies text into different categories, such as positive, negative, or neutral sentiments, based on the emotional tone expressed in the text.
   - **Emotion Detection**: This method identifies specific emotions, such as happiness, sadness, anger, or fear, conveyed in the text.

2. **Methodology**:
   - **Natural Language Processing (NLP)**: Uses computational techniques to process and analyze large volumes of textual data.
   - **Text Mining**: Applies algorithms and machine learning models to extract sentiment-related features from text, including sentiment scores, key phrases, and context.

3. **Applications in Pharmaceutical Marketing**:
   - **Brand Perception**: Assesses how pharmaceutical brands are perceived by stakeholders based on online reviews, social media conversations, patient forums, and healthcare provider feedback.

- **Product Feedback**: Monitors and analyzes patient and healthcare provider feedback on pharmaceutical products, including efficacy, safety, side effects, and treatment experiences.

- **Campaign Effectiveness**: Evaluates the impact of marketing campaigns, educational initiatives, and promotional activities on audience and brand sentiment.

- **Competitor Analysis**: Compares sentiment trends between pharmaceutical products or brands to identify competitive strengths and weaknesses.

- **Patient Insights**: Provides insights into patient sentiments, preferences, and satisfaction levels regarding treatment options, healthcare services, and disease management.

4. **Tools and Technologies**:

- **Sentiment Analysis Tools**: Utilize specialized software and platforms with sentiment analysis capabilities, such as sentiment lexicons, machine learning models, and sentiment scoring algorithms.

- **Social Media Monitoring Tools**: These tools track and analyze sentiment expressed on social media platforms (e.g., Twitter, Facebook, LinkedIn) through real-time monitoring and sentiment visualization dashboards.

- **Text Analytics Platforms**: Integrates sentiment analysis with text mining topic modeling and predictive analytics to derive actionable insights from unstructured textual data.

Benefits of Sentiment Analysis in Pharmaceutical Marketing

- **Insight Generation**: Provides actionable insights into stakeholder opinions, preferences, and perceptions, enabling pharmaceutical companies to tailor marketing strategies and communication approaches.

- **Real-Time Monitoring**: This system tracks sentiment trends in real-time, allowing timely responses to emerging issues, crises, or opportunities in the market.

- **Customer Experience Improvement**: Identifies areas for improvement in patient care, customer service, and product development based on sentiment-driven feedback.

- **Competitive Intelligence**: Benchmarks sentiment metrics against competitors to identify market gaps, differentiation opportunities, and competitive positioning strategies.

- **Regulatory Compliance**: Monitors and addresses compliance-related concerns by identifying potential misperceptions or misinformation online about pharmaceutical products.

Challenges of Sentiment Analysis

- **Contextual Understanding**: This requires context-aware algorithms to accurately interpret nuances, sarcasm, and cultural variations in sentiment expressed in text.

- **Data Privacy**: Ensures compliance with data privacy regulations (e.g., GDPR, HIPAA) when analyzing patient-related data or sensitive healthcare information.

- **Data Volume and Noise**: This function handles large volumes of unstructured data and filters irrelevant noise to extract meaningful sentiment insights.

In conclusion, sentiment analysis is a valuable tool in pharmaceutical marketing for understanding stakeholder sentiments, enhancing brand perception, and informing strategic decision-making. By leveraging advanced NLP techniques and sentiment analysis tools, pharmaceutical companies can gain actionable insights that drive customer engagement, brand loyalty, and market competitiveness in an increasingly digital and data-driven healthcare landscape.

15.3   Understanding Patient and HCP Sentiments

Understanding patient and healthcare provider (HCP) sentiment is crucial in pharmaceutical marketing and healthcare management, as it provides insights into their attitudes, emotions, and opinions towards treatments, medications, healthcare services, and overall patient care experiences. Here is a deeper exploration of how patient and HCP sentiment can be understood and leveraged in pharmaceutical marketing:

Understanding Patient Sentiment

1. **Patient Experience and Satisfaction**:
   - **Feedback and Reviews**: Analyzing patient reviews, feedback surveys, and online discussions on social media platforms and patient forums to gauge satisfaction levels and identify areas for improvement.
   - **Patient Journey Mapping**: This involves mapping the journey from diagnosis to treatment and follow-up care to understand emotional highs and lows, pain points, and moments of satisfaction or dissatisfaction.
   - **Patient-Reported Outcomes (PROs)**: Using PRO measures to capture patient-reported symptoms, quality of life impacts, treatment preferences, and satisfaction with treatment outcomes.

2. **Patient Preferences and Needs**:
   - **Treatment Preferences**: Assessing patient preferences for treatment modalities (e.g., oral medications, injectables), dosing frequency, side effect profiles, and convenience factors influencing adherence.
   - **Healthcare Delivery**: Evaluating patient sentiments towards telemedicine, remote monitoring, digital health tools, and patient support programs to improve healthcare access and convenience.

3. **Emotional and Psychosocial Factors**:
   - **Emotional Impact of Illness**: Understanding the emotional toll of chronic illnesses, disease progression, and the

psychological aspects influencing treatment adherence and quality of life.

- **Psychosocial Support**: Identifying patient sentiments toward support services, peer networks, counseling, and educational resources that address emotional and social needs.

Understanding Healthcare Provider (HCP) Sentiment

1. **Clinical Decision-Making**:

   - **Treatment Decision Drivers**: Investigating HCP sentiments towards clinical guidelines, evidence-based medicine, therapeutic efficacy, safety profiles, and cost-effectiveness of pharmaceutical products.

   - **Clinical Outcomes**: Assessing HCP perceptions of treatment outcomes, patient response rates, disease management challenges, and real-world effectiveness in clinical practice.

2. **Educational and Informational Needs**:

   - **Continuing Medical Education (CME)**: Identifying HCP sentiments towards CME programs, medical conferences, scientific publications, and updates on emerging therapies and treatment guidelines.

   - **Access to Information**: Evaluating HCP preferences for access to medical literature, clinical trial data, drug information resources, and digital platforms supporting evidence-based practice.

3. **Professional Relationships and Collaboration**:

   - **Industry Interactions**: Monitor HCP sentiments toward pharmaceutical industry interactions, promotional activities, detailing visits, and engagement with medical representatives.

   - **Interdisciplinary Collaboration**: Understanding sentiments towards multidisciplinary care teams, collaboration with specialists, referral patterns, and patient care coordination.

Leveraging Sentiment Analysis in Pharmaceutical Marketing

- **Segmentation and Targeting**: Segmenting patient and HCP populations based on sentiment profiles to tailor marketing strategies, educational initiatives, and communication approaches that resonate with their needs and preferences.

- **Brand Perception Management**: Monitoring and managing brand perception through sentiment analysis of online reviews, social media mentions, and healthcare provider feedback to address concerns, mitigate negative sentiment, and amplify positive experiences.

- **Patient Engagement and Adherence**: Designing patient-centric programs, digital health solutions, and patient support services aligned with sentiment-driven insights to enhance treatment adherence, patient satisfaction, and health outcomes.

- **HCP Engagement Strategies** Include developing targeted educational content, clinical support tools, and value-added services based on HCP sentiment analysis to foster trust, enhance professional relationships, and support evidence-based decision-making.

Challenges and Considerations

- **Privacy and Ethics**: Ensuring compliance with patient privacy regulations (e.g., GDPR, HIPAA) and ethical guidelines when collecting and analyzing sensitive healthcare data.

- **Data Quality and Interpretation**: Addressing challenges related to data noise, bias, and the need for context-aware analysis to interpret sentiment in healthcare-related data accurately.

- **Integration with Clinical Insights**: Aligning sentiment analysis findings with clinical insights, real-world evidence, and patient-reported outcomes to drive holistic healthcare strategies and patient-centered care.

In conclusion, understanding patient and HCP sentiment through advanced sentiment analysis techniques empowers pharmaceutical companies to deliver personalized, empathetic, and effective healthcare solutions. By leveraging these insights, pharmaceutical marketers can optimize engagement strategies, improve healthcare outcomes, and foster meaningful relationships with patients and providers in an evolving healthcare landscape.

15.4 Tools and Techniques for Sentiment Analysis

Sentiment analysis, or opinion mining, employs various tools and techniques from natural language processing (NLP), machine learning, and text analytics to extract and analyze sentiments expressed in textual data. In pharmaceutical marketing and healthcare, these tools and techniques are crucial in understanding patient, healthcare provider (HCP), and stakeholder sentiments towards medications, treatments, healthcare services, and brand perceptions. Here is an overview of the tools and techniques commonly used for sentiment analysis:

Tools for Sentiment Analysis

1. **Python Libraries**:
   - **NLTK (Natural Language Toolkit)** is a comprehensive library for NLP tasks in Python, including sentiment analysis, tokenization, stemming, and part-of-speech tagging.
   - **TextBlob**: Simplifies text processing tasks such as sentiment analysis and language transition with an easy-to-use API.
   - **spaCy**: Provides efficient tokenization, part-of-speech tagging, named entity recognition, and dependency parsing, suitable for large-scale text processing and sentiment analysis.
   - **VADER (Valence-Aware Dictionary and Sentiment Reasoner)** is a lexicon and rule-based sentiment analysis tool specifically tuned for social media texts. It provides sentiment scores based on predefined lexicons.

2. **Machine Learning Frameworks**:
   - **Scikit-learn** offers machine learning algorithms for text classification tasks, including sentiment analysis, using models like Support Vector Machines (SVM) and Naive Bayes.

- **TensorFlow and Keras** are deep learning frameworks for building and training neural networks. They are suitable for advanced sentiment analysis tasks requiring context-aware understanding.

- **PyTorch** provides flexible and efficient deep-learning capabilities for sentiment analysis, particularly in researching and developing custom models.

3. **Cloud-Based Services**:

   - **Google Cloud Natural Language API** offers sentiment analysis, scores, and entity recognition capabilities.

   - **Amazon Comprehend** is a fully managed NLP service by AWS (Amazon Web Services), providing sentiment analysis, entity recognition, and language detection through API calls.

   - **Microsoft Azure Text Analytics**: Includes sentiment analysis and other NLP capabilities, such as key phrase extraction and language detection, accessible via API.

Techniques for Sentiment Analysis

1. **Lexicon-Based Approaches**:

   - **Sentiment Lexicons** are predefined dictionaries or lexicons containing sentiment scores for words and phrases. They facilitate sentiment classification based on predefined polarity (positive, negative, neutral).

   - **Rule-based systems** apply linguistic rules and heuristics to analyze text sentiment based on patterns, grammatical structures, and context clues.

2. **Machine Learning Approaches**:

   - **Supervised Learning**: Trains models on labeled datasets (e.g., sentiment-labeled reviews) to predict sentiment polarity (positive, negative) of unseen text using algorithms like SVM, Naive Bayes, or neural networks.

- **Unsupervised Learning**: Clusters text data based on the similarity of sentiment expressions without pre-labeled data, using techniques like clustering or topic modeling.

3. **Hybrid Approaches**:

   - **Combining Lexicon and Machine Learning**: This technique integrates lexicon-based sentiment scoring with machine learning models to improve accuracy and handle nuanced sentiments.

   - **Aspect-Based Sentiment Analysis**: This technique analyzes sentiments toward specific aspects or entities within the text (e.g., sentiment toward drug efficacy, side effects, and pricing).

4. **Contextual and Emotion Detection**:

   - **Context-Aware Analysis**: This technique enhances accuracy by considering contextual clues, sarcasm, irony, and cultural nuances in sentiment interpretation.

   - **Emotion Detection**: This technique identifies specific emotions (e.g., joy, anger, sadness) expressed in the text beyond simple positive/negative sentiment classification.

Considerations and Challenges

- **Data Quality**: Ensuring high-quality and representative datasets for training models and lexicons, addressing bias and noise in sentiment analysis results.

- **Multi-Lingual Support**: Handling sentiments expressed in different languages and cultural contexts, requiring language-specific models and lexicons.

- **Privacy and Ethics**: Adhering to data privacy regulations (e.g., GDPR, HIPAA) when handling sensitive healthcare-related texts and patient information.

- **Real-Time Analysis**: Implementing scalable and efficient sentiment analysis pipelines for real-time monitoring and actionable insights in dynamic healthcare environments.

In summary, leveraging tools and techniques for sentiment analysis enables pharmaceutical marketers, healthcare providers, and researchers to extract valuable insights from textual data, understand stakeholder sentiments, and optimize strategies for patient engagement, healthcare delivery, and brand management in the pharmaceutical industry.

Part V. Strategic Planning and Brand Management

Strategic planning and brand management are critical components of pharmaceutical product management, ensuring that a product meets market needs and achieves long-term success and sustainability. Here is a detailed description of these areas:

Strategic Planning

1. **Definition and Importance**
   - **Definition:** Strategic planning involves defining a pharmaceutical product's long-term goals and objectives, determining the strategies to achieve them, and aligning resources and actions accordingly. It is crucial for navigating the competitive landscape and adapting to market changes.
   - **Importance:**
     - **Alignment:** Ensures all activities and resources align with the company's vision and goals.
     - **Adaptability:** Helps anticipate and respond to market shifts and emerging trends.
     - **Resource Optimization:** Allocates resources effectively to maximize ROI.

2. **Key Components**

   2.1. Market Analysis
   - **Approach:** Conduct a thorough analysis of market trends, competitive landscape, and customer needs.
   - **Tools:** Market research reports, SWOT analysis, PESTEL analysis.
   - **Outcome:** Identifies opportunities and threats, guiding strategic decisions.

   2.2. Goal Setting
   - **Approach:** Define clear, measurable objectives for the product's performance and market position.
   - **Tools:** SMART criteria (Specific, Measurable, Achievable, Relevant, Time-bound).
   - **Outcome:** Provides direction and benchmarks for success.

2.3. Strategy Formulation

- **Approach:** Develop strategies to achieve the defined goals, including market entry strategies, pricing, and distribution plans.
- **Tools:** Strategic frameworks such as Porter's Five Forces and Ansoff Matrix.
- **Outcome:** Creates a roadmap for achieving objectives.

2.4. Implementation and Planning

- **Approach:** Outline the steps and resources needed to execute the strategies.
- **Tools:** Project management tools, Gantt charts.
- **Outcome:** Ensures that strategies are actionable and achievable.

2.5. Monitoring and Evaluation

- **Approach:** Regularly review performance against goals and adjust strategies as needed.
- **Tools:** Key Performance Indicators (KPIs) and performance dashboards.
- **Outcome:** Keeps the plan on track and responsive to changes.

Brand Management

1. Definition and Importance

- **Definition:** Brand management involves creating and maintaining a positive perception of a product in the market, differentiating it from competitors, and building long-term brand equity. It is essential for driving product success and customer loyalty.
- **Importance:**
    - **Differentiation:** Distinguishes the product from competitors and establishes a unique value proposition.
    - **Customer Loyalty:** Builds trust and loyalty among healthcare professionals (HCPs) and patients.

- **Market Positioning:** Enhances the product's position in the market and drives demand.

2. Key Components

2.1. Brand Identity

- **Approach:** Develop a clear and compelling brand identity that reflects the product's values and benefits.
- **Tools:** Brand guidelines, visual identity systems (logos, color schemes).
- **Outcome:** Creates a consistent and recognizable brand image.

2.2. Brand Positioning

- **Approach:** Define the product's position in the market relative to competitors and target segments.
- **Tools:** Positioning statements, perceptual mapping.
- **Outcome:** Establishes how the product is perceived and valued by customers.

2.3. Brand Messaging

- **Approach:** Craft messaging that communicates the brand's value proposition and resonates with target audiences.
- **Tools:** Messaging frameworks and communication strategies.
- **Outcome:** Ensures that all marketing communications are aligned and impactful.

2.4. Brand Promotion

- **Approach:** Implement promotional strategies to enhance brand visibility and reputation.
- **Tools:** Advertising campaigns, digital marketing, public relations.
- **Outcome:** Increases brand awareness and drives engagement.

2.5. Brand Monitoring

- **Approach:** Track and analyze brand performance and perception in the market.

- **Tools:** Brand tracking studies, social media monitoring.
- **Outcome:** Provides insights into brand health and areas for improvement.

2.6. Brand Equity Management

- **Approach:** Build and maintain brand equity by delivering consistent value and positive experiences.
- **Tools:** Customer satisfaction surveys and loyalty programs.
- **Outcome:** Enhances long-term brand value and market position.

Integration of Strategic Planning and Brand Management

- Effective strategic planning and brand management are interconnected as strategic goals guide brand management efforts, and strong brand management supports achieving strategic objectives. For example:
- Market Analysis from strategic planning informs brand positioning and messaging.
- Goal Setting helps define brand objectives and key performance indicators.
- Strategy Formulation aligns with brand promotion and communication strategies.
- Implementation Planning ensures that brand initiatives are executed effectively.
- Monitoring and Evaluation of both strategic and brand performance enables continuous improvement.

By integrating strategic planning with brand management, pharmaceutical product managers can create a cohesive approach that drives product success, enhances market positioning, and builds lasting brand equity.

Pharmaceutical Product Launch and Lifecycle Management

Pharmaceutical product launch and lifecycle management are integral processes in ensuring a product's success from development through its entire market presence. Here is a detailed look into each:

Pharmaceutical Product Launch

1. Definition and Importance

- **Definition:** A product launch is the process of introducing a new pharmaceutical product to the market. It encompasses strategic planning, execution, and post-launch activities to ensure the product's successful entry and growth.

- **Importance:**

  - **Market Penetration:** Establishes the product's presence and drives initial adoption.

  - **Competitive Advantage:** Differentiates the product from competitors and secures market share.

  - **Revenue Generation:** Creates revenue streams and contributes to financial performance.

2. Key Phases

2.1. Pre-Launch Planning

- **Objective:** Prepare for a successful launch by defining strategy and tactics.

- **Activities:** Conduct market research, identify target segments, develop positioning strategies, and prepare marketing materials.
- **Deliverables:** Market analysis reports, strategic plans, and promotional materials.

2.2. Regulatory Approvals

- **Objective:** Obtain necessary approvals from regulatory bodies to market the product.
- **Activities:** Submit regulatory submissions, address queries from regulatory agencies, and ensure compliance with guidelines.
- **Deliverables:** Approved product labels and marketing authorizations.

2.3. Launch Preparation

- **Objective:** Execute launch strategies and ensure all systems are in place.
- **Activities:** Coordinate with supply chain, train sales teams, launch promotional campaigns, and set up distribution channels.
- **Deliverables:** Training programs, marketing campaigns, and distribution plans.

2.4. Product Launch

- **Objective:** Introduce the product to the market and drive initial adoption.
- **Activities:** Execute launch events, initiate marketing campaigns, and engage with key stakeholders (e.g., healthcare professionals and patients).
- **Deliverables:** Launch events, promotional materials, and initial sales data.

2.5. Post-Launch Evaluation

- **Objective:** Assess the launch's success and make necessary adjustments.

- **Activities:** Monitor sales performance, collect feedback, evaluate marketing effectiveness, and adjust strategies as needed.
- **Deliverables:** Performance reports, feedback analysis, revised strategies.

Lifecycle Management

1. Definition and Importance

- **Definition:** Lifecycle management involves managing a pharmaceutical product throughout its entire lifecycle, from development through decline. This includes optimizing the product's market performance, extending its lifecycle, and adapting to changes in the market environment.
- **Importance:**
  - **Maximizing Value:** Enhances the product's profitability and market position.
  - **Adapting to Changes:** Responds to market dynamics, competition, and regulatory changes.
  - **Sustaining Market Presence:** Extends the product's relevance and success over time.

2. Key Phases

2.1   Product Introduction

- **Objective:** Launch the product and establish it in the market.
- **Activities:** Implement launch strategies, build brand awareness, and generate initial sales.
- **Deliverables:** Market presence, initial sales data, brand recognition.

2.2   Growth Phase

- **Objective:** Increase market share and drive revenue growth.
- **Activities:** Expand marketing efforts, enhance distribution, explore new markets, and engage in promotional activities.
- **Deliverables:** Increased sales, market share growth, expanded customer base.

2.3 Maturity Phase

- **Objective:** Maintain market position and optimize performance.
- **Activities:** Focus on customer retention, optimize pricing strategies, and explore new applications or indications.
- **Deliverables:** Stable revenue, high customer loyalty, optimized pricing.

2.4 Decline Phase

- **Objective:** Manage the product's decline and transition out of the market.
- **Activities:** Evaluate opportunities for product modifications (e.g., new formulations), manage inventory, and phase out marketing efforts.
- **Deliverables:** Managed product phase, transition plans, and inventory reduction.

2.5 End-of-Life Management

- **Objective:** Officially withdraw the product from the market and manage remaining assets.
- **Activities:** Ensure compliance with regulatory requirements for product discontinuation, manage remaining inventory, and communicate with stakeholders.
- **Deliverables:** Product withdrawal notices, compliance documentation, and stakeholder communication.

Integration of Launch and Lifecycle Management

- Effective product launch and lifecycle management require seamless integration to ensure that a product enters the market successfully and remains competitive and profitable throughout its lifecycle.
- **Launch Preparation and Lifecycle Planning:** Strategic launch plans should align with lifecycle management goals to ensure long-term success.

- **Ongoing Evaluation:** Continuous monitoring and evaluation during the lifecycle inform adjustments and improvements in the launch and post-launch phases.

- **Adaptation to Market Changes:** Flexibility in launch and lifecycle strategies helps respond to evolving market conditions, competitor actions, and regulatory changes.

- By effectively managing a pharmaceutical product's launch and lifecycle, product managers can drive long-term success, optimize market performance, and sustain competitive advantage.

16.1 Introduction: Overview of Product Launch and Lifecycle Management

In the dynamic pharmaceutical industry, successfully launching and managing a product's lifecycle is pivotal to achieving sustained success and profitability. The pharmaceutical product launch and lifecycle management process are crucial for ensuring the product reaches the market effectively and remains competitive throughout its time in the marketplace.

Pharmaceutical Product Launch is the gateway through which new drugs debut. It is a multifaceted process that involves strategic planning, regulatory compliance, market preparation, and promotional efforts. A successful launch sets the foundation for a product's future success, establishing its presence, generating initial market traction, and positioning it against competitors. It is a period of intense activity and coordination across various departments, including regulatory affairs, marketing, sales, and supply chain management.

Lifecycle Management, on the other hand, encompasses the ongoing management of a product from its introduction through to its eventual decline. This process involves maximizing the product's value over time, adapting to market changes, extending its market life, and optimizing its performance. Effective life cycle management requires continuous monitoring and strategic adjustments to address evolving market dynamics, competition, and regulatory environments.

Both launch and lifecycle management are interconnected, with the strategies and decisions made during the launch phase influencing the effectiveness of lifecycle management efforts. For product managers, understanding and mastering these processes is essential for ensuring that a product enters the market successfully and maintains its relevance and profitability throughout its lifecycle.

In this context, Pharmaceutical Product Launch and Lifecycle Management serve as a product manager's core elements, demanding a comprehensive approach to planning, execution, and strategic adaptation. This introduction sets the stage for exploring the detailed phases, challenges, and best practices of effectively launching and managing pharmaceutical products.

Overview of Pharmaceutical Product Launch and Lifecycle Management

Pharmaceutical Product Launch and Lifecycle Management are integral to ensuring the success and sustainability of pharmaceutical products in a competitive market. Here is a comprehensive overview of each:

Pharmaceutical Product Launch

1. **Pre-Launch Preparation:**
   - **Market Research:** Conduct in-depth research to understand market needs, patient demographics, competitor landscape, and potential market size. This helps identify opportunities and tailor strategies.
   - **Regulatory Approvals:** Secure all necessary regulatory approvals from agencies like the FDA or EMA. This includes filing for New Drug applications (NDAs) or Marketing Authorization Applications (MAAs) and ensuring compliance with local regulations.
   - **Stakeholder Engagement:** Build awareness and support by engaging with key stakeholders, including healthcare professionals (HCPs), payers, and patient advocacy groups.

2. **Launch Strategy:**
   - **Market Positioning:** Develop a clear value proposition and positioning strategy based on the product's benefits, unique selling points (USPs), and competitive advantage.
   - **Marketing and Promotion:** Create and execute a comprehensive marketing plan that includes promotional

campaigns, educational content, digital marketing, and sales materials. Tailor messages to different audiences, such as HCPs and patients.

- **Sales Strategy:** Prepare sales teams with training and tools to effectively communicate the product's benefits and address potential questions or objections.

- **Distribution and Supply Chain:** Ensure the product is available in the right quantities and locations, with a robust distribution and supply chain plan to meet market demand.

3. **Launch Execution:**

- **Monitoring and Adjustment:** Track the launch progress through key performance indicators (KPIs) and adjust strategies as needed based on market feedback and performance metrics.

- **Customer Feedback:** Collect and analyze feedback from HCPs and patients to refine marketing tactics and address any issues promptly.

Pharmaceutical Lifecycle Management

1. **Post-Launch Activities:**

- **Ongoing Market Analysis:** Continuously monitor market trends, competitive landscape, and patient needs. Use this information to adapt strategies and maintain product relevance.

- **Lifecycle Strategies:** Implement strategies to extend the product's lifecycle, such as line extensions, new indications, or improved formulations. Evaluate opportunities for product differentiation and market expansion.

2. **Product Optimization:**

- **Performance Monitoring:** Regularly assess the product's performance using sales data, market analysis, and patient outcomes. Identify areas for improvement and implement changes to optimize performance.

- **Regulatory Compliance:** Ensure ongoing compliance with regulatory requirements, including updating labeling and managing post-marketing commitments.

3. **Phase-Out and Transition:**

- **End-of-Life Planning:** Plan for the product's eventual decline, including strategies for phase-out and transition. This might involve managing inventory levels, informing stakeholders, and planning for successor products.

- **Knowledge Transfer:** Document and transfer knowledge gained throughout the product's lifecycle to support future product development and management efforts.

4. **Maximizing Value:**

- **Market Adaptation:** Adapt to evolving market conditions and emerging trends to continue leveraging the product's value. Consider partnerships, licensing opportunities, or new market segments.

Conclusion

Managing a pharmaceutical product's launch and lifecycle requires a strategic approach and ongoing adaptation. Successful product management ensures a strong market entry and maximizes the product's value over time. Pharmaceutical product managers can drive success and achieve long-term product growth by understanding and implementing the best launch and lifecycle management practices.

16.2 Importance of Pharmaceutical Product Management

Pharmaceutical product management is crucial for the success and sustainability of pharmaceutical products throughout their lifecycle. Here is a detailed look at why it is so important:

1. **Maximizing Market Potential**

   - **Market Entry and Positioning:** Effective product management ensures that pharmaceutical products are launched with a clear market strategy, addressing unmet needs and positioning the product to gain a competitive advantage.

   - **Market Analysis:** By understanding market dynamics, patient needs, and competitor actions, product managers can identify opportunities for growth and make informed decisions to maximize market potential.

2. **Ensuring Compliance and Regulatory Success**

   - **Regulatory Navigation:** Pharmaceutical product managers oversee the regulatory process, ensuring that products meet all necessary legal requirements and obtain the required approvals from regulatory bodies.

   - **Compliance Management:** They ensure ongoing compliance with regulations, which is crucial for avoiding legal issues and maintaining market access.

3. **Driving Strategic Planning and Execution**

   - **Strategic Vision:** Product managers develop and execute strategic plans for the product's lifecycle, including market entry, growth strategies, and eventual phase-out.

   - **Resource Allocation:** They manage resources effectively, including budget, personnel, and marketing efforts, to ensure successful product launch and sustainability.

4. Enhancing Cross-Functional Collaboration

- **Coordination:** Product managers act as the central point of contact between various departments, such as R&D, marketing, sales, and regulatory affairs, ensuring alignment and effective strategy execution.

- **Team Leadership:** They lead cross-functional teams, fostering collaboration and driving efforts towards common goals.

5. Optimizing Product Performance

- **Market Monitoring:** Continuous monitoring of product performance, including sales data, market trends, and feedback, allows product managers to make data-driven adjustments and improvements.

- **Lifecycle Management:** They implement strategies to extend the product's lifecycle, such as exploring new indications, improving formulations, or managing product variations.

6. Managing Financial Performance

- **Budgeting and Forecasting:** Product managers are responsible for financial planning, including budgeting, forecasting, and managing expenditures to ensure the product's profitability.

- **Financial Metrics:** They analyze key financial metrics, such as return on investment (ROI) and market share, to gauge the product's financial success and make informed decisions.

7. Driving Innovation and Differentiation

- **Innovation:** Product managers identify opportunities for innovation, such as new formulations, delivery methods, or complimentary products, to meet evolving market needs and differentiate the product.

- **Competitive Edge:** Product managers help the company maintain a competitive edge by staying ahead of industry trends and leveraging technological advancements.

8. Enhancing Customer Engagement

- **Stakeholder Relations:** Product managers build relationships with key stakeholders, including healthcare professionals, patients, and payers, ensuring that the product meets their needs and expectations.

- **Customer Feedback:** They gather and analyze customer feedback to refine product offerings and improve customer satisfaction.

9. Supporting Global Expansion

- **Market Entry:** For multinational companies, product managers play a key role in developing strategies for global market entry and expansion, addressing local regulatory requirements, and adapting marketing strategies to different regions.

- **Cultural Adaptation:** They ensure that the product and marketing strategies are culturally and regionally appropriate, enhancing acceptance and success in diverse markets.

Conclusion

Pharmaceutical product management is essential for navigating the complexities of the pharmaceutical industry, from development through commercialization. It involves strategic planning, regulatory compliance, cross-functional collaboration, financial oversight, and market analysis. By effectively managing these aspects, product managers ensure that pharmaceutical products are successful, profitable, and aligned with both market needs and company goals.

16.3 Pharmaceutical Product Launch

A pharmaceutical product launch is a critical phase in the product lifecycle where a new drug or therapy is introduced to the market. This process involves strategic, regulatory, and operational activities to ensure the product's successful entry and acceptance in the marketplace. Here is a comprehensive description of the key aspects involved in a pharmaceutical product launch:

1. Pre-Launch Planning

- **Market Research and Analysis:** Conduct thorough research to understand market needs, target patient populations, competitive landscape, and potential barriers to entry. This includes assessing unmet medical needs, evaluating market size, and identifying key opinion leaders (KOLs).

- **Regulatory Strategy:** Develop a regulatory strategy that includes planning submission timelines, understanding regulatory requirements, and preparing for interactions with regulatory agencies such as the FDA, EMA, or other national bodies.

- **Pricing and Reimbursement:** Determine the pricing strategy and reimbursement approach, which involves assessing cost-effectiveness, negotiating with payers, and understanding pricing regulations in different markets.

- **Stakeholder Engagement:** Identify and engage key stakeholders, including healthcare professionals (HCPs), patients, advocacy groups, and payers. Building relationships with these stakeholders is crucial for gaining support and ensuring successful product adoption.

2. Launch Preparation

- **Marketing Strategy:** Develop a comprehensive marketing plan that includes positioning, messaging, branding, and promotional activities. This plan should outline how the product will be differentiated from competitors and communicated to target audiences.

- **Sales Training:** Train the sales team and field representatives on the product's benefits, features, and competitive positioning. Provide them with tools, materials, and strategies to effectively communicate the product's value to healthcare professionals.

- **Launch Materials:** Prepare and produce all necessary materials, including promotional brochures, educational content, digital assets, and patient information. Ensure that these materials comply with regulatory guidelines and are tailored to the needs of different stakeholders.

- **Distribution and Supply Chain:** Coordinate with supply chain partners to ensure the product is available in the required quantities and locations. This includes managing inventory and logistics and ensuring compliance with distribution regulations.

3. **Execution of the Launch**

- **Launch Event:** Organize a launch event or series of events to introduce the product to key stakeholders. This could include medical conferences, seminars, webinars, or promotional events. Use these events to generate excitement, provide education, and build relationships.

- **Marketing and Promotion:** Implement the marketing plan by deploying advertising campaigns, digital marketing, social media strategies, and other promotional activities. Monitor the effectiveness of these activities and adjust strategies as needed.

- **Sales Execution:** Activate the sales force to promote the product to healthcare professionals. Monitor their activities, provide support, and ensure they effectively address customer needs and objections.

4. **Post-Launch Activities**

- **Performance Monitoring:** Track the product's performance through sales data, market feedback, and key performance indicators (KPIs). Analyze this data to assess the product's market penetration, effectiveness of marketing strategies, and overall success.

- **Ongoing Support:** Support healthcare professionals, patients, and other stakeholders. Address any issues, answer quesions, and other additional training or resources needed.

- **Feedback and Improvement:** Gather feedback from users and stakeholders to identify areas for improvement. Use this feedback to refine marketing strategies, address concerns, and enhance the product or its messaging.

- **Lifecycle Management:** Begin planning for the product's lifecycle management, including strategies for managing product updates, extensions, or potential market challenges.

Conclusion

A successful pharmaceutical product launch requires meticulous planning, strategic execution, and ongoing management. From pre-launch activities to post-launch support, every phase ensures that the product meets market needs, complies with regulations, and achieves commercial success. Effective product management throughout this process helps position the product for success, optimize its impact in the market, and achieve long-term growth sustainability.

16.4 Pre-Launch Planning

Pre-launch planning is a crucial phase in the pharmaceutical product launch process, settling the foundation for a successful market introduction. This phase involves comprehensive preparation and strategic activities that ensure the product is ready for market entry and poised for success. Here is an in-depth look at the key components of pre-launch planning:

1. Market Research and Analysis

- **Market Needs Assessment:** Identify unmet medical needs and determine how the new product addresses these needs. This involves understanding the disease or condition the product targets, patient demographics, and current treatment options.

- **Competitive Analysis:** Evaluate the competitive landscape to understand existing products, their strengths and weaknesses, and market share. Identify direct and indirect competitors and assess their marketing strategies, pricing, and positioning.

- **Stakeholder Mapping:** Identify key stakeholders, such as healthcare professionals (HCPs), patients, payers, and regulatory bodies. Understand their needs, preferences, and concerns to tailor your strategy effectively.

- **Market Segmentation:** Segment the market based on geography, demographics, and patient characteristics. This helps in targeting specific groups with customized strategies and messaging.

2. Regulatory Strategy

- **Regulatory Pathway:** Develop a clear regulatory strategy outlining the requirements for approval in different markets. This includes understanding the regulatory agencies' guidelines (e.g., FDA, EMA) and submission planning.

- **Submission Timelines:** Create a timeline for regulatory submissions, including pre-IND (Investigational New Drug) meetings, clinical trial applications, and final product approval. Factor in potential delays and prepare contingency plans.

- **Compliance:** Ensure all documentation, clinical data, and marketing materials comply with regulatory standards. Develop strategies for addressing any regulatory challenges or queries.

3. Pricing and Reimbursement Strategy

- **Pricing Strategy:** Develop a pricing strategy based on market research, cost of production, and perceived value. Consider factors such as competitive pricing, payer expectations, and pricing regulations.

- **Reimbursement Strategy:** Understand the reimbursement landscape and develop strategies for securing reimbursement from payers. This involves preparing Health Technology Assessments (HTAs) and cost-effectiveness analyses to support the product's value proposition.

- **Market Access:** Engage with payers and formulary committees early to facilitate market access. Build relationships and provide evidence of the product's benefits and cost-effectiveness.

4. Stakeholder Engagement

- **Key Opinion Leaders (KOLs):** Identify and engage with KOLs and influencers. These experts can provide valuable insights, support clinical trial recruitment, and advocate for the product.

- **Patient Advocacy Groups:** Collaborate with advocacy organizations to understand patient needs and gain support. These groups can also assist in education and awareness campaigns.

- **Healthcare Professionals:** Establish relationships with HCPs who will be key in prescribing the product. Provide them with information, training, and resources to facilitate adoption.

5. Marketing Strategy Development

- **Positioning and Messaging:** Develop a clear and compelling product positioning that highlights its unique benefits and differentiates it from competitors. Create messaging that resonates with target audiences and aligns with their needs.

- **Brand Development:** Create a strong brand identity, including the product name, logo, and visual elements. Ensure consistency across all materials and channels.

- **Communication Plan:** Develop a communication plan that outlines how the product will be introduced to the market. This plan should include advertising, public relations, digital marketing, and promotional activities.

6. Sales and Training Preparation

- **Sales Strategy:** Develop a sales strategy that includes targeting and segmentation, sales tactics, and performance metrics. Plan for the product's distribution and availability.

- **Sales Training:** Prepare training programs for the sales team to ensure they know the product's features, benefits, and competitive positioning. Equip them with tools and materials to engage effectively with HCPs.

- **Marketing Materials:** Create marketing materials, including brochures, presentations, and digital assets, that support the sales efforts and align with the overall marketing strategy.

7. Distribution and Supply Chain Planning

- **Supply Chain Management:** Develop a supply chain plan that ensures the product is available in the required quantities and locations. This includes managing inventory, production schedules, and logistics.

- **Distribution Channels:** Identify and establish relationships with distribution partners. Ensure that distribution processes comply with regulatory requirements and meet market demands.

Conclusion

Pre-launch planning is a comprehensive process that requires meticulous attention to detail and strategic foresight. By addressing market needs, regulatory requirements, pricing strategies, stakeholder engagement, and marketing preparation, pharmaceutical product managers can lay a solid foundation for a successful product launch. Effective pre-launch planning helps mitigate risks, optimize market entry, and maximize the product's potential impact.

16.5 Launch Execution

Launch execution is when all the preparatory work culminates in introducing a new pharmaceutical product. This stage involves coordinating various activities to ensure a successful launch and establishing the product's presence in the market. Here is a detailed look at the key components of launch execution:

1. **Final Preparation**

   - **Regulatory Approval Confirmation:** Ensure that all regulatory approvals are in place and that the product meets all legal and compliance requirements. Verify that labeling, packaging, and promotional materials comply with regulations.

   - **Distribution Readiness:** Confirm that the supply chain and distribution channels are fully operational. Ensure that the product is available in the intended markets and that distribution partners are prepared to handle the product.

   - **Sales and Marketing Materials:** Finalize and distribute marketing and sales materials, including brochures, presentations, and digital assets. Ensure that all materials are aligned with the product's branding and messaging.

   - **Training Programs:** Complete training for the sales force, healthcare professionals (HCPs), and internal teams. Ensure they are well-informed about the product, its benefits, and key selling points.

2. **Market Introduction**

   - **Launch Event:** Organize a launch event or series of events to introduce the product to key stakeholders, including HCPs, industry leaders, and media. The event should highlight the product's key benefits, clinical data, and market positioning.

   - **Media and Public Relations:** Implement a media and public relations strategy to generate buzz and awareness. This may include press releases, media interviews, and articles in industry publications.

- **Digital Campaigns:** Launch digital marketing campaigns across various channels, including social media, email, and online advertising. Use targeted messaging to reach specific segments of the audience and drive engagement.

- **Sales Activation:** Deploy the sales team to engage with HCPs, hospitals, and clinics. Focus on building relationships, providing product information, and addressing questions or concerns.

3. Monitoring and Support

- **Sales Tracking:** Monitor sales performance and track key metrics, such as market share, sales volume, and revenue. Use this data to assess the launch strategy's effectiveness and identify areas for improvement.

- **Market Feedback:** Collect feedback from HCPs, patients, and other stakeholders to gauge the product's reception and identify any issues. Use surveys, focus groups, and direct interactions to gather insights.

- **Customer Support:** Provide ongoing support to healthcare professionals and patients. This includes addressing product-related issues, answering questions, and offering additional resources or training.

- **Adverse Event Reporting:** Implement a system for reporting and managing adverse events or side effects. Ensure that all reports are handled in compliance with regulatory requirements and that appropriate actions are taken.

4. Adjustments and Optimization

- **Performance Review:** Regularly review the launch performance against predefined goals and objectives. Analyze sales data, market penetration, and feedback to evaluate the launch's success.

- **Strategic Adjustments:** Based on the performance review, make necessary marketing and sales strategy adjustments. This may involve refining messaging, adjusting pricing, or targeting new market segments.

- **Ongoing Marketing Efforts:** Continue marketing and promotional activities to maintain product visibility and drive ongoing engagement. This may include follow-up campaigns, continued media outreach, and participation in industry events.

5. Long-term Planning

- **Lifecycle Management:** Develop a plan for the product's lifecycle management, including strategies for product enhancements, line extensions, and market expansion. Ensure that the product remains competitive and continues to meet market needs.

- **Post-Launch Evaluation:** Conduct a comprehensive post-launch evaluation to assess the overall impact of the launch. Identify lessons learned and best practices to inform future product launches.

- **Stakeholder Relationships:** Maintain and strengthen relationships with key stakeholders, including HCPs, patients, and payers. Continue to engage with them to gather insights and foster loyalty.

Conclusion

Launch execution is a critical phase that requires careful coordination and strategic implementation. By effectively managing the introduction of a pharmaceutical product, ensuring regulatory compliance, and engaging with stakeholders, pharmaceutical product managers can maximize the product's impact and achieve long-term success. The ability to monitor performance, make adjustments, and plan for the future ensures that the product remains competitive and continues to meet the market's needs.

16.6 Post-Launch Monitoring

Post-launch monitoring is a crucial phase in the lifecycle of a pharmaceutical product, involving the continuous assessment and evaluation of the product's performance and impact after it has been introduced to the market. This phase helps ensure the product meets its market goals, adheres to regulatory requirements, and addresses emerging issues. Here is a detailed overview of post-launch monitoring:

1. **Sales Performance Tracking**

   - **Sales Metrics:** Monitor key sales metrics such as sales volume, revenue, market share, and growth rates. Compare these metrics against predefined targets and benchmarks to assess the product's performance.

   - **Geographic and Demographic Analysis:** Analyze sales data by region, demographic segment, and sales channel. This helps identify areas of strength and opportunities for further market penetration.

   - **Sale Trends:** Track sales trends over time to identify patterns, seasonal variations, and deviations from expected performance.

2. **Market and Competitive Analysis**

   - **Market Feedback:** Collect feedback from healthcare professionals (HCPs), patients, and other stakeholders regarding the product's efficacy, safety, and overall satisfaction. Use surveys, interviews, and focus groups to gather insights.

   - **Competitive Landscape:** Monitor the competitive landscape to assess the performance of competing products. Analyze competitor strategies, market share, and any new product developments.

   - **Market Dynamics:** Stay informed about changes in market dynamics, such as shifts in patient needs, emerging trends, and regulatory updates that could impact the product's success.

3. Adverse Event Monitoring

- **Adverse Event Reporting:** Implement a system for collecting and reporting adverse events or side effects related to the product. Ensure all reports are documented, analyzed, and communicated to relevant regulatory authorities.

- **Risk Management:** Assess and manage any risks associated with the product. Develop strategies to address adverse events, improve product safety, and mitigate potential issues.

- **Regulatory Compliance:** Ensure ongoing compliance with regulatory requirements for adverse event reporting and risk management. Update safety information and product labeling as needed.

4. Customer Support and Engagement

- **Support Channels:** Provide robust channels for healthcare professionals and patients to address questions, concerns, and issues related to the product.

- **Educational Resources:** Continue to offer educational resources and training for HCPs and patients to ensure proper use of the product and enhance their understanding.

- **Patient Feedback:** Collect and analyze patient feedback to assess their experience with the product. Use this information to improve patient support and address any challenges.

5. Marketing and Communication

- **Marketing Effectiveness:** Evaluate the effectiveness of ongoing marketing and promotional activities. Assess the impact of campaigns, advertisements, and digital marketing efforts on product awareness and engagement.

- **Adjustments:** Based on performance data and market feedback, make necessary adjustments to marketing strategies. This may include refining messaging, targeting new segments, or enhancing promotional activities.

- **Public Relations:** Manage public relations efforts to maintain a positive image of the product and address any negative publicity or concerns.

6. Financial and Operational Metrics

- **Cost Management:** Monitor the financial performance of the product, including production costs, marketing expenses, and profitability. Ensure that the product remains financially viable.

- **Operational Efficiency:** Assess the efficiency of supply chain and distribution operations. Identify and address any bottlenecks or issues impacting product availability and delivery.

- **Budget Adherence:** Track adherence to the budget and financial forecasts. Make adjustments as needed to stay within budget and optimize resource allocation.

7. Strategic Planning

- **Long-Term Strategy:** Develop and implement a long-term strategy for managing the product's lifecycle. This may include product enhancements, line extensions, and market expansion plans.

- **Product Evolution:** Evaluate opportunities for product evolution based on market needs and technological advancements. Consider potential improvements or new indications that could enhance the product's value.

- **Future Planning:** Plan for future phases of the product lifecycle, including potential challenges and opportunities. Prepare for shifts in market conditions and evolving stakeholder needs.

Conclusion

Post-launch monitoring is essential for ensuring the ongoing success of the pharmaceutical product. By continuously tracking performance, managing risks, engaging with stakeholders, and adapting strategies, pharmaceutical product managers can address emerging issues, optimize the product's impact, and ensure long-term success. Effective post-launch monitoring helps maintain product quality, meet market demands, and uphold regulatory compliance, contributing to the product's overall success.

Lifecycle Management

Lifecycle management (LCM) in pharmaceutical product management is a strategic approach that maximizes a drug's commercial potential throughout its various stages—from inception and market introduction to maturity and eventual decline. Effective lifecycle management ensures a product remains competitive, profitable, and relevant in an ever-changing market. It's not just about the initial launch but the continuous efforts to sustain, grow, and extend the product's market presence.

In this chapter, we will explore the foundational concepts of lifecycle management (LCM), its significance in pharmaceutical marketing, and how pharmaceutical companies can strategically plan and execute LCM to navigate the complexities of the market.

The Importance of Lifecycle Management

Every pharmaceutical product has a finite lifecycle that typically follows a familiar curve: development, introduction, growth, maturity, and decline. Successful management of this cycle can help pharmaceutical companies:

- **Maximize Revenue**: By extending the profitable phases and delaying the decline of the drug.
- **Adapt to Market Changes**: Through strategic responses to competition, patent expirations, and regulatory changes.

- **Maintain Market Relevance**: By offering value through innovations such as new formulations or expanded indications.

When done well, lifecycle management transforms a static product into a dynamic asset that can evolve with market demands and regulatory landscapes.

The Lifecycle Stages

1. **Development and Pre-Launch Planning**:
   - Focuses on clinical trials, regulatory approval, and market research to ensure a successful launch. Decisions made here have long-lasting impacts on the product's lifecycle.

2. **Product Launch**:
   - The introduction of the product into the market, where brand strategy, pricing, positioning, and marketing campaigns play crucial roles.

3. **Growth Stage**:
   - This is the phase where the product gains market share, and sales increase rapidly. Strategies focus on expanding market penetration and maximizing product adoption.

4. **Maturity Stage**:
   - Sales begin to plateau. This is a critical period when companies introduce lifecycle extension strategies such as new indications, combinations, or geographic expansion.

5. **Decline Stage**:
   - Sales fall due to patent expiration, competition, or market saturation. The focus is often on maintaining profitability through cost management and transitioning to newer products.

Strategic Elements of Lifecycle Management

1. **Product Differentiation**:
   - Establishing a unique value proposition to stand out from the competition is essential for the initial launch and sustaining the product in its later stages.

2. **Post-Launch Surveillance and Adaptation**:
   - Continuous monitoring of product performance, patient outcomes, and market conditions, allowing for quick adjustments in marketing strategies or formulation.

3. **Patent Strategy and Intellectual Property Management**:
   - Protecting and extending market exclusivity through intellectual property strategies like patent extensions, new formulations, or combination therapies.

4. **Lifecycle Extension**:
   - Implementing reformulations, line extensions, or geographic expansion extends the product lifecycle beyond its original scope.

5. **Managing Product Decline**:
   - Preparing for the decline phase involves optimizing profitability through cost-effective strategies while planning to transition to newer products.

In the following sections, we will dive deeper into each phase of the lifecycle, explore strategies for successful lifecycle management, and analyze real-world case studies highlighting successful and challenging LCM scenarios. The goal is to provide pharma product managers with a comprehensive toolkit for maximizing the value of their products throughout their lifecycle.

17.1 Introduction to Lifecycle Management

Lifecycle management (LCM) in pharmaceutical product management is a strategic approach that involves overseeing a product from its initial development and market launch through its entire commercial life until its phase-out or discontinuation. Effective lifecycle management ensures that a pharmaceutical product remains competitive, meets market needs, and maximizes its value and profitability throughout its entire lifecycle.

What is Lifecycle Management?

Lifecycle management is a comprehensive strategy encompassing all stages of a product's life, from inception through development, market introduction, growth, maturity, and eventual decline. It involves coordinating various activities and strategically optimizing the product's performance and value at each stage. The goal is to extend the product's market presence, adapt to changing market conditions, and ensure continued commercial success.

Why is Lifecycle Management Important?

1. **Maximizing Product Value**: Effective LCM helps maximize a product's value by identifying growth opportunities, optimizing market strategies, and extending its life.

2. **Adapting to Market Changes**: The pharmaceutical market is dynamic, with evolving treatment guidelines, competitive pressures, and regulatory changes. LCM allows product managers to adapt to these changes and keep the product relevant.

3. **Strategic Resources Allocation**: By managing the product's lifecycle strategically, pharmaceutical companies can allocate resources more efficiently, focusing on high-impact activities and areas with the greatest potential for return on investment.

4. **Risk Mitigation**: LCM involves identifying and mitigating risks associated with the product, such as declining sales, regulatory issues, or market competition. This proactive approach helps minimize potential negative impacts.

5. **Enhancing Patient Outcomes**: By continually evaluating and improving the product, LCM contributes to better patient outcomes by making effective treatments and innovations available.

Phases of Lifecycle Management

1. **Introduction**: The phase where the product is launched into the market. Key activities include market entry strategies, promotional campaigns, and initial sales efforts.

2. **Growth**: During this phase, the product experiences increasing market acceptance and sales. Strategic focus on market expansion, product optimization, and building a strong market presence.

3. **Maturity**: The product reaches peak market penetration and sales. The focus shifts to maintaining market position, managing competition, and exploring product enhancements or line extension opportunities.

4. **Decline**: Sales decline due to market saturation, competition, or changing treatment needs. Strategies during this phase involve managing the decline, optimizing costs, and planning for product phase-out.

5. **Post-Lifecycle**: After the product is discontinued, activities include managing the remaining inventory, communicating with stakeholders, and conducting post-mortem analyses to derive lessons learned.

Role of the Product Manager In Lifecycle Management

Pharmaceutical product managers play a crucial role in lifecycle management. Their responsibilities include:

- **Strategic Planning**: Developing and executing strategies for each phase of the product lifecycle to achieve business objectives and market goals.

- **Cross-Functional Coordination**: Working with various teams, including marketing, sales, R&D, and regulatory affairs, to ensure cohesive and effective lifecycle management.

- **Market Analysis**: Monitor market trends, competitor activities, and stakeholder feedback to inform decision-making and strategy adjustments.

- **Risk Management**: Identifying potential risks and implementing mitigation strategies to address challenges and minimize impacts on the product's success.

- **Stakeholder Engagement**: Engaging with healthcare professionals, patients, and other stakeholders to gather insights, address concerns, and ensure the product remains aligned with market needs.

Post-Lifecycle Activities

- **Lessons Learned**: Conduct a post-mortem analysis to identify lessons learned from the product's lifecycle. Evaluate what worked well and what could be improved for future product management efforts.

- **Intellectual Property and Data Management**: Manage any remaining intellectual property rights and data associated with the product. Consider opportunities to leverage intellectual property in future developments.

- **Customer Transition**: Facilitate customers' transition to alternative treatments or products. Provide support and communication to ensure a smooth transition for patients and healthcare providers.

Conclusion

Lifecycle management is a critical aspect of pharmaceutical product management that requires a strategic and proactive approach to oversee a product's entire life. By effectively managing each phase, pharmaceutical companies can enhance the product's value, adapt to market changes, and achieve long-term commercial success. The role of the product manager is essential in executing these strategies and ensuring the product's continued relevance and profitability.

17.2 Strategies for Lifecycle Extension

Extending the lifecycle of a pharmaceutical product involves implementing strategies that prolong its market presence and maintain or enhance its profitability. Lifecycle extension is crucial for maximizing the return on investment and ensuring sustained success despite evolving market conditions and competition. Here are some effective strategies for extending the lifecycle of a pharmaceutical product:

1. **Product Line Extensions and Variations**

   - **Formulation Changes**: Introduce new formulations, such as extended-release or combination products, to meet patient needs or preferences.

   - **New Indications**: Obtain regulatory approval for new indications or uses of the product to expand its market and reach new patient populations.

   - **Dosage Forms**: Develop alternative dosage forms, such as tablets, injectables, or topical solutions, to offer more convenience or cater to different patient preferences.

2. **Market Expansion**

   - **Geographic Expansion**: Launch the product in new geographical markets where it is unavailable. This can include entering emerging markets with growing healthcare needs.

   - **New Patient Segments**: Target new patient demographics or subpopulations that may benefit from the product, such as specific age groups or disease stages.

3. **Enhancing Product Value**

   - **Improved Formulations**: Invest in research and development to enhance the product's efficacy, safety, or patient compliance, thereby increasing its value proposition.

   - **Packaging Innovations**: Update packaging to improve usability, safety, or convenience, which can positively impact patient adherence and market appeal.

4. Strategic Pricing and Reimbursement

- **Pricing Strategies**: Implement pricing strategies that optimize the product's market position, such as value-based or tiered pricing models.

- **Reimbursement Negotiations**: Work with payers to secure favorable reimbursement terms, which can increase product accessibility and adoption.

5. Marketing and Promotion

- **Targeted Marketing Campaigns**: Develop targeted marketing campaigns to address new patient populations, emerging trends, or shifts in healthcare priorities.

- **Educational Initiatives**: Launch educational programs for healthcare professionals and patients to increase awareness and understanding of the product's benefits and uses.

6. Lifecycle Management Innovations

- **Clinical Trials and Research**: Conduct additional clinical trials to gather new evidence on the product's effectiveness or safety, which can support new claims or indications.

- **Patient Support Programs**: Implement patient support programs to enhance adherence, manage side effects, and provide additional resources for patients using the product.

7. Strategic Partnerships

- **Collaborations and Licensing**: Partner with other companies or organizations to explore new applications, co-market the product, or leverage complementary technologies and expertise.

- **Academic and Research Collaborations**: Collaborate with academic institutions or research organizations to generate new data, explore innovative uses, or enhance product development.

8. Regulatory and Compliance Strategies

- **Regulatory Updates**: Stay abreast of regulatory changes and adapt the product's labeling, indications, or marketing materials to comply with new guidelines or requirements.

- **Regulatory Submissions**: To extend the product's lifecycle, pursue additional regulatory approvals for new indications, formulations, or markets.

9. Customer Feedback and Insights

- **Patient and HCP Feedback**: Collect and analyze feedback from patients and healthcare professionals to identify areas for improvement and address unmet needs.

- **Market Research**: Conduct ongoing market research to understand evolving trends, competitive dynamics, and opportunities for product enhancement.

10. Post-Market Surveillance and Quality Improvement

- **Monitoring and Evaluation**: Implement robust post-market surveillance to monitor product performance, identify potential issues, and ensure ongoing quality and safety.

- **Continuous Improvement**: Use insights from post-market data to continuously improve the product, address emerging concerns, and enhance its market position.

Conclusion

Lifecycle extension strategies are essential for maximizing the value and success of a pharmaceutical product throughout its commercial life. By leveraging product innovations, market expansions, strategic pricing, and effective marketing, pharmaceutical companies can extend the product's lifecycle, adapt to changing market conditions, and sustain profitability. These strategies require product managers' proactive and strategic approach to ensure the product remains relevant and competitive in the ever-evolving healthcare landscape.

17.3 Product Maturity and Decline

Understanding the phases of product maturity and decline is essential for effective pharmaceutical product management. These phases present unique challenges and opportunities that require strategic planning to maximize product value and manage transitions smoothly.

Product Maturity

1. **Characteristics of Product Maturity**

   - **Stable Sales and Market Share**: During maturity, the product's sales growth stabilizes as it has captured a significant market share. The product has reached widespread acceptance and is well-established in the market.

   - **Intense Competition**: The product faces intense competition from established brands and new entrants. Competition may offer similar or improved versions of the product.

   - **High Market Penetration**: The product is commonly used by the target patient population, and its benefits are well understood by healthcare professionals (HCPs) and patients.

   - **Steady Revenue**: Revenue growth slows as the market becomes saturated and the focus shifts to maintaining market share and profitability.

2. **Strategies During Product Maturity**

   - **Product Differentiation**: Enhance the product's differentiation through improvements, such as new formulations, packaging innovations, or additional indications, to maintain a competitive edge.

   - **Cost Management**: Optimize production and operational costs to maintain profitability despite slower revenue growth. Implement efficiencies in manufacturing and supply chain management.

- **Enhanced Marketing**: Focus on targeted marketing and promotional activities to reinforce the product's benefits, maintain brand loyalty, and address emerging market needs.

- **Customer Engagement**: Strengthen relationships with HCPs and patients through education, support programs, and personalized communication to ensure continued product usage and adherence.

- **Market Expansion**: Explore opportunities to expand into new geographic markets or patient segments to sustain growth and capture additional revenue.

Product Decline

1. Characteristics of Product Decline

- **Decreasing Sales and Market Share**: Sales begin to decline due to market saturation, loss of patent protection, or competitors' introduction of superior alternatives.

- **Reduced Market Demand**: Demand for the product decreases as new treatments become available or changes in medical guidelines reduce the product's relevance.

- **Increasing Competition**: Competitive pressure intensifies as new or improved products enter the market, offering better efficacy, safety, or convenience.

- **Eroding Profit Margins**: Profit margins may shrink due to declining sales volumes, increased competition, or rising costs associated with maintaining the product in the market.

2. Strategies During Product Decline

- **Cost Reduction**: Implement cost-cutting measures to manage declining revenues. This may involve streamlining operations, reducing marketing expenditures, and optimizing supply chain processes.

- **Product Rationalization**: Evaluate the product's performance and consider discontinuation if it is no longer viable or profitable. Focus on reallocating resources to more promising products or new opportunities.

- **Focus on Niche Markets**: Identify and target niche markets or specific patient segments where the product still holds value. Tailor marketing efforts to these segments to maximize revenue.

- **Lifecycle Management**: Implement strategies to extend the product's lifecycle, such as pursuing additional indications, conducting new research, or exploring alternative uses.

- **Transition Planning**: Develop a transition plan to manage the product exit from the market. This includes communicating with stakeholders, managing inventory levels, and ensuring continuity of patient care.

Conclusion

Managing the phases of product maturity and decline requires a strategic approach to adapt to changing market dynamics and ensure continued success. During Maturity, the focus is on maintaining market share, optimizing costs, and enhancing differentiation. In decline, the emphasis shifts to cost management, rationalization, and exploring niche opportunities. Effective management of these phases is crucial for maximizing the value of pharmaceutical products and ensuring a smooth transition as market conditions evolve.

The Role of the Product Manager

The role of a pharmaceutical product manager is multifaceted and critical to the success of a pharmaceutical product throughout its lifecycle. This position involves overseeing various product development, marketing, and market performance aspects. Here is an in-depth look at the key responsibilities and functions of a pharmaceutical product manager:

1. **Strategic Planning**

   - **Market Analysis**: Conduct thorough market research to understand industry trends, competitive landscape, and unmet medical needs. Use this information to develop strategic plans for product positioning and differentiation.

   - **Product Strategy**: Define the product vision and strategy in alignment with the company's objectives. Create long-term plans that outline goals, target markets, and key performance indicators (KPIs).

   - **Lifecycle Management**: Develop and implement strategies for each phase of the product lifecycle, from development to decline, to maximize the product's value and adapt to market changes.

2. **Cross-Functional Coordination**

   - **Team Leadership**: Lead and coordinate cross-functional teams, including research and development (R&D), marketing, sales, regulatory affairs, and medical affairs, to ensure the successful execution of product strategies.

- **Communication**: Facilitate effective communication between teams and stakeholders. Ensure that all departments are aligned with the product strategy and objectives.

- **Project Management**: Oversee the progress of product development projects, ensuring that timelines, budgets, and milestones are met. Address any issues or roadblocks that arise during the process.

3. **Market Access and Regulatory Compliance**

- **Regulatory Affairs**: Work with regulatory teams to navigate the approval process and ensure compliance with regulatory requirements. Prepare and review regulatory submissions and interact with regulatory agencies as needed.

- **Market Access**: Develop strategies to gain market access and reimbursement for the product. Collaborate with payer and health economics teams to demonstrate the product's value and negotiate pricing and reimbursement terms.

4. **Marketing and Sales**

- **Marketing Strategy**: Create and implement marketing plans that include positioning, messaging, and promotional activities. Develop marketing materials and campaigns to support product launches and ongoing promotions.

- **Sales Enablement**: Develop training materials, tools, and strategies to support the sales team. Monitor sales performance and adjust marketing tactics as needed to achieve sales targets.

- **Customer Engagement**: Build relationships with healthcare professionals (HCPs), patients, and other stakeholders. Gather feedback and insights to inform product improvements and marketing strategies.

5. Financial Management

- **Budgeting**: Develop and manage the product budget, including marketing, R&D, and operational expenses. Ensure that all expenditures are aligned with the product's financial goals.

- **Forecasting**: Create sales forecasts and financial projections based on market data, historical performance, and strategic plans. Monitor actual performance against forecasts and adjust strategies as needed.

- **Cost Management**: Optimize product development, production, and marketing costs to maximize profitability and return on investment.

6. Innovation and Problem-Solving

- **Innovation**: Identify opportunities for product innovation and improvement. Stay informed about emerging trends, technologies, and scientific advancements that could impact the product or the market.

- **Problem-Solving**: Address challenges and issues that arise during the product lifecycle. Develop and implement solutions to overcome obstacles and achieve product objectives.

7. Monitoring and Evaluation

- **Performance Tracking**: Monitor performance through KPIs, sales data, market share analysis, and other metrics. Evaluate the effectiveness of marketing and sales strategies and make data-driven decisions to optimize performance.

- **Market Feedback**: Gather and analyze feedback from customers, HCPs, and other stakeholders. Use this feedback to make informed decisions about product improvements and strategic adjustments.

Conclusion

A pharmaceutical product manager's role is integral to a product's success in the highly competitive and regulated pharmaceutical industry. It requires strategic thinking, cross-functional leadership, financial acumen, and market insight. By effectively managing all aspects of the product lifecycle, from development to market exit, the product manager ensures that the product delivers value to patients, meets business objectives, and achieves long-term success.

18.1 Strategic Planning and Cross-Functional Leadership

Strategic Planning

1. Market Analysis and Research

- **Objective**: Understand the market landscape, identify opportunities, and recognize threats.
- **Activities**:
  - **Competitive Analysis**: Assess competitors' products, strategies, strengths, and weaknesses.
  - **Customer Insights**: Gather data on customer needs, preferences, and behaviors through surveys, interviews, and market reports.
  - **Trend Identification**: Monitor industry trends, regulatory changes, and technological advancements that may impact the product.

2. Product Vision and Strategy Development

- **Objective**: Define the long-term goals and strategic direction for the product.
- **Activities**:
  - **Vision Statement**: Develop a clear and compelling vision for the product that aligns with the company's overall mission and objectives.
  - **Strategic Objectives**: Set specific, measurable, achievable, relevant, and time-bound (SMART) goals for the product.
  - **Positioning and Differentiation**: Determine how the product will be positioned in the market and what makes it unique compared to competitors.

3. Business Case and Financial Planning

- **Objective**: Justify the investment in the product and ensure its financial viability.

- **Activities:**
  - **Business Case Development**: Create a detailed business case that outlines the product's value proposition, target market, competitive advantage, and expected return on investment (ROI).
  - **Budgeting and Forecasting**: Develop budgets and financial forecasts to plan for development, marketing, and operational costs. Adjust forecasts based on market conditions and performance.

4. Strategic Roadmap and Implementation Plan

- **Objective**: Develop a detailed plan to execute the product strategy and achieve objectives.
- **Activities:**
  - **Roadmap Creation**: Outline key milestones, timelines, and deliverables for each phase of the product lifecycle.
  - **Action Plans**: Develop actionable plans for product development, marketing, sales, and post-launch activities. Assign responsibilities and establish deadlines.

5. Performance Measurement and Adaptation

- **Objective**: Monitor progress and make adjustments as needed to stay on track.
- **Activities:**
  - **Key Performance Indicators (KPIs)**: Define KPIs to track progress towards strategic goals and measure success.
  - **Review and Adaptation**: Regularly review performance data, market feedback, and industry trends. Adapt strategies and plans based on insights and changing conditions.

Cross-Functional Leadership

1. Team Coordination and Collaboration

- **Objective**: Ensure effective collaboration among different functional teams to achieve product goals.

- **Activities**:
  - **Team Leadership**: Lead and motivate cross-functional teams, including R&D, marketing, sales, and medical affairs.
  - **Facilitation of Communication**: Establish regular communication channels and meetings to keep all teams informed and aligned with the product strategy.
  - **Conflict Resolution**: Address and resolve conflicts or issues among team members or between departments.

2. Stakeholder Management

- **Objective**: Build and maintain strong relationships with internal and external stakeholders.
- **Activities**:
  - **Internal Stakeholders**: Work closely with executives, department heads, and other key internal stakeholders to gain support and ensure alignment with the product strategy.
  - **External Stakeholders**: Engage with external stakeholders such as healthcare professionals (HCPs), patients, regulatory bodies, and partners to gather feedback and build relationships.

3. Project Management and Execution

- **Objective**: Oversee the execution of strategic plans and ensure timely delivery of milestones.
- **Activities**:
  - **Project Planning**: Develop detailed project plans, including timelines, resource allocation, and deliverables.
  - **Monitoring and Reporting**: Track project progress, manage risks, and report on performance to stakeholders. Ensure that projects stay on schedule and within budget.
  - **Resource Management**: Allocate resources effectively and manage budgets to support project execution.

4. Decision-Making and Problem-Solving

- **Objective**: Make informed decisions and address challenges that impact the product's success.
- **Activities**:
  - **Data-Driven Decisions**: Use data and insights to make strategic decisions and solve problems.
  - **Problem Identification**: Identify potential issues or obstacles early and develop strategies to address them.
  - **Solution Implementation**: Implement solutions and monitor their effectiveness. Make adjustments as needed to achieve desired outcomes.

5. Change Management

- **Objective**: Manage change effectively and ensure smooth transitions during strategic shifts.
- **Activities**:
  - **Change Planning**: Develop and communicate plans for implementing changes in strategy, process, or product direction.
  - **Impact Assessment**: Assess the impact of changes on teams, stakeholders, and project outcomes.
  - **Support and Training**: Support and train teams to help them adapt to changes and maintain productivity.

Conclusion

Strategic planning and cross-functional leadership are crucial components of pharmaceutical product management. Effective strategic planning ensures that the product is well-positioned to succeed in the market, while cross-functional leadership ensures that all teams work together harmoniously to achieve the product's goals. By mastering these areas, pharmaceutical product managers can drive product success, adapt to changing market conditions, and deliver value to patients and stakeholders.

18.2 Continuous Market Assessment and Adaptation

1. Continuous Market Assessment

- **Objective**: Ensure the product remains relevant and competitive in a dynamic market environment.

- **Key Activities**:

  - **Market Monitoring**: Regularly track market trends, emerging technologies, and competitor activities to stay informed about changes that could impact the product. This involves subscribing to industry reports, monitoring news sources, and attending relevant conferences and webinars.

  - **Customer Feedback Collection**: Continuously gather customer feedback, including healthcare professionals (HCPs), patients, and other stakeholders. Use surveys, interviews, focus groups, and online reviews to understand their needs, preferences, and experiences with the product.

  - **Competitive Analysis**: Perform ongoing analysis of competitors' products, marketing strategies, pricing, and market positioning. This helps identify strengths and weaknesses relative to competitors and spot potential opportunities and threats.

  - **Sales and Performance Data Analysis**: Analyze sales data, market share, and other performance metrics to assess the product's performance. Look for trends, anomalies, and patterns that could provide insights into market dynamics or product performance.

  - **Regulatory and Policy Changes**: Keep abreast of healthcare regulations, policies, and reimbursement landscape changes that could affect the product. This involves tracking legislative developments and understanding their potential impact on the product's market access and commercial viability.

2. Adaptation Strategies

- **Objective**: Adjust strategies and tactics based on market insights to optimize product performance and maintain competitive advantage.

- **Key Activities**:

  - **Strategy Review and Adjustment**: Regularly review and update the product's strategic plan based on market assessment findings. This might include revising the product's positioning, target audience, or marketing strategies to align with current market conditions.

  - **Product Improvements**: Implement changes or enhancements to the product based on customer feedback and market needs. This could involve modifying the product's formulation, packaging, or labeling to address unmet needs or improve user experience.

  - **Marketing and Sales Tactics**: Adjust marketing and sales tactics in response to market feedback and competitive dynamics. This could include changing promotional strategies, refining messaging, or adjusting pricing strategies to better meet market demands.

  - **Operational Flexibility**: Ensure that the product's supply chain, manufacturing processes, and distribution channels are flexible enough to adapt to market demand, disruptions, and changes. This may involve establishing contingency plans and building relationships with alternative suppliers.

  - **Stakeholder Engagement**: Maintain strong relationships with key stakeholders, including HCPs, patients, and payers. Engage them through ongoing communication, educational initiatives, and collaborations to ensure continued support and alignment with the product's value proposition.

  - **Innovation and Technology Integration**: Explore and integrate new technologies and innovations that could enhance the product's performance or create additional value for customers. This could involve adopting digital health tools, leveraging data analytics, or incorporating advancements in drug delivery systems.

Here is a case study that shows how a pharmaceutical company adapted to a new regulatory requirement.

Case Study #13: Adapting to a New Regulatory Requirement

Scenario:

A pharmaceutical company's product was initially launched with an effective and well-tolerated dosing regimen. However, a new regulatory requirement mandated changes to the labeling and dosing guidelines.

Adaption Process:

1. **Market Assessment**: The company thoroughly reviewed new regulatory requirements and assessed their potential impact on product usage and market positioning.

2. **Stakeholder Consultation**: Engaged with regulatory bodies, healthcare professionals, and patients to understand the implications of the new requirements and gather feedback on the proposed changes.

3. **Strategic Adjustment**: The product's labeling and dosing instructions were revised to comply with the new regulations, and marketing materials and educational resources were updated to reflect the changes.

4. **Communication and Training**: The company trained sales representatives, healthcare professionals, and other stakeholders on the new dosing guidelines and updated labeling.

5. **Monitoring and Feedback**: Continuously monitored the product's performance and stakeholder feedback following the adaptation to ensure compliance and address any emerging issues.

Conclusion

Pharmaceutical product managers must continuously assess and adapt to the market to ensure their products remain competitive and meet evolving needs. By actively monitoring market trends, gathering customer feedback, and adjusting strategies and tactics accordingly, they can enhance product performance, maintain market relevance, and drive long-term success.

18.3 Leveraging Digital Tools for Launch and Lifecycle Management

1. Digital Tools for Product Launch

- **Objective**: Enhance the efficiency and effectiveness of product launches through digital technologies and platforms.

- **Key Tools and Techniques**:

  - **Digital Project Management Platforms**: Tools like Asana, Trello, and Microsoft Project help streamline project management tasks, facilitate team collaboration, and track progress. These platforms allow for real-time updates, task assignments, and milestone tracking, ensuring that all aspects of the launch are on schedule.

  - **Customer Relationship Management (CRM) Systems**: CRM systems like Salesforce and HubSpot manage interactions with healthcare professionals (HCPs), patients, and other stakeholders. They support targeted outreach, lead management, and engagement metrics tracking, helping build and maintain relationships throughout the launch process.

  - **Marketing Automation Tools**: Platforms like Marketo, Eloqua, and Mailchimp automate marketing activities, including email campaigns, social media posts, and digital advertisements. These tools enable personalized communication, track campaign performance, and optimize marketing strategies based on data-driven insights.

  - **Content Management Systems (CMS)**: CMS platforms like WordPress and Drupal facilitate digital content creation, management, and optimization. They are crucial for developing and updating product websites, educational materials, and promotional content.

  - **Data Analytics and Business Intelligence (BI) Tools**: Tools like Google Analytics, Tableau, and Power BI provide insights into market trends, customer behavior, and campaign performance. These tools enable data-driven decision-making and help assess the effectiveness of launch strategies.

- **Virtual Collaboration Platforms**: Solutions like Zoom, Microsoft Teams, and Slack support remote collaboration and communication among cross-functional teams. They are essential for coordinating efforts, conducting virtual meetings, and sharing information during the launch phase.

2. Digital Tools for Lifecycle Management

- **Objective**: Optimize the ongoing management of the product through digital tools that support continuous improvement and adaptation.

- **Key Tools and Techniques**:

  - **Product Lifecycle Management (PLM) Systems**: PLM tools like PTC Windchill and Siemens Teamcenter manage the entire product lifecycle, from development to discontinuation. They support version control, regulatory compliance, and documentation management.

  - **Market Intelligence Platforms**: Tools such as IQVIA and Symphony Health provide real-time insights into market dynamics, competitor activities, and healthcare trends. These platforms help product managers identify opportunities and threats throughout the product lifecycle.

  - **Patient and HCP Engagement Platforms**: Digital tools like MedPage Today and HealthTap facilitate ongoing engagement with patients and HCPs. They support patient education, gather feedback, and enhance communication, which is crucial for maintaining product relevance and addressing evolving needs.

  - **Digital Analytics for Performance Monitoring**: Tools like Mixpanel and Adobe Analytics track and analyze user interactions with digital assets, such as websites and mobile apps. They provide insights into user behavior, product performance, and areas for improvement.

- **Regulatory Compliance and Risk Management Tools**: Platforms like Veeva Vault and MasterControl manage regulatory documents, compliance processes, and risk assessments. They ensure adherence to regulatory requirements and help mitigate risks throughout the product lifecycle.

- **Customer Feedback and Support Tools**: Solutions such as Zendesk and Freshdesk handle customer support inquiries, manage feedback, and track issue resolution. These tools are essential for addressing concerns, improving product quality, and enhancing customer satisfaction.

The following case study demonstrates how you can optimize product launches with digital tools.

Case Study #14: Optimizing a Product Launch with Digital Tools

Scenario:

A pharmaceutical company is launching a new medication and aims to ensure successful market entry through digital tools.

Implementation:

1. **Project Management**: The company uses Asana to manage launch tasks, track progress, and coordinate efforts across teams. This helps ensure all launch activities are completed on time and within budget.

2. **CRM and Marketing Automation**: Salesforce CRM manages HCP interactions, while Marketo automates email campaigns and social media promotions. This approach allows for personalized outreach and targeted marketing.

3. **Data Analytics**: Google Analytics and Tableau monitor website traffic, campaign performance, and market trends. The insights from these tools guide decision-making and strategy adjustments.

4. **Virtual Collaboration**: Zoom and Slack facilitate remote meetings and team collaboration, ensuring seamless communication and coordination during the launch phase.

5. **Content Management**: WordPress is utilized to update the product website with relevant information, educational resources, and promotional content, ensuring that stakeholders have access to accurate and up-to-date details.

Enhancing Lifecycle Management with Digital Tools

The same pharmaceutical company needs to manage the product throughout its lifecycle, focusing on performance optimization and market adaptation.

Implementation:

1. **PLM System**: PTC Windchill is used to manage product documentation, regulatory compliance, and version control, ensuring smooth operations and adherence to standards.

2. **Market Intelligence**: IQVIA provides insights into competitor activities and market trends, helping the company adjust its strategies and identify new opportunities.

3. **Engagement Platforms**: MedPage Today and HealthTap are used to engage with HCPs and patients, gather feedback, and address evolving needs.

4. **Digital Analytics**: Mixpanel tracks user interactions with the product's digital assets, providing insights into user behavior and identifying areas for improvement.

5. **Customer Support**: Zendesk manages customer support inquiries and feedback, helping to resolve issues and improve customer satisfaction.

Conclusion

Leveraging digital tools is essential for optimizing product launch and lifecycle management in the pharmaceutical industry. By incorporating project management platforms, CRM systems, marketing automation tools, and other digital solutions, product managers can enhance efficiency, improve decision-making, and drive success throughout the product's lifecycle.

Real-World Case Studies

The following five case studies illustrate how pharmaceutical companies can leverage digital tools, artificial intelligence, and machine learning to optimize product launches, lifecycle and supply chain management, and treatment personalization.

Case Study #15: Launching a New Drug with Digital Tools by a Pharma MNC!

Background:

A leading multinational pharmaceutical company was preparing to launch a new medication for chronic disease. The company aimed to streamline the launch process, maximize market impact, and ensure effective stakeholder engagement.

Implementation:

- **Project Management**: The company used Asana for comprehensive project management. Asana facilitated fast-tracking, team collaboration, and milestone monitoring, ensuring all aspects of the launch were coordinated and completed on schedule.

- **CRM and marketing Automation**: Salesforce CRM managed interactions with healthcare professionals (HCPs) and tracked engagement metrics. Marketo was used for automated email campaigns and targeted digital advertisements, allowing for personalized communication and optimized marketing efforts.

- **Data Analytics**: Google Analytics and Tableau provided real-time insights into website traffic, campaign performance, and market trends. These analytics tools helped the company make data-driven decisions, adjust strategies as needed, and evaluate the effectiveness of the launch.

- **Virtual Collaboration**: Tools like Zoom and Slack supported remote meetings and collaboration among cross-functional teams, ensuring seamless communication and coordination throughout the launch phase.

- **Content Management**: WordPress was used to update and manage the product website, providing accurate and up-to-date information to stakeholders, including educational resources and promotional content.

Outcome:

Digital tools enabled the company to efficiently manage the launch process, engage effectively with stakeholders, and monitor real-time performance. The result was a successful market entry, with well-coordinated efforts and optimized marketing strategies.

Case Study #16: Managing Product Lifecycle with Digital Tools at Pharma Company X!

Background:

Pharma Company X sought to enhance the management of an established product throughout its lifecycle, focusing on performance optimization, regulatory compliance, and market adaptation.

Implementation:

- **PLM System**: Company X implemented PTC Windchill for product lifecycle management. The system managed documentation, regulatory compliance, and version control, ensuring efficient operations and adherence to standards.

- **Market Intelligence**: IQVIA provided insights into competitor activities, market trends, and emerging opportunities. These insights guided Company X's strategic adjustments and helped identify new growth opportunities.

- **Patient and HCP Engagement**: Company X utilized MedPage Today and HealthTap to engage with patients and HCPs. These platforms facilitated ongoing communication, gathered feedback, and addressed evolving needs.

- **Digital Analytics**: The company used Mixpanel to track user interactions with the product's digital assets. The data collected provided insights into user behavior and identified areas for improvement.

- **Customer Support**: Zendesk managed customer support inquiries and feedback. This tool helped resolve issues promptly, improving customer satisfaction and ensuring a positive experience for users.

Outcome:

By leveraging digital tools, Company X effectively managed the product's lifecycle, optimizing performance and maintaining relevance in the market. The company's ability to adapt to market changes and engage with stakeholders contributed to sustained product success.

Case Study #17: Enhancing Engagement with Digital Tools at Company Y!

Background:

Company Y aimed to improve engagement with healthcare professionals (HCPs) and patients for respiratory treatments, focusing on targeted marketing and personalized communication.

Implementation:

- **Digital Advertising**: Company Y used AI-powered tools to target online ads for respiratory treatments, focusing on regions with high prevalence rates. This approach ensured that the advertising efforts were directed toward areas with greater need.

- **CRM and Marketing Automation**: The company used Salesforce CRM to manage HCP interactions and track engagement. Marketo facilitated personalized email campaigns and digital outreach, enhancing communication with HCPs and patients.

- **Data Analytics**: Google Analytics and Tableau provided insights into campaign performance and market dynamics. These tools helped the company adjust strategies and optimize marketing efforts based on real-time data.

- **Virtual Collaboration**: Zoom and Slack supported remote coordination among marketing teams, enabling efficient communication and collaboration.

Outcome:

Company Y's use of digital tools led to improved targeting of advertising efforts, enhanced engagement with HCPs and patients, and more effective marketing strategies. The company addressed regional needs more effectively and optimized outreach efforts.

Case Study #18: Optimizing Supply Chain Management at Company Z.

Background:

Company Z sought to optimize its supply chain operations to ensure the timely delivery of medications and improve overall efficiency.

Implementation:

- **Supply Chain Management Tools**: Company Z utilized AI-enhanced supply chain management tools to predict demand, optimize inventory levels, and manage logistics. These tools provided real-time data and insights, helping Company Z to make informed decisions and proactively address challenges.
- **Digital Collaboration**: Platforms like Microsoft Teams facilitated communication and collaboration among supply chain teams, ensuring seamless coordination and timely decision-making.

Outcome:

Implementing AI and data analytics tools significantly improved supply chain efficiency. Company Z optimized inventory levels, reduced lead times, and enhanced overall operational performance.

Case Study #19: Personalizing Cancer Treatment at OncoWell Pharma.

Background:

OncoWell Pharma aimed to enhance personalized cancer treatment by leveraging AI and digital tools to analyze patient data and tailor treatment plans.

Implementation:

- **AI and Machine Learning**: OncoWell Pharma used AI algorithms to analyze large datasets of patient information, including genetic profiles and treatment outcomes. These insights were used to develop personalized treatment plans based on individual patient needs.

- **Data Integration**: OncoWell Pharma employed digital tools to integrate data from various sources, including clinical trials and electronic health records (EHRs). This integration provided a comprehensive view of patient data, supporting more informed treatment decisions.

- **Patient Engagement**: Digital platforms facilitated communication between patients and healthcare providers, allowing real-time updates and personalized support.

Outcome:

Using AI and digital tools enabled Oncowell Pharma to develop more effective and personalized treatment plans for cancer patients. Integrating data and insights led to improved patient outcomes and a more tailored approach to treatment.

These case studies demonstrate the significant impact of digital tools on pharmaceutical product management, from optimizing launches and lifecycle management to enhancing supply chain operations and personalized treatment. The successful implementation of these tools underscores the importance of embracing digital technologies in the pharmaceutical industry.

19.1 Successful Product Launches and Lifecycle Extensions

Product Launches

Key Elements of A Successful Product Launch

1. **Market Research and Planning:**

   - **Example: *Pfizer's Launch of Ibrance (Palbociclib)*:** Before launching Ibrance, Pfizer conducted extensive market research to understand the needs of oncologists and breast cancer patients. The company identified a gap in effective treatments for HR-positive breast cancer and planned its launch strategy to address this need with targeted messaging and education.

2. **Regulatory Approvals and Compliance:**

   - **Example: *Moderna's COVID-19 Vaccine*:** Moderna navigated a complex regulatory landscape to gain the FDA and other regulatory bodies' emergency use authorization (EUA). The company's swift regulatory approach was crucial in the rapid development and deployment of the vaccine.

3. **Marketing and Sales Strategy:**

   - **Example: *Gilead's Launch of Truvada for PrEP (Pre-Exposure Prophylaxis)*:** Gilead's comprehensive marketing strategy included educational campaigns targeting healthcare providers and high-risk populations, emphasizing the importance of PrEP in HIV prevention.

4. **Distribution and Supply Chain Management:**

   - **Example: *AstraZeneca's Launch of Tagrisso*:** AstraZeneca ensured robust distribution channels and supply chain logistics to meet the high demand for Tagrisso, a treatment for non-small cell lung cancer (NSCLC).

5. Post-Launch Monitoring and Feedback:

- **Example: *Boehringer Ingelheim's Launch of Jardiance*:** After launching Jardiance, Boehringer Ingelheim used real-world data and feedback to monitor the drug's performance and gather insights for continuous improvement in marketing and patient support.

Lifecycle Extensions

A. Strategies for Extending Product Life:

1. Line Extensions and New Indications:

- **Example: *Johnson & Johnson's Stelara*:** Johnson & Johnson extended Stelara's lifecycle by gaining approval for additional indications beyond its initial use for psoriasis, including Crohn's disease and ulcerative colitis.

2. Formulation Changes and New Delivery Methods:

- **Example: *AbbVie's Humira*:** AbbVie extended Humira's lifecycle by developing new formulations and delivery methods, such as Humira citrate-free, to address patient preferences and reduce injection pain.

3. Geographic Expansion:

- **Example: *Roche's Herceptin*:** Roche expanded Herceptin's market by launching it in new geographic regions, increasing its patient base, and maintaining its relevance in the oncology market.

4. Repositioning and Rebranding:

- **Example: *Eli Lilly's Cialis*:** Eli Lilly successfully repositioned Cialis with a new indication for benign prostatic hyperplasia (BPH), extending the product's lifecycle and maintaining market interest.

5. Patient Support and Education Programs:

- **Example: *Novartis's Gilenya*:** Novartis implemented comprehensive patient support programs for Gilenya, including educational resources and adherence tools, to enhance patient engagement and ensure long-term use.

B. Real-World Examples of Lifecycle Management:

1. Pfizer's Lipitor:

- **Lifecycle Management Strategy**: Pfizer managed Lipitor's lifecycle by aggressively defending its market position against generic competitors through patent extensions and continued marketing efforts.

- **Outcome**: Lipitor remained a top-selling drug for many years, generating substantial revenue even as generic versions became available.

2. Merck's Keytruda:

- **Lifecycle Management Strategy**: Merck extended Keytruda's lifecycle by continuously adding new indications and expanding its use in combination therapies for various cancers.

- **Outcome**: Keytruda became a leading immunotherapy with many applications, contributing significantly to Merck's oncology portfolio.

3. Sanofi's Lantus:

- **Lifecycle Management Strategy**: Sanofi extended Lantus's lifecycle by developing a follow-on product, Lantus SoloStar, and, later, Toujeo, a more concentrated version of insulin.

- **Outcome**: These extensions allowed Sanofi to capture continued market share and address evolving patient needs.

These examples illustrate how pharmaceutical companies effectively launch new products and extend the lifecycle of existing ones through strategic planning, market analysis, and continuous innovation.

19.2 Lessons Learned from Market Failures

1. Inadequate Market Research and Understanding

Real-World Example #1: Theranos

- **Overview**: Theranos, founded by Elizabeth Holmes, aimed to revolutionize blood testing with a technology that promised to perform numerous tests from a single drop of blood. However, the technology was never proven reliable, and the company faced scrutiny and legal action.

- **Analysis**:

  - **Lack of Validation**: Theranos needed to provide credible data validating its technology. It conducted limited testing and did not publish peer-reviewed research, which is crucial for gaining credibility in the scientific community.

  - **Impact**: The company's misleading claims and eventual fraud charges led to its downfall. Patients received incorrect diagnoses, and investors lost substantial amounts of money.

  - **Lesson**: Rigorous validation and transparent communication are essential. Thorough market research and reliable evidence are crucial to building trust and credibility.

Real-world Example #2: Neurontin (Gabapentin) and Off-Label Marketing

- **Overview**: Pfizer faced allegations for marketing Neurontin, an anti-seizure drug, for off-label uses not approved by the FDA. The company was fined $430 million for these practices.

- **Analysis**:

  - **Regulatory Overreach**: Pfizer's promotion of Neurontin for uses beyond its approved indications violated FDA regulations and led to legal consequences.

  - **Impact**: The legal penalties and damages to Pfizer's reputation highlighted the risks of non-compliance and the importance of adhering to regulatory guidelines.

- **Lesson**: Strict adherence to regulatory guidelines is crucial to avoid legal repercussions and protect the company's reputation

2. Regulatory Hurdles and Compliance Issues

Real-World Example #3: Avandia (Rosiglitazone)

- **Overview**: Avandia, a diabetes drug by GlaxoSmithKline (GSK), was linked to increased cardiovascular risks. Following mounting evidence and public concern, the drug faced severe restrictions and was withdrawn in some markets.

- **Analysis**:
  - **Risk Management**: The late recognition of cardiovascular risks highlighted gaps in risk management and post-market surveillance.
  - **Impact**: The drug's withdrawal and restrictions affected patient access and resulted in financial losses for GSK.
  - **Lesson**: Effective risk assessment and proactive communication with regulatory bodies are essential for managing potential adverse effects and ensuring patient safety.

Real-World Example #4: Genzyme's Fabry Disease Drug Approval

- **Overview**: Genzyme faced delays in approving Fabry disease treatment due to issues with its manufacturing process, leading to FDA scrutiny.

- **Analysis**:
  - **Manufacturing Issues**: Genzyme's challenges in meeting manufacturing standards delayed the drug's approval and impacted market entry.
  - **Impact**: The delays affected patient access to the drug and had financial repercussions for the company.
  - **Lesson**: Ensuring robust manufacturing processes and regulatory compliance from the beginning can prevent costly delays and obstacles.

3. Inadequate Customer and Market Fit

Real-World Example #5: Merck's Vioxx

- **Overview**: Vioxx, a painkiller developed by Merck, was withdrawn from the market due to its association with increased cardiovascular risk.

- **Analysis**:

  - **Lack of Long-Term Data**: The failure to identify long-term cardiovascular risk before the drug's widespread use led to significant issues.

  - **Impact**: The withdrawal had severe financial and reputational consequences for Merck.

  - **Lesson**: Comprehensive pre-market testing and continuous post-market monitoring are vital for ensuring patient safety and maintaining product viability.

Real-World Example #6: Arena Pharmaceuticals' Lorcaserin (Belviq)

- **Overview**: Lorcaserin, a weight-loss drug by Arena Pharmaceuticals, was withdrawn due to cancer risks identified after approval.

- **Analysis**:

  - **Post-Market Surveillance**: The drug's withdrawal due to safety concerns underlined the importance of long-term surveillance and risk management.

  - **Impact**: The withdrawal affected Arena Pharmaceuticals' market position and financial performance.

  - **Lesson**: Continuous monitoring of long-term effects and robust risk-benefit analysis are crucial for sustaining market presence.

4. Product Development and Launch Missteps

Real-World Example #7: Novartis's Exjade

- **Overview**: Exjade, used for treating iron overload, faced market challenges due to poor patient adherence and side effects.

- **Analysis**:
  - **Patient Adherence**: The drug's complexity and side effects led to poor patient adherence and market uptake.
  - **Impact**: The market adoption and patient management challenges affected Novartis's sales and market position.
  - **Lesson**: Addressing patient adherence through user-friendly formulations and support programs is key to product success.

Real-World Example #8: Eli Lilly's Evista (Raloxifene)

- **Overview**: Evista, a drug for osteoporosis, faced market challenges due to patent expirations and generic competition.
- **Analysis**:
  - **Patents and Competition**: The expiration of patents and the emergence of generics led to reduced market share and revenue.
  - **Impact**: Eli Lilly had to navigate increased competition and shifting market dynamics.
  - **Lesson**: Developing patent protection and differentiation strategies is essential for maintaining market share.

5. **Strategic Misalignment and Execution Issues**

Real-World Example #9: Johnson & Johnson's DePuy Synthes

- **Overview**: The recall of DePuy Synthes hip implants due to high failure rates affected J&J's reputation and market performance.
- **Analysis**:
  - **Quality Control**: The recall highlighted quality control issues and product defects management.
  - **Impact**: The recall led to financial losses and damaged J&J's brand reputation.
  - **Lesson**: Effective quality control and strategic management of recalls are vital for maintaining brand integrity.

Real-World Example #10: AstraZeneca's Crestor

- **Overview:** Crestor, a cholesterol-lowering drug, faced challenges due to patent expirations and generic competition.

- **Analysis:**

  - **Patent Expiration:** The expiration of Crestor's patents led to increased competition from generics and reduced revenue.

  - **Impact:** AstraZeneca needed to adapt strategies to maintain product relevance and market position.

  - **Lesson:** Developing strategies for patient protection, lifecycle management, and market adaptation is crucial for sustaining competitive advantage.

Key Takeaways

1. **Thorough Validation:** Comprehensive validation and market research are crucial for successful product development and market entry.

2. **Regulatory Compliance:** Adhering to regulatory requirements and maintaining transparency is essential to avoid legal and financial repercussions.

3. **Customer-Centric Approach:** Understanding and addressing patient needs and market dynamics are key to product success.

4. **Continuous Monitoring:** Implementing robust post-market surveillance and risk management strategies helps identify and mitigate issues early.

5. **Strategic Adaptation:** Adapting strategies based on market changes and competitive pressures is vital for maintaining product relevance and success.

19.3 The Evolving Role of Pharma Product Manager

In the dynamic field of pharmaceutical product management, navigating challenges and seizing the opportunities presented by traditional practices and modern digital innovations is crucial. As the industry evolves, product managers play a pivotal role in driving the success of pharmaceutical products through various stages of their lifecycle. Here is a summary of key insights and takeaways:

1. **The Evolving Role of Pharma Product Managers**
   - Pharmaceutical product managers are central to the success of drug products, from conception through to market success. Their role has expanded from basic product portfolio management to strategic leadership, data-driven decision-making, and digital transformation. This evolution reflects broader industry changes, where understanding complex market dynamics and leveraging new technologies are essential.

2. **Importance of Core Competencies**
   - Effective product management requires a blend of scientific knowledge, business acumen, and regulatory expertise. Understanding drug development processes, pharmacology, epidemiology, and biostatistics ensures that product managers can make informed decisions and communicate product benefits and risks effectively. Simultaneously, financial management skills, project management expertise, and regulatory knowledge are crucial for navigating the complexities of the pharmaceutical industry.

3. **Harnessing Digital Tools and Technologies**
   - The digital age has introduced informative tools and technologies, including artificial intelligence, machine learning, and digital marketing strategies. Pharma product managers must embrace these innovations to enhance decision-making, optimize marketing strategies, and

improve patient engagement. Effectively leveraging digital tools can lead to more personalized marketing, efficient operations, and better management of product lifecycles.

4. Navigating Challenges and Seizing Opportunities

- The pharmaceutical landscape is fraught with challenges, such as regulatory hurdles, market competition, and the need for continuous innovation. However, these challenges also present opportunities for product managers to drive growth and success. By staying abreast of industry trends, employing strategic planning, and adapting to new market demands, product managers can position their products for long-term success.

5. Case Studies and Real-world Insights

- Examining real-world case studies provides valuable lessons in both successes and failures. From effective product launches and lifecycle management to learning from market failures and strategic missteps, these examples underscore the importance of thorough planning, rigorous validation, and adaptive strategies.

Looking Ahead

Product managers must remain agile and forward-thinking as the pharmaceutical industry evolves. Embracing emerging technologies, understanding evolving market dynamics, and maintaining a patient-centered approach will drive future success. The ability to integrate digital innovations with traditional practices will define the next generation of pharmaceutical product management.

In conclusion, the role of a pharmaceutical product manager is both challenging and rewarding. By mastering core competencies, leveraging digital advancements, and learning from real-world experiences, product managers can effectively navigate the industry's complexities and contribute to developing successful and impactful pharmaceutical products.

19.4 Key Takeaways and Best Practices

1. Embrace Digital Transformation

- **Takeaway**: The digital age has revolutionized pharmaceutical product management. Embracing digital tools and technologies is no longer optional but essential for success.

- **Best Practices**:

  - **Leverage AI and Data Analytics**: Use AI for data-driven insights, predictive analytics, and personalized marketing. Implement machine learning algorithms to optimize decision-making and forecasting.

  - **Utilize Digital Marketing Strategies**: Incorporate digital channels like social media, email, and content marketing to effectively engage with healthcare professionals (HCPs) and patients.

  - **Implement Digital Therapeutics**: Explore and integrate digital therapeutics into your product offerings to enhance patient outcomes and provide comprehensive care.

2. Prioritize Scientific Knowledge

- **Takeaway**: A deep understanding of scientific concepts, including pharmacology, epidemiology, and biostatistics, is crucial for effective product management.

- **Best Practices**:

  - **Stay Updated on Scientific Advances**: Stay abreast of the latest research and developments in drug discovery, clinical trials, and therapeutic areas.

  - **Understand Drug Mechanisms and Therapeutic Categories**: Ensure thorough knowledge of how drugs work and their therapeutic uses to communicate effectively with stakeholders and design relevant marketing strategies.

  - **Apply Epidemiological Insights**: Use epidemiological data to understand disease prevalence, patient demographics, and market needs, guiding strategic decisions.

3. Master Financial Management

- **Takeaway**: Financial acumen is vital for managing budgets, forecasting, and ensuring the economic viability of pharmaceutical products.

- **Best Practices**:

  - **Develop Accurate Budgets and Forecasts**: Create detailed financial plans to track expenditures, revenue, and profitability. Adjust forecasts based on market dynamics and performance data.

  - **Monitor Financial Metrics**: Regularly review key financial metrics such as ROI, cost-per-acquisition, and profitability to ensure alignment with strategic goals.

4. Optimize Project Management

- **Takeaway**: Effective project management ensures timely and successful product launches and lifecycle management.

- **Best Practices**:

  - **Implement Agile Methodologies**: Use agile project management techniques to adapt to changes and streamline processes. Regularly review and adjust plans based on project progress and feedback.

  - **Manage Cross-Functional Teams**: Coordinate efforts across different departments to ensure cohesive execution. Foster clear communication and collaboration among team members.

5. Navigate Regulatory Affairs

- **Takeaway**: Understanding and adhering to regulatory requirements is crucial for successful product development and market entry.

- **Best Practices**:

  - **Stay Compliant**: Ensure all product documentation, marketing materials, and promotional activities comply with regulatory guidelines and standards.

- **Engage with Regulatory Authorities**: Maintain open lines of communication with regulatory agencies to address any concerns and expedite approval processes.

6. Integrate Marketing and Sales Strategies

- **Takeaway**: Effective integration of marketing and sales strategies maximizes product impact and market penetration.
- **Best Practices**:
  - **Develop Coordinated Campaigns**: Align marketing efforts with sales objectives to create cohesive campaigns that address the needs of HCPs and patients.
  - **Measure and Optimize**: Continuously track the performance of marketing and sales activities. Use insights to refine strategies and improve outcomes.

7. Apply Social Science Insights

- **Takeaway**: Insights from sociology, anthropology, and behavioral economics enhance understanding of patient behaviors and market dynamics.
- **Best Practices**:
  - **Utilize Sociological and Anthropological Research**: Apply social and cultural studies findings to tailor marketing messages and product offerings to diverse patient populations.
  - **Implement Behavioral Economics Principles**: Use principles like nudging to positively influence patient and HCP behaviors, guiding them towards better health outcomes.

8. Learn from Real-World Case Studies

- **Takeaway**: Analyzing successful and unsuccessful case studies provides valuable lessons for future endeavors.
- **Best Practices**:
  - **Study Successful Launches**: Review case studies of successful product launches to identify best practices and strategies contributing to their success.

- **Analyze Failures**: Examine market failures to understand what went wrong and avoid similar pitfalls in future projects.

By following these key takeaways and best practices, pharmaceutical product managers can more effectively navigate the industry's complexities, drive product success, and contribute to improved patient outcomes and business growth.

Competitive Analysis and Scenario Planning

Competitive Analysis and Scenario Planning are critical strategic tools in pharmaceutical product management. They help managers understand the market landscape and prepare for future uncertainties.

Competitive Analysis

Definition: Competitive analysis involves assessing competitors' strengths, weaknesses, strategies, and market positioning. This process helps identify opportunities and threats within the industry, enabling product managers to develop more effective strategies.

Key Components:

1. **Market Positioning**: Evaluate your competitors' positions relative to your product. Analyze their market share, geographic reach, and target demographics.

2. **Product Offerings**: Review competitors' products, including their features, benefits, pricing, and lifecycle stages, to understand how they compare to your product.

3. **Sales and Marketing Strategies**: Investigate competitors' marketing tactics, sales channels, and promotional activities. Identify their key messages and how they engage with HCPs and patients.

4. **Strengths and Weaknesses**: Assess competitors' strengths (e.g., brand reputation, advanced technology) and weaknesses (e.g., gaps, product portfolio, customer complaints).

5. **Market Trends**: Monitor industry trends and competitor responses to these trends. Look for emerging technologies, regulatory changes, and evolving patient needs.

6. **SWOT Analysis**: Conduct a SWOT (Strengths, Weaknesses, Opportunities, Threats) analysis for your product and competitors. This will help you identify areas where you can gain a competitive edge.

Case Study

The following case study illustrates the importance of competitive analysis and how its insights into the market landscape help pharmaceutical companies refine their product strategies.

Case Study #20: Eli Lilly's Competitive Analysis for Diabetes Treatments

Background:

Eli Lilly is a major player in the diabetes market, known for its innovative insulin therapies and other diabetes management solutions. To maintain and strengthen its market position, Eli Lilly conducts rigorous competitive analysis to understand its competitors and adapt its strategies accordingly.

Competitive Analysis Process:

1. **Market Positioning:**
   - Eli Lilly analyzes the market positioning of competitors such as Novo Nordisk and Sanofi. This involves assessing their market share in various regions, product portfolio, and target demographics. For instance, they evaluate how Novo Nordisk's GLP-1 receptor agonists compare with Lilly's offerings regarding efficacy and patient preference.

2. **Product Offerings:**
   - The company reviews the features and benefits of competitors' products. This includes examining clinical trial results, product efficacy, side effects, and patient adherence rates. For example, Lilly compares the long-acting insulins offered by competitors with its long-acting insulin formulations to identify unique selling propositions.

3. **Sales and Marketing Strategies:**
   - Eli Lilly studies the marketing tactics used by competitors, including their digital marketing campaigns, promotional events, and partnerships with healthcare professionals (HCPs). They analyze how competitors position their products in advertisements and how they engage with HCPs through continuing education and advisory boards.

4. Strengths and Weaknesses:

- Through SWOT analysis, Lilly identifies key strengths and weaknesses of competitors. For instance, they might find that a competitor has a strong brand presence but lacks comprehensive patient support programs, which could be an opportunity for Lilly to differentiate itself.

5. Market Trends:

- The company tracks emerging trends in diabetes care, such as advancements in continuous glucose monitoring and digital health tools. They evaluate how competitors respond to these trends and identify opportunities for innovation in their product line.

Impact:

By conducting this thorough competitive analysis, Eli Lilly can make informed decisions about product development, marketing strategies, and market positioning. For instance, if they identify a growing preference for personalized diabetes management, they might accelerate the development of new tools or therapies that cater to this trend.

Scenario Planning

Definition: Scenario planning is a strategic method used to anticipate and prepare for various potential future scenarios. It involves creating and analyzing different plausible futures to develop flexible strategies that adapt to changing circumstances.

Key Components:

1. **Identify Key Drivers**: Determine the critical factors that could impact the market, such as regulatory changes, technological advancements, and economic shifts.

2. **Develop Scenarios**: Create a range of possible future scenarios based on different combinations of key drivers. Each scenario should represent a plausible future state.

3. **Analyze Impact**: Assess the potential impact of each scenario on your product and business strategy. Consider how each scenario might affect market conditions, competitor actions, and consumer behavior.

4. **Develop strategic Responses**: Formulate strategic responses for each scenario. Identify actions that can be taken to capitalize on opportunities or mitigate risks associated with different scenarios.

5. **Monitor and Adapt**: Regularly review and update scenarios based on new information and changing conditions. Adapt strategies as needed to stay responsive to evolving market dynamics.

Case Study

The following case study demonstrates how scenario planning can help prepare companies for future uncertainties, enabling them to thrive in dynamic environments.

Case Study #21: Merck's Scenario Planning for Emerging Markets

Background:

Merck operates in various emerging markets with diverse economic, regulatory, and competitive environments. Scenario planning helps Merck navigate these complexities and adapt its strategies to future uncertainties.

Scenario Planning Process:

1. **Identify Key Drivers:**
   - Merck identifies critical drivers that could impact its operations in emerging markets, such as changes in healthcare regulations, economic fluctuations, and technological advancements. For example, it considers how potential changes in government healthcare policies might affect drug pricing and access.

2. **Develop Scenarios:**
   - The company creates different scenarios based on the identified drivers. For instance, they might develop scenarios that include:
     - **Scenario 1**: A favorable regulatory environment with reduced barriers to market entry.
     - **Scenario 2**: Increased regulatory scrutiny and stricter compliance requirements.
     - **Scenario 3**: Economic downturn leading to reduced healthcare spending.

3. **Analyze Impact:**
   - Merck evaluates how each scenario could affect its market strategy, product pricing, distribution channels, and overall business operations. For example, in a scenario with increased regulatory scrutiny, it might need to invest in regulatory compliance and documentation.

4. **Develop Strategic Responses**:

- Merck formulates strategic responses based on the scenarios to address potential challenges and seize opportunities. For example, if they anticipate a regulatory tightening scenario, they might plan to strengthen their regulatory affairs team and enhance their compliance processes.

5. **Monitor and Adapt**:

- Merck continuously monitors market changes and updates its scenarios and strategies accordingly. It uses market intelligence and feedback to adjust its approaches as needed, ensuring it remains agile and responsive to evolving conditions.

Impact:

Through scenario planning, Merck can proactively manage risks and capitalize on opportunities in emerging markets. Preparing for various regulatory scenarios can ensure smoother market entry and better regulatory compliance, enhancing market presence and competitiveness.

Summary

Competitive Analysis provides insights into the competitive landscape, helping product managers identify strategic opportunities and threats.

Scenario Planning prepares managers for future uncertainties by developing flexible strategies based on various plausible scenarios. These tools enable pharma product managers to navigate complex markets, adapt to changes, and make informed strategic decisions.

20.1 Conducting a Comprehensive Competitive Analysis

Conducting a comprehensive competitive analysis involves several key steps designed to gather and analyze information about competitors to inform strategic decisions. Here is a detailed guide on how to conduct a comprehensive competitive analysis:

1. Define Objectives and Scope

- **Objectives**:
  - Identify what you need to learn about your competitors (e.g., market position, strengths, weaknesses, strategic initiatives).
  - Determine how this information will be used (e.g., to refine your marketing strategy, develop new products, improve market positioning).
- **Scope**:
  - Decide which competitors to analyze (e.g., direct, indirect, new entrants).
  - Determine the time frame for the analysis (e.g., historical performance, current state, future outlook).

2. Identify Competitors

- **Direct Competitors**:
  - Companies offering similar products or services in the same market.
- **Indirect Competitors**
  - Companies offering alternative solutions that address the same customer needs.
- **Emerging Competitors**
  - New entrants or startups that might disrupt the market.

3. Gather Information

- **Market Position**:
  - **Market Share**: Estimate the competitor's market share using industry reports and market research.

- **Geographic Presence**: Understand the regions where competitors operate and their market penetration.
- **Product and Service Offerings**:
  - **Product Portfolio**: List and describe the competitor's products or services.
  - **Features and Benefits**: Analyze their offerings' features, benefits, and differentiators.
  - **Pricing**: Review pricing strategies and compare them with your own.
- **Sales and Marketing Strategies**:
  - **Promotion**: Examine advertising campaigns, digital marketing strategies, and promotional activities.
  - **Sales Channels**: Identify the sales channels used by competitors (e.g., direct sales, distributors, online platforms).
  - **Customer Engagement**: Assess how competitors engage with their customers (e.g., loyalty programs, customer service).
- **Financial Performance**:
  - **Revenue and Profitability**: Look at financial reports, annual reports, and industry benchmarks for revenue, profit margins, and growth rates.
  - **Investment and Funding**: Analyze recent funding rounds, investments, or partnerships.
- **SWOT Analysis**:
  - **Strengths**: Identify the competitor's key strengths (e.g., strong brand, advanced technology).
  - **Weaknesses**: Determine their weaknesses (e.g., limited product range, poor customer service).
  - **Opportunities**: Spot potential opportunities that competitors might leverage.
  - **Threats**: Recognize external threats impacting their performance (e.g., regulatory changes, economic downturn).

4. Analyze Competitor Strategies

- **Strategic Initiatives:**
  - Review strategic moves such as mergers and acquisitions, alliances, or new market entries.
- **Innovation:**
  - Investigate their approach to innovation, including R&D activities and new product development.
- **Regulatory Compliance:**
  - Understand how competitors handle regulatory and compliance issues, including recent changes or challenges.
- **Customer Feedback:**
  - Analyze customer reviews, ratings, and feedback to gauge customer satisfaction and product performance.

5. Compare and Benchmark

- **Benchmarking:**
  - Compare your findings with your company's performance and strategies. Identify gaps and areas for improvement.
- **Competitive Positioning:**
  - Map out where you and your competitors stand regarding market share, product features, and customer satisfaction.

6. Develop Strategic Insights

- **Opportunity Identification:**
  - Determine areas where you can gain a competitive advantage or capitalize on competitor's weaknesses.
- **Strategic Recommendations:**
  - Formulate actionable recommendations based on your analysis. These could include adjusting pricing strategies, enhancing product features, or exploring new market segments.

7. Implement and Monitor

- **Action Plan**:
  - Develop an action plan based on the insights gained from the competitive analysis. This plan should include specific actions, timelines, and responsible parties.
- **Ongoing Monitoring**:
  - Monitor competitors and update your analysis regularly to stay informed about market changes and emerging trends.

Conclusion

A comprehensive competitive analysis provides valuable insights into the market landscape, helping pharmaceutical product managers make informed strategic decisions. By understanding competitors' strengths, weaknesses, strategies, and market positions, product managers can better position their products, enhance their marketing strategies, and drive business success.

20.2 SWOT Analysis

SWOT analysis is a strategic planning tool used to identify and evaluate the strengths, Weaknesses, Opportunities, and Threats of a business venture or product. In pharmaceutical product management, SWOT analysis helps assess the internal and external factors that can impact the success of a pharmaceutical product, brand, or company. Here is a breakdown of each component:

SWOT Analysis Components

1. **Strengths**:
   - **Internal Factors**: These are the positive attributes and resources within the organization that provide a competitive advantage. In pharmaceuticals, strengths might include:
     - Strong research and development capabilities leading to innovative products.
     - Established brand reputation and recognition in specific therapeutic areas.
     - Intellectual property rights (patents) protect proprietary technologies or formulations.
     - Efficient manufacturing processes ensure quality and cost-effectiveness.
     - Skilled and experienced workforce, including scientific experts and regulatory specialists.

2. **Weaknesses**:
   - **Internal Factors**: These are areas where the organization may be lacking or vulnerable compared to competitors. Weaknesses in pharmaceuticals could include:
     - Limited pipeline of new drug candidates or dependence on a few products.
     - High production costs impact profitability.
     - Regulatory challenges delaying product approvals or market entry.
     - Weak brand awareness or perceived gaps in clinical efficacy or safety data.

- Challenges in adapting to rapidly changing market trends or technological advancements.

3. **Opportunities**:

- **External Factors**: These are favorable conditions or trends in the external environment that the organization could exploit to its advantage. Opportunities in pharmaceuticals include:

  - Growing demand for innovative treatments in specific therapeutic areas (e.g., oncology, rare diseases).
  - Expanding global healthcare markets and emerging economies offering market expansion opportunities.
  - Advances in biotechnology, genomics, or personalized medicine open new avenues for drug development.
  - Strategic partnerships with academic institutions, research organizations, or biotech startups.
  - Increasing emphasis on digital health technologies and telemedicine enhancing patient engagement and healthcare delivery.

4. **Threats**:

- **External Factors**: These are factors outside the organization's control that could pose risks or challenges to its success. Threats in pharmaceuticals may include:

  - Intense competition from generic drugs or biosimilars impacts market share and pricing strategies.
  - Stringent regulatory requirements and compliance costs affect product development timelines and profitability.
  - Pricing pressures from payers, healthcare providers, or government regulations.
  - Patent expirations lead to loss of exclusivity and revenue erosion.
  - Shifting healthcare policies, trade tariffs, or geopolitical instability impact market access and operations.

Application of SWOT Analysis in Pharmaceutical Product Management

1. **Strategic Planning**: SWOT analysis helps pharmaceutical product managers formulate strategies that capitalize on strengths, address weaknesses, exploit opportunities, and mitigate threats. It informs decision-making processes for product development, marketing strategies, market expansion, and competitive positioning.

2. **Risk Management**: SWOT analysis enables proactive risk mitigation strategies by identifying potential threats and weaknesses. This includes contingency planning, regulatory compliance measures, and adapting to market dynamics to minimize vulnerabilities and optimize operational resilience.

3. **Market Differentiation**: Understanding strengths and weaknesses relative to competitors helps pharmaceutical companies differentiate their products in the marketplace. Leveraging strengths (e.g., scientific expertise, brand reputation) and addressing weaknesses (e.g., regulatory challenges, product portfolio gaps) can enhance competitive advantage and market penetration.

4. **Resource Allocation**: SWOT analysis guides resource allocation decisions, including investments in research and development, marketing campaigns, talent acquisition, and infrastructure development. It effectively allocates resources to capitalize on opportunities and address critical weaknesses.

5. **Monitoring and Adaptation**: SWOT analysis is not a one-time exercise but an ongoing process. Pharmaceutical product managers continually monitor internal and external factors, update SWOT analyses as conditions change, and adapt strategies to maintain competitiveness and achieve long-term success.

By systematically evaluating strengths, weaknesses, opportunities, and threats, pharmaceutical product managers can develop informed strategies, mitigate risks, and effectively navigate the complex landscape of pharmaceutical product management.

20.3   Scenario Planning and Risk Management

Scenario planning and risk management are crucial processes in strategic planning, particularly in complex and dynamic fields such as pharmaceutical product management. Here is a detailed look at each:

Scenario Planning

1. **Definition**:
   - Scenario planning is a strategic method for envisioning and preparing for various future scenarios by analyzing potential changes in external factors that could impact the business. It helps organizations anticipate uncertainties and develop flexible strategies.

2. **Steps in Scenario Planning**:

   A. Identify Key Drivers:
   - **External Factors**: Consider factors such as regulatory changes, technological advancements, market trends, economic conditions, and competitive dynamics.
   - **Internal Factors**: Evaluate internal drivers such as company capabilities, resource availability, and strategic goals

   B. Develop Scenarios:
   - **High-Impact Scenarios**: Create multiple scenarios that explore a range of possible futures, including best-case, worst-case, and most likely scenarios.
   - **Scenario Narrative**: Develop detailed narratives for each scenario, describing how these external factors could evolve and impact the business.

   C. Analyze Implications:
   - **Strategic Impact**: Assess how each scenario affects your business operations, market position, and objectives.

- **Opportunities and Threats**: Identify potential opportunities and threats arising from each scenario.

D. Develop Response Strategies:

- **Action Plans**: Formulate specific action plans and contingency strategies for each scenario to address potential challenges and capitalize on opportunities.
- **Flexibility**: Ensure that strategies are adaptable to different scenarios and can be modified as new information emerges.

E. Monitor and Update:

- **Ongoing Monitoring**: Regularly track relevant indicators and external factors to identify signs of changes that could affect your scenarios.
- **Scenario Updates**: Update scenarios and strategies based on new data and changing conditions.

3. Benefits of Scenario Planning

- **Enhanced Preparedness**: Improves the organization's readiness for unexpected changes and uncertainties.
- **Informed Decision-Making**: Provides a framework for making informed decisions and setting priorities.
- **Strategic Flexibility**: Allows for flexible and adaptable strategies that can be adjusted as solutions evolve.

Risk Management

1. Definition:

- Risk management identifies, assesses, and mitigates risks that could impact an organization's objectives. It involves proactive measures to minimize the negative effects of uncertainties and threats.

2. Steps in Risk Management:

A. Risk Identification:

- **Internal Risks**: Identify risks arising from internal sources such as operational inefficiencies, regulatory non-compliance, or resource constraints.

- **External Risks**: Identify risks from external factors such as market volatility, competitive pressures, or geopolitical issues

B. Risk Assessment:

- **Likelihood and Impact**: Evaluate the likelihood of each risk occurring and the potential impact on the organization. This helps in prioritizing risks based on their severity.

- **Risk Matrix**: Use a risk matrix or assessment tools to visualize and categorize risks.

C. Risk Mitigation:

- **Mitigation Strategies**: Develop and implement strategies to mitigate or reduce the impact of identified risks. This may include risk avoidance, reduction, transfer (e.g., insurance), or acceptance.

- **Contingency Planning**: Create contingency plans and response strategies to address risks if they materialize.

D. Risk Monitoring and Review:

- **Ongoing Monitoring**: Continuously monitor risks and their indicators to detect changes in risk levels or emerging new risks.

- **Review and Update**: Regularly review and update risk management strategies and contingency plans based on new information or changes in the risk environment.

3. **Benefits of Risk Management**:

- **Reduced Uncertainty**: Helps in minimizing the impact of uncertainties and potential disruptions.

- **Enhanced Decision-Making**: Provides a structured approach to assess and address risks, improving decision-making.

- **Improved Resilience**: Strengthens the organization's ability to withstand and recover from adverse events.

Conclusion

Scenario Planning and **Risk Management** are integral to effective strategic planning in pharmaceutical product management. Scenario planning prepares organizations for various possible futures, enabling them to remain agile and responsive to changes. Risk management systematically identifies, assesses, and mitigates risks, proactively managing potential threats. Together, these processes help product managers make informed decisions, develop robust strategies, and navigate the complexities of the pharmaceutical industry effectively.

20.4 Scenario Planning Techniques

Scenario planning is a strategic technique to anticipate and prepare for future uncertainties by creating plausible scenarios and exploring their potential implications. In pharmaceutical product management, scenario planning helps decision-makers navigate dynamic market conditions, regulatory changes, technological advancements, and other external factors impacting product development, market access, and commercialization strategies. Here are key techniques and steps involved in scenario planning:

Scenario Planning Techniques

1. **Environmental Scanning**:
   - **Identify Key Drivers**: Conduct thorough environmental analysis to identify critical factors influencing the pharmaceutical industry, such as regulatory trends, technological innovations, healthcare policies, and market dynamics.
   - **Trend Analysis**: Monitor emerging trends, demographic shifts, patient preferences, and competitive developments to anticipate future scenarios and their potential impacts on pharmaceutical products and markets.

2. **Scenario Generation**:
   - **Diverse Scenarios**: Develop multiple scenarios representing a range of plausible futures based on different combinations of key drivers and uncertainties. Common scenarios may include optimistic, pessimistic, and moderate outcomes.
   - **Storytelling**: Craft narratives or stories for each scenario to illustrate how external factors (e.g., regulatory changes and competitive actions) could unfold over time and affect the pharmaceutical landscape.

3. **Quantitative Modeling**:
   - **Data Analysis**: Use quantitative models, statistical techniques, and predictive analytics to simulate scenarios

and quantify potential outcomes. This may involve forecasting market demand, pricing dynamics, revenue projections, and profitability under various scenarios.

- **Sensitivity Analysis**: Conduct sensitivity analysis to assess the impact of critical variables (e.g., pricing, market share, regulatory approvals) on scenario outcomes and strategic decisions.

4. **Stakeholder Engagement**:

- **Cross-functional Collaboration**: Engage stakeholders from different departments (e.g., R&D, marketing, regulatory affairs) to gather diverse perspectives, insights, and expertise. Collaborative workshops or brainstorming sessions can foster creativity and consensus-building around scenario planning.

5. **Scenario Evaluation**:

- **Scenario Testing**: Evaluate each scenario based on its plausibility, coherence, and potential strategic implications for pharmaceutical products, market strategies, and organizational capabilities.

- **Risk Assessment**: Assess risks and uncertainties associated with each scenario, including operational risks, regulatory hurdles, competitive threats, and market volatility.

6. **Decision-Making and Strategy Development**:

- **Strategy Formulation**: Develop adaptive strategies and contingency plans tailored to each scenario's potential outcomes. Prioritize strategic initiatives, resource allocation, and operational adjustments to capitalize on opportunities and mitigate risks identified through scenario planning.

- **Iterative Process**: Continuously review and update scenario plans based on new information, changing market conditions, and evolving stakeholder priorities to maintain strategic agility and responsiveness.

Benefits of Scenario Planning in Pharmaceutical Product Management

- **Enhanced Strategic Preparedness**: Scenario planning enables pharmaceutical product managers to proactively anticipate future challenges and opportunities, reducing reactive decision-making and enhancing strategic preparedness.

- **Risk Mitigation**: Scenario planning facilitates proactive risk mitigation strategies and contingency planning to safeguard business operations and investments by identifying potential risks and uncertainties early.

- **Innovation and Adaptation**: Scenario planning fosters innovation by encouraging creative thinking, exploring alternative futures, and identifying new growth opportunities that align with emerging market trends and patient needs.

- **Stakeholder Alignment**: Engaging stakeholders in scenario planning fosters cross-functional collaboration, consensus-building, and alignment around strategic priorities, enhancing organizational agility and decision-making effectiveness.

In summary, scenario planning equips pharmaceutical product managers with the foresight and flexibility to navigate uncertainty, capitalize on opportunities, and sustain competitive advantage in an evolving healthcare landscape.

20.5 Strategic Responses to Market Dynamics

Strategic responses to market dynamics involve proactive and adaptive actions that a company takes to address changes in the market environment. These dynamics can include shifts in consumer behavior, competitive actions, regulatory changes, technological advancements, and economic fluctuations. Effective strategic responses ensure that a company remains competitive, meets market demands, and achieves its business objectives. Here is a detailed description of how to strategically respond to market dynamics:

1. Understanding Market Dynamics

A. Market Analysis:

- **Market Trends**: Analyze market trends regularly to identify emerging opportunities and threats. This includes examining consumer preferences, technological advancements, and regulatory changes.

- **Competitive Landscape**: Evaluate the competitive environment to understand competitors' strategies and actions. Identify their strengths, weaknesses, and market positioning.

B. Consumer Insights:

- **Customer Needs**: Gather and analyze data on customer needs and preferences. Use surveys, focus groups, and market research to understand what drives consumer behavior.

- **Feedback Mechanisms**: Implement mechanisms to collect feedback from customers and stakeholders to gain insights into their experiences and expectations.

2. Developing Strategic Responses

A. Strategic Adaptation:

- **Product Innovation** involves modifying or developing new products to meet changing consumer needs or address gaps in the market. This may involve enhancing existing products or creating entirely new offerings.

- **Market Segmentation**: Reassess and refine market segmentation strategies to target specific customer groups effectively. Tailor marketing messages and product offerings to meet the needs of different segments.

B. Competitive Positioning:

- **Differentiation**: Enhance your product's unique value proposition to stand out in the market. This could involve differentiating on features, quality, price, or customer service.

- **Strategic Alliances**: From strategic partnerships or alliances to leverage complementary strengths and expand market reach. Collaborations can enhance competitive positioning and provide access to new markets.

C. Operational Efficiency:

- **Process Optimization**: Streamline operations to improve efficiency and reduce costs. This can include optimizing supply chain processes, enhancing manufacturing capabilities, and implementing lean practices.

- **Technology Integration**: Invest in technology and digital tools to improve operational processes and productivity. This can include adopting new software solutions, automating processes, and utilizing data analytics.

3. Implementing Strategic Responses

A. Action Plans:

- **Execution Strategies**: Develop detailed action plans outlining the steps needed to implement strategic responses. Assign responsibilities, set timelines, and allocate resources for each initiative.

- **Change Management**: Effectively manage the change process to ensure smooth implementation. Communicate changes to stakeholders, provide training, and address any resistance.

B. Monitoring and Evaluation:

- **Performance Metrics**: Establish key performance indicators (KPIs) to measure the success of strategic responses. Monitor progress against these metrics to assess effectiveness.

- **Continuous Improvement**: Regularly review and adjust strategies based on performance data and feedback. Make iterative improvements to enhance outcomes and adapt to evolving market conditions.

Examples

A. Market Expansion:

- Apple Inc. responded to the growing demand for wearable technology by expanding its product line to include the Apple Watch. This strategic move allowed Apple to capture a significant share of the wearable market and reinforce its position as a leader in consumer electronics.

B. Product Innovation:

- Tesla Inc. continuously innovates its electric vehicle lineup to meet changing consumer preferences and regulatory requirements. The introduction of new models and advancements in battery technology have positioned Tesla as a pioneer in the electric vehicle industry.

C. Operational Efficiency

- Amazon leverages advanced logistics and automation technologies to optimize its supply chain and delivery processes. This focus on operational efficiency has enabled Amazon to offer fast shipping and maintain a competitive edge in the e-commerce market.

Conclusion

Strategic responses to market dynamics involve a comprehensive approach to understanding market changes, developing and implementing adaptive strategies, and continuously monitoring and refining those strategies. By staying attuned to market trends and consumer needs, companies can navigate challenges, seize opportunities, and achieve sustained success.

Strategic Marketing Planning

Strategic marketing planning is a systematic process that involves developing and implementing strategies to achieve a company's marketing objectives and align them with overall business goals. It encompasses assessing the market environment, setting goals, formulating strategies, and executing plans to drive growth and competitive advantage. Here is a detailed breakdown of the strategic marketing planning process:

1. Market Analysis

A. Market Research

- **Objective**: Gather and analyze data on market conditions, customer needs, and competitor activities.
- **Techniques**: Use quantitative and qualitative research methods such as surveys, interviews, focus groups, and market reports.
- **Outcome**: Identify market opportunities, threats, and trends.

B. Competitive Analysis

- **Objective**: Assess the strengths, weaknesses, strategies, and market positioning of key competitors.
- **Techniques**: Analyze competitors' product offerings, pricing, marketing strategies, and market share.
- **Outcome**: Determine competitive advantages and areas for differentiation.

C. Customer Segmentation

- **Objective**: Segment the market into distinct groups based on demographics, psychographics, behaviors, and needs.

- **Techniques**: Use segmentation criteria such as age, income, lifestyle, and purchasing behavior.

- **Outcome**: Tailor marketing strategies to target specific customer segments effectively.

2. Setting Objectives

A. Goal Setting:

- **Objective**: Define clear, measurable, and time-bound marketing goals that align with overall business objectives.

- **Types of Goals**: Include sales targets, market share growth, brand awareness, customer acquisition, and retention.

B. Key Performance Indicators:

- **Objective**: Establish metrics to measure progress toward achieving marketing objectives.

- **Examples**: Sales revenues, conversion rates, customer satisfaction scores, and return on investment (ROI).

3. Strategy Formulation

A. Positioning:

- **Objective**: Develop a unique value proposition differentiating your product or brand from competitors.

- **Techniques**: Identify key benefits, features, and values that resonate with target customers.

B. Marketing Mix (4Ps):

- **Product**: Define the product features, quality, branding, and packaging.

- **Price**: Set pricing strategies based on cost, competition, and perceived value.

- **Place**: Determine distribution channels and logistics for product delivery.
- **Promotion**: Plan promotional activities such as advertising, public relations, sales promotions, and digital marketing.

C. Strategic Initiatives:

- **Objective**: Develop specific initiatives to achieve marketing goals.
- **Examples**: Launching new products, entering new markets, or enhancing customer service.

4. Implementation

A. Action Plans:

- **Objective**: Create detailed plans outlining the steps, resources, and timelines required to execute marketing strategies.
- **Components**: Include responsibilities, deadlines, budget allocation, and task prioritization.

B. Resource Allocation:

- **Objective**: Allocate budget and resources effectively to support marketing initiatives.
- **Components**: Include marketing budget, human resources, technology, and external agencies.

5. Monitoring and Evaluation

A. Performance Tracking:

- **Objective**: Monitor progress against KPIs and marketing objectives.
- **Techniques**: Track performance metrics using dashboards, reports, and analytics tools.

B. Feedback and Adjustments:

- **Objective**: Gather feedback from customers and stakeholders to assess the effectiveness of marketing strategies.

- **Techniques**: Conduct surveys, analyze customer feedback, and review campaign results.
- **Outcome**: Make data-driven adjustments to improve marketing strategies and tactics.

Examples

A. Successful Product Launch

- **Example**: **Apple Inc**. successfully launched the iPhone through a strategic marketing plan that included innovative product positioning, a well-defined target audience, and a multi-channel promotional strategy.

B. Market Expansion

- **Example**: **Netflix** expanded its market presence by developing a global content strategy tailored to regional preferences, leveraging data analytics to drive content creation and acquisition.

C. Brand Revitalization

- **Example**: **Old Spice** revitalized its brand through a strategic marketing campaign that included a memorable advertising series, targeted social media engagement, and a refreshed brand image.

Conclusion

Strategic marketing planning is a crucial process that enables companies to align their marketing efforts with business goals, effectively target their audience, and drive growth. By conducting market analysis, setting clear objectives, formulating and implementing strategies, and continuously monitoring performance, organizations can achieve a competitive edge and deliver value to their customers.

21.1 Building a Strategic Marketing Plan

Building a strategic marketing plan involves a systematic approach to creating a roadmap for achieving marketing goals and objectives aligned with the overall business strategy. Here is a comprehensive guide to building a strategic marketing plan:

1. **Executive Summary**
   - **Objective**: Provide a concise marketing plan overview summarizing key elements and goals.
   - **Components**:
     - **Purpose and Scope**: Outline the purpose of the plan and its scope.
     - **Key Objectives**: Highlight the primary marketing objectives and expected outcomes.
     - **Summary of Strategies**: Briefly describe the main strategies and tactics.

2. **Situation Analysis**
   - **Objective**: Analyze the current market environment to understand the context in which the marketing plan will be executed.
   - **Components**:
   A. **Market Research**:
     - **Data Collection**: Gather data on market size, growth trends, customer demographics, and buying behavior.
     - **Analysis**: To assess the market landscape, use tools such as SWOT analysis (Strengths, Weaknesses, Opportunities, and Threats).
   B. **Competitive Analysis**:
     - **Competitor Identification**: Identify key competitors and their market position.
     - **SWOT Analysis**: Analyze competitors' strengths, weaknesses, opportunities, and threats.

- **Market Positioning**: Determine how your product or service differs from competitors.

C. Customer Insights:

- **Segmentation**: Segment the market based on demographics, psychographics, and behavioral criteria.
- **Buyer Personas**: Develop detailed profiles of target customers to tailor marketing efforts.

3. Marketing Objectives

- **Objective**: Define clear, measurable goals the marketing plan aims to achieve.
- **Components**:
  - **SMART Goals**: Ensure objectives are Specific, Measurable, Achievable, Relevant, and Time-bound.
  - **Examples**: Increase market share by 10% in the next 12 months or boost online sales by 20% within six months.

4. Strategy Development

- **Objective**: Formulate strategies to achieve the marketing objectives based on the insights gained from situation analysis.
- **Components**:

A. Positioning:

- **Value Proposition**: Define what makes your product or service unique and valuable to the target market.
- **Brand Messaging**: Develop key messages that communicate the value proposition effectively.

B. Marketing Mix (4Ps):

- **Product**: Define product features, benefits, and unique selling points.
- **Price**: Set pricing strategies based on market conditions, cost, and perceived value.
- **Place**: Determine distribution channels and logistics.
- **Promotion**: Plan promotional activities, including advertising, public relations, and digital marketing.

5. **Action Plan**
   - **Objective**: Develop a detailed plan of action for executing the marketing strategies.
   - **Components**:

   A. **Tactical Initiatives**:
   - **Campaigns**: Outline specific marketing campaigns and activities.
   - **Channels**: Select marketing channels such as social media, email, content marketing, etc.

   B. **Timeline**:
   - **Schedule**: Create a timeline for implementing marketing activities, including key milestones and deadlines.

   C. **Budget**:
   - **Allocation**: Allocate budget across different marketing activities and channels.
   - **Monitoring**: Track expenditures and ensure alignment with the budget.

6. **Implementation**
   - **Objective**: Execute the marketing plan according to the action plan.
   - **Components**:

   A. **Resource Management**:
   - **Team Roles**: Assign responsibilities to team members.
   - **Tools**: Utilize marketing tools and platforms to execute campaigns.

   B. **Execution**:
   - **Campaign Launch**: Roll out marketing campaigns and activities as planned.
   - **Coordination**: Ensure coordination among team members and departments.

7. **Monitoring and Evaluation**
   - **Objective**: Track the performance of the marketing plan and make data-driven adjustments.

- **Components**:

A. Performance Metrics:

- **KPIs**: Define key performance indicators (KPIs) to measure the success of marketing activities.
- **Tools**: Use analytical tools to track metrics such as website traffic, conversion rates, and sales.

B. Reporting:

- **Regular Updates**: Provide regular reports on marketing performance to stakeholders.
- **Analysis**: Analyze data to assess the effectiveness of strategies and tactics.

C. Adjustments:

- **Feedback Loop**: Gather feedback from customers and stakeholders.
- **Optimization**: Make necessary adjustments to improve marketing effectiveness.

8. Review and Refinement

- **Objective**: Continuously review and refine the marketing plan based on performance and changing market conditions.
- **Components**:

A. Post-Campaign Analysis:

- **Success Evaluation**: Assess the success of marketing campaigns and activities.
- **Lessons Learned**: Identify lessons learned and areas for improvement.

B. Strategic Adjustments:

- **Market Trends**: Stay updated on market trends and adjust strategies accordingly.
- **Continuous Improvement**: Implement changes to enhance the marketing plan.

Case Studies

Here are case study examples that illustrate successful strategic marketing planning in the pharmaceutical industry.

Case Study #22: Pfizer's COVID-19 Vaccine Launch

Overview:

Pfizer's strategic marketing plan for its COVID-19 vaccine involved a well-coordinated global launch strategy that addressed urgent public health needs and leveraged digital and traditional marketing channels.

Key Elements:

A. Situation Analysis:

- **Market Research**: Pfizer conducted extensive research to understand the pandemic's impact and the urgent need for a vaccine.

- **Competitive Analysis**: The company evaluated competitors and collaborated with BioNTech to develop a unique mRNA vaccine.

B. Marketing Objectives:

- **Objective**: Ensure rapid vaccine distribution and widespread adoption to combat the epidemic.

- **SMART Goal**: Distribute 2 billion doses by the end of 2021.

C. Strategy Development:

- **Positioning**: Position the vaccine as a groundbreaking solution based on cutting-edge technology and rigorous testing.

- **Marketing Mix**:
  - **Product**: Emphasized the vaccine's efficacy and safety based on clinical trials.
  - **Price**: Offered the vaccine at a cost to ensure global accessibility.
  - **Place**: Coordinated with governments and health organizations for distribution.
  - **Promotion**: Utilized a mix of media channels, including TV, social media, and public service announcements.

D. Action Plan:

- **Tactical Initiatives**: Launched a global awareness campaign highlighting the vaccine's benefits and safety.
- **Timeline**: Managed a phased rollout, starting with high-risk populations and expanding to broader demographics.

E. Monitoring and Evaluation:

- **Performance Metrics**: Tracked vaccination rates, public perception, and distribution efficiency.
- **Adjustments**: Adjusted messaging and distribution strategies based on real-time feedback and challenges.

F. Outcomes:

- **Success**: Achieved rapid vaccine distribution and significant impact on public health, contributing to global efforts in controlling the pandemic.

Case Study #23: Merck's Keytruda

Overview:

Merck's launch of Keytruda, an immune checkpoint inhibitor for cancer treatment, involved a comprehensive marketing strategy highlighting groundbreaking efficacy in oncology.

Key Elements

A. Situation Analysis:

- **Market Research**: Identified the growing need for effective cancer immunotherapies.
- **Competitive Analysis**: Analyzed competing immuno-therapies and differentiated Keytruda based on its clinical benefits.

B. Marketing Objectives:

- **Objective**: Establish Keytruda as a leading treatment option for various cancers.
- **SMART Goals**: Increase market share and drive adoption among oncologists.

C. Strategy Development:

- **Positioning**: Positioned Keytruda as a novel and effective treatment option with proven results in clinical trials.
- **Marketing Mix**:
  - **Product**: Highlighted Keytruda's clinical success and patient outcomes.
  - **Price**: Developed a pricing strategy that included patient assistance programs.
  - **Place**: Focused on establishing partnerships with oncology centers and hospitals.
  - **Promotion**: Implemented a targeted campaign involving medical conferences, peer-reviewed publications, and digital marketing.

D. Action Plan:

- **Tactical Initiatives**: Conducted educational webinars for oncologists and supported clinical trial data dissemination.
- **Timeline**: Rolled out the campaign in phases, targeting key oncology centers first.

E. Monitoring and Evaluation:

- **Performance Metrics**: Monitored adoption rates, sales data, and feedback from healthcare professionals.
- **Adjustments**: Adapted strategies based on competitive landscape changes and market feedback.

F. Outcomes:

- **Success**: Keytruda became one of the leading treatments for various cancers, significantly increasing Merck's oncology portfolio.

Case Study #24: Novartis' Kymriah

Overview:

Novartis' launch of Kymriah, the first FDA-approved CAR-T cell therapy for cancer, involved a strategic marketing plan focused on educating healthcare professionals and securing market access.

Key Elements:

A. Situation Analysis

- **Market Research**: Identified the need for innovative cancer treatments and the potential of CAR-T therapies.
- **Competitive Analysis**: Assessed the competitive landscape for cell and gene therapies.

B. Marketing Objectives

- **Objective:** Position Kymriah as a leading CAR-T therapy and gain market acceptance.
- **SMART Goal**: Achieve brand adoption in major cancer treatment centers.

C. Strategy Development

- **Positioning**: Emphasized Kymriah's innovative approach and efficacy in treating relapsed/refractory B-cell lymphoma.
- **Marketing Mix**:
  - **Product**: Focused on the unique aspects of CAR-T technology and patient outcomes.
  - **Price**: Developed pricing models considering the high cost of CAR-T therapy and reimbursement strategies.
  - **Place**: Partnered with treatment centers and provided infrastructure support.
  - **Promotion**: Launched an educational campaign for hematologists and oncologists, including clinical trial results and patient testimonials.

D. Action Plan

- **Tactical Initiatives**: Conducted workshops and informational sessions for healthcare providers.
- **Timeline**: Coordinated with treatment centers for launch readiness and patient onboarding.

E. Monitoring and Evaluation

- **Performance Metrics**: Tracked adoption rates, patient outcomes, and market feedback.
- **Adjustments**: Made adjustments to marketing strategies based on market reception and feedback.

F. Outcomes:

- **Success**: Kymriah established itself as a leading CAR-T therapy, significantly impacting treatment options for specific cancers.

These case studies demonstrate how pharmaceutical companies successfully implement strategic marketing plans to launch and manage products in a competitive market.

21.2 Aligning Marketing Strategies with Business Objectives

Aligning marketing strategies with business objectives is crucial for ensuring that marketing efforts contribute to a company's overall goals in the pharmaceutical industry. This alignment helps to drive product success, optimize resources, and achieve desired outcomes. Here are the details:

1. Understanding Business Objectives

A. Define Company Goals:

- **Revenue Targets**: Achieving specific financial targets, such as sales growth or profitability.
- **Market Share**: Expanding the company's presence in existing or new markets.
- **Patient Outcomes**: Improving patient outcomes and treatment adherence.
- **Innovation**: Advancing the development and introduction of new products or therapies.
- **Regulatory Compliance**: Ensuring adherence to regulatory standards and requirements.

B. Align with Strategic Vision:

- **Long-term Vision**: Ensure marketing strategies support the company's long-term vision and mission.
- **Strategic Priorities**: Focus on areas identified as strategic priorities, such as expanding into new therapeutic areas or geographic regions.

2. Developing Marketing Strategies

A. Conduct Market Analysis:

- **Market Needs**: Identify the needs and preferences of targeted customers (e.g., healthcare professionals and patients).
- **Competitive Landscape**: Analyze competitors to understand their strategic and market positioning.

- **Trends and Insights**: Utilize market research to identify trends, opportunities, and challenges.

B. Set Marketing Objectives:

- **SMART Goals**: Develop marketing objectives that are Specific, Measurable, Achievable, Relevant, and Time-bound.

- **KPIs**: Define key performance indicators (KPIs) to track progress and measure success.

C. Craft Marketing Strategies:

- **Positioning**: Define how the product will be positioned in the market relative to competitors.

- **Messaging**: Develop key messages that resonate with the target audience and align with business goals.

- **Channels**: Select marketing channels (e.g., digital, print, events) that are most effective for reaching the target audience.

3. Implementing and Integrating Marketing Strategies

A. Action Plans:

- **Tactical Initiatives**: Create detailed action plans that outline specific activities, timelines, and responsibilities.

- **Resource Allocation**: Allocate resources effectively to support the execution of marketing strategies.

B. Cross-functional Collaboration:

- **Internal Alignment**: Work closely with other departments (e.g., sales, R&D, regulatory affairs) to ensure alignment and integration.

- **Feedback Loop**: Establish a feedback loop to gather input from stakeholders and adjust strategies as needed.

4. Monitoring and Evaluating Performance

A. Track Performance:

- **Metrics and KPIs**: Monitor performance against KPIs and metrics to evaluate the effectiveness of marketing strategies.

- **Campaign Analysis**: Asses the success of individual campaigns and initiatives.

B. Adjust Strategies:

- **Continuous Improvement**: Use insights and feedback to make data-driven adjustments to marketing strategies.
- **Adapt to Changes**: Be flexible and responsive to market or business environment changes.

Case Studies

The following case studies illustrate how pharmaceutical companies align their marketing strategies with business objectives.

Case Study #25: Pfizer's COVID-19 Vaccine Launch

Business Objective: Rapidly develop and distribute a vaccine to combat the COVID-19 pandemic.

Marketing Strategy and Alignment:

- **Development and Approval**: Pfizer, in partnership with BioNTech, focused on the rapid development of the mRNA vaccine. They prioritized expedited clinical trials and regulatory approvals.

- **Positioning**: Pfizer Positioned the vaccine as a groundbreaking solution in the fight against COVID-19, emphasizing its high efficacy rate and safety profile.

- **Global Coordination**: The company coordinated with global health organizations, such as the World Health Organization (WHO) and national health agencies, to ensure swift distribution and equitable access.

- **Communication Channels**: Utilized multiple channels for communication, including press releases, social media, and public service announcements. Pfizer also engaged with healthcare professionals through webinars and educational resources.

- **Market Impact**: Pfizer's strategic approach led to the rapid rollout of the vaccine, which became one of the first widely administered vaccines globally. Effective communication and coordination helped build public trust and facilitated widespread vaccine uptake.

Case Study #26: Merck's Keytruda

Business Objective: Establish Keytruda as a leading cancer treatment.

Marketing Strategy and Alignment:

- **Clinical Efficacy**: Merck focused on demonstrating Keytruda's clinical efficacy through rigorous clinical trials and published results in leading medical journals.

- **Targeting Key Oncology Centers**: The company targeted top oncology centers and key opinion leaders (KOLs) to advocate using Keytruda in cancer treatment.

- **Medical Conferences**: Leveraged conferences to present new data, engage with healthcare professionals, and build credibility in the oncology community.

- **Digital and Traditional Marketing**: Used a mix of digital marketing, including targeted ads, educational webinars, and traditional marketing strategies such as print advertisements and medical journal articles.

- **Patient Support Programs**: Implemented patient support programs to assist with access to the medication and address any concerns related to its use.

Case Study #27: Novartis' Kymriah

Business Objective: Gain market acceptance for Kymriah as a CAR-T therapy.

Marketing Strategy Alignment:

- **Education and Awareness**: Novartis focused on educating healthcare professionals (HCPs) about CAR-T therapy and Kymriah's benefits through detialed presentations, training programs, and informational materials.

- **Treatment Centers**: Developed partnerships with treatment centers and hospitals specializing in advanced therapies to facilitate the administration of Kymriah.

- **Pricing Models**: Addressed high-cost concerns by developing flexible pricing models and providing financial assistance programs for patients.

- **Regulatory and Market Access**: Worked on gaining regulatory approvals and negotiating with payers to secure reimbursement for Kymriah.

- **Patient Advocacy**: Engaged with patient advocacy groups to raise awareness about the treatment options and support patient access.

Summary

These case studies illustrate how pharmaceutical companies align their marketing strategies with business objectives:

- **Pfizer's COVID-19 Vaccine Launch**: Focused on rapid development, global coordination, and effective communication to achieve a critical public health goal.
- **Merck's Keytruda**: Emphasized clinical efficacy, targeted key stakeholders, and utilized a mix of marketing strategies to establish Keytruda as a leading cancer treatment.
- **Novartis' Kymriah**: Focused on education, partnerships with treatment centers, flexible pricing models, and patient advocacy to gain acceptance for CAR-T therapy.

These approaches demonstrate the importance of aligning marketing efforts with strategic business goals to drive product success and achieve desired outcomes.

21.3 Measuring and Optimizing Marketing Performance

Measuring and optimizing marketing performance is crucial to ensure that pharmaceutical marketing strategies are effective, efficient, and aligned with business goals. Here is a detailed look at the process:

1. Defining Key Performance Indicators (KPIs)

- KPIs are metrics that help assess the success of marketing activities. In pharmaceutical marketing, KPIs may include:
- **Sales Metrics**:
  - ***Revenue Growth***: Track increases in sales revenue over specific periods.
  - ***Market Share***: Measures the product's share of the total market compared to competitors.
- **Engagement Metrics**:
  - ***Click-Through Rate (CTR)***: The percentage of people who click on an advertisement or email link compared to those who see it.
  - ***Engagement Rate***: Measures interactions (likes, shares, comments) on social media or other platforms.
- **Conversion Metrics**:
  - ***Conversion Rate***: The percentage of leads or prospects who take the desired action, such as completing a form or making a purchase.
  - ***Cost per Acquisition (CPA)***: The cost of acquiring a new customer or lead.
- **Awareness Metrics**:
  - ***Brand Recognition***: Assesses the target audience's ability to recognize or recall a brand.
  - ***Share of Voice***: Measures the brand's presence in industry conversations compared to competitors.
- **Compliance Metrics**:
  - ***Regulatory Compliance***: Tracks adherence to regulatory guidelines and standards.

2. Data Collection and Analysis

- **Data Collection**: Gather data from various sources to assess marketing performance, including:
  - *Sales Data*: Sales reports, revenue figures, and market share information.
  - *Digital Analytics*: Web traffic, email open rates, click-through rates, and social media engagement.
  - *Surveys and Feedback*: Customer and healthcare professional feedback, market research surveys, and focus group data.
  - *Compliance Reports*: Records related to regulatory adherence and marketing practices.
- **Data Analysis**: Analyze the collected data to identify trends, patterns, and areas for improvement:
  - *Trend Analysis*: Examine historical data to understand performance trends over time.
  - *Benchmarking*: Compare performance against industry standards and competitor benchmarks.
  - *Segmentation Analysis*: Breakdown data by different customer segments or market regions to identify specific performance drivers.

3. Optimization Strategies

- **Optimization** involves refining marketing strategies based on performance insights:
  - *A/B Testing*: Conduct experiments by comparing different versions of marketing materials (e.g., emails, ads) to determine which performs better.
  - *Content Refinement*: Adjust content based on engagement metrics and feedback to improve relevance and effectiveness.
  - *Targeting and Segmentation*: Refine audience targeting and segmentation based on performance data to enhance engagement and conversion rates.

- ***Budget Allocation:*** Reallocate marketing budgets based on the performance of different channels or campaigns to maximize ROI.

4. Reporting and Communication

- **Regular Reporting**: Create detailed reports summarizing performance metrics, insights, and recommendations. Share these reports with key stakeholders to ensure transparency and alignment.

- **Continuous Improvement**: Use performance data to refine and improve marketing strategies continuously. This involves:

  - ***Reviewing Results***: Regularly review marketing performance results to identify successes and improvement areas.

  - ***Adjusting Strategies***: Make data-driven adjustments to marketing strategies and tactics based on performance insights.

  - ***Feedback Loop***: Establish a feedback loop where insights from performance measurement are used to inform future marketing decisions.

Case Studies

Three case studies show how companies can leverage data and analytics to improve campaign performance.

Case Study #28: Company X's Digital Campaign Optimization

Background:

Company X, a global pharmaceutical company, launched a series of digital advertising campaigns to promote a new drug. The initial phase of the campaign involved broad targeting and standard ad formats.

Approach:

- **Data Collection**: Company X utilized digital analytics tools to track performance metrics such as click-through rates (CTR), conversion rates, and engagement levels.

- **Analysis**: Company X conducted a detailed data analysis to identify which ad creatives and audience segments performed best. Key performance indicators (KPIs) included CTR and coversion rates.

- **Optimization**: The company identified high-performing ad creatives and audience segments based on the analysis. The team then adjusted the campaign by:

  - *Reallocating Budget*: Shifting budget towards the best-performing creatives and audience segments.

  - *Refining Ad Creatives*: Modifying underperforming ads to align better with audience preferences.

  - *Targeting*: Narrowing down the target audience based on successful segments identified during the analysis.

Results:

- **Increased CTR**: Company X observed a significant increase in click-through rates for the optimized ads.

- **Higher Conversion Rates**: The conversion rates improved due to more relevant and targeted ad creatives.

- **Enhanced ROI**: The campaign's overall return on investment (ROI) increased due to optimized budget allocation and more effective targeting.

Key Takeaways:

- **Importance of Data-Driven Decisions**: Analyzing performance metrics helped Company X make informed decisions to optimize their campaign.

- **Role of A/B Testing**: Testing different ad creatives and targeting strategies was crucial in identifying what worked best.

Case Study #29: Company Y's Content Marketing Enhancement

Background:

Company Y, a leading biopharmaceutical company, targeted healthcare professionals (HCPs) to promote a new treatment option.

Approach:

- **Data Collection**: Company Y tracked key metrics such as email open rates, click-through rates, and engagement with the content.
- **Analysis**: The team performed A/B testing to compare email subject lines and content formats. The goal was to identify which version generated higher engagement.
- **Optimization**: Based on the results of the A/B tests, Company Y:
  - *Refined Content*: Enhanced the content and design of emails that performed well.
  - *Optimized Subject Lines*: Used successful subject lines from A/B tests to increase open rates in future campaigns.
  - *Segmented Audience*: Tailored content to different segments of the HCP audience based on their preferences and behaviors.

Results:

- *Improved Open Rates*: The changes led to higher open rates for email campaigns.
- *Enhanced Engagement*: Higher engagement levels were observed with the optimized content and subject lines.
- *Increased Lead Generation*: More leads were generated due to improved email performance.

Key Takeaways:

- *Value of A/B Testing*: Testing different versions of content helps identify what resonates with the target audience.

- ***Importance of Continuous Optimization***: Regular content refinement based on performance data can significantly enhance campaign results.

Case Study #30: Company Z's Social Media Adjustment

Background:

Company Z, a multinational pharmaceutical company, launched a disease awareness campaign on social media platforms to educate the public and healthcare professionals about a chronic condition.

Approach:

- **Data Collection**: Company Z used social media analytics to monitor engagement metrics, sentiment analysis, and audience interactions.

- **Analysis**: The team analyzed social media data to assess the campaign's effectiveness and understand the audience's sentiment and engagement patterns.

- **Optimization**: Based on the insights:

  - *Content Adjustments*: Company Z modified the content to better address audience concerns and interests.

  - *Enhanced Targeting*: Adjusted targeting strategies to focus on regions and demographics with higher engagement and positive sentiment.

  - *Engagement Strategies*: Implemented new engagement tactics like interactive posts and Q&A sessions to foster greater interaction.

Results:

- *Increased Engagement*: The campaign saw higher engagement rates due to more relevant and targeted content.

- *Positive Sentiment*: The sentiment analysis indicated a more positive response from the audience after content adjustments.

- *Greater Reach*: The refined targeting strategies led to a broader and more effective reach within the target audience.

Key Takeaways:

- ***Impact on Social Media Analytics*:** Analyzing engagement and sentiment data is crucial to optimizing social media campaigns.
- ***Need for Adaptive Strategies*:** Adjusting content and targeting based on real-time feedback can enhance the effectiveness of social media campaigns.

Conclusion

These case studies highlight the importance of leveraging data and analytics to measure and optimize marketing performance. Pharmaceutical companies can make data-driven decisions using tools and techniques such as A/B testing, digital analytics, and social media monitoring, improving campaign effectiveness and ROI. The key takeaways from these examples emphasize the role of continuous optimization, the value of understanding audience behavior, and the need for adaptive strategies to respond to market dynamics.

21.4 Example of a Marketing Plan for a Hypothetical Diabetes Medication Brand Glucotrol

Glucotrol Annual Marketing Plan

Brand Name: Glucotrol

Therapeutic Area: Diabetes Management

Product Attributes: Superior blood sugar control, minimal side effects, convenient once-daily dosage

Executive Summary:

The annual marketing plan for Glucotrol outlines strategic initiatives to launch and establish Glucotrol as a leading diabetes medication. The plan targets healthcare professionals (HCPs) and patients, aiming to build brand awareness, drive adoption, and ensure patient adherence through comprehensive marketing and engagement activities.

1. Market Objectives

- Achieve a 5 percent market share within the first year.
- Increase brand awareness to 70 percent among target HCPs and 50 percent among patients.
- Achieve an 80 percent patient adherence rate.

2. Target Audience

2.1. Primary Target Audience

- **Healthcare Professionals (HCPs)**:
  - Endocrinologists
  - General Practitioners
  - Diabetologists

2.2. Secondary Target Audience

- **Patients**:
  - Type 2 Diabetes Patients

- Patients experiencing side effects from current medications
- Patients seeking more convenient treatment options

3. Market Analysis

3.1. Market Overview

- **Global Diabetes Market Size**: $XX billion (2023)
- **Growth Rate**: CAGR of X% (2023-2028)

3.2. Competitive Landscape:

- Competitor analysis highlighting strengths and weaknesses of key players
- SWOT analysis for Glucotrol

4. Marketing Strategy

4.1. Brand Positioning

- "Glucotrol offers superior blood sugar control with minimal side effects in a convenient once-daily dosage, empowering patients to manage their diabetes effectively and easily."

4.2. Marketing Mix (4Ps)

Product:

- Once-daily oral tablet
- Easy-to-open blister packs
- Patient information leaflet with detailed dosage instructions

Price:

- Competitive pricing with tiered discounts for bulk purchases by healthcare institutions.
- Work with insurance providers to ensure coverage.

Place:

- Distribution through pharmacies, hospitals, and online platforms.
- Focus on initial launch in major markets (US, Europe), followed by expansion in emerging markets.

Promotion:

HCP Engagement:

- Educational seminars and webinars
- Detailing and sampling through sales representatives
- Peer-to-peer interactions and endorsements

Patient Engagement:

- Digital marketing campaigns (social media, email marketing)
- Patient support programs (hotlines, mobile apps)
- Educational content (blogs, videos, brochures)

5. **Detailed Action Plan**

   Q1: January - March

   - **Objective**: Build initial awareness and educate HCPs
   - **Marketing Activities**:
     - Finalize branding and packaging
     - Train sales and marketing teams
     - Launch HCP educational campaigns via webinars and seminars
     - Develop and distribute detailing kits and samples to HCPs
     - Launch the Glucotrol website with detailed product information

   Budget Allocation: $X million

   Q2: April - June

   - **Objective**: Begin patient-centric marketing and expand reach
   - **Marketing Activities**:
     - Launch digital marketing campaigns targeting patients (social media, search engine, email marketing).
     - Initiate patient support programs, including hotlines and mobile apps.
     - Distribute educational content (blogs, videos, brochures) online and offline.

* Begin distribution in key markets
* **Budget Allocation**: $X million

Q3: July - September

* **Objective**: Gather feedback, optimize strategies, and expand distribution
* **Marketing Activities**:
* Conduct market research and gather feedback from HCPs and patients.
* Adjust marketing strategies based on feedback.
* Expand distribution to additional markets.
* Continue digital and traditional marketing efforts.
* Increase HCP engagement through peer-to-peer interactions and endorsements.
* **Budget Allocation**: $X million

Q4: October - December

* **Objective**: Evaluate performance and plan for the next year
* **Marketing Activities**:
* Evaluate year-end performance using key performance indicators (KPIs).
* Conduct surveys and focus groups to gather in-depth feedback.
* Plan for next year's marketing initiatives based on performance data.
* Introduce new patient support features and enhancements to existing programs.
* **Budget Allocation**: $X million

6. Monitoring and Evaluation

6.1. Key Performance Indicators (KPIs):

* **Sales Metrics**: Market share, prescription share, sales volume, revenue
* **Awareness Metrics**: Brand recognition, website traffic, social media engagement

- **Engagement Metrics**: HCP participation in educational programs, patient enrollment in support programs
- **Adherence Metrics**: Patient feedback, prescription refill rates

6.2. Feedback and Adjustments

- Regularly review KPIs and adjust strategies as needed.
- Conduct surveys and focus groups to gather feedback from HCPs and patients.
- Implement improvements based on feedback and performance data.

7. Budget

Total Budget: $XX million

- **Q1**: $X million
- **Q2**: $X million
- **Q3**: X million
- **Q4**: X million

Conclusion

By executing this comprehensive annual marketing plan, Gluctotrol aims to establish itself as a leading brand in the diabetes medication market. The focus on superior efficacy, minimal side effects, and convenience will drive adoption among HCPs and patients, improving diabetes management and enhancing patient lives.

Brand Planning and Execution

Brand Planning

Brand Planning is a strategic process that involves defining a brand's identity, positioning, and marketing strategy to achieve long-term success. This involves creating a strong brand identity for pharmaceutical brands and addressing regulatory considerations, market needs, and competitive dynamics.

1. **Brand Identity Development**

   - **Brand Vision and Mission**: Define the brand's overarching purpose and direction. This includes understanding the brand's role in improving patient outcomes and aligning with the company's overall mission.

   - **Core Values**: Establish the core principles that guide the brand's actions and communications. This often includes values such as innovation, trust, and patient-centricity for pharmaceutical brands.

   - **Brand Personality** determines the tone, style, and characteristics the brand will convey in its messaging. This helps create a consistent and relatable brand image.

2. **Market Analysis**

   - **Market Research**: Conduct thorough research to understand market dynamics, patient needs, healthcare provider preferences, and competitive landscape. Use quantitative (e.g., surveys, sales data) and qualitative (e.g., focus groups, interviews) methods.

- **Segmentation**: Identify distinct segments within the market based on factors such as demographics, disease states, and treatment preferences. Tailor marketing strategies to address the specific needs of each segment.

- **Competitive Analysis**: Analyze competitors' strengths, weaknesses, strategies, and market positions. This helps identify opportunities for differentiation and understand competitive pressures.

3. Brand Positioning

- **Unique Value Proposition (UVP)**: Clearly define what differentiates the brand from competitors. This could be based on efficacy, safety profile, patient support programs, or other unique features.

- **Positioning Statement**: Develop a concise statement communicating the brand's unique value and target audience. This statement should guide all marketing and communication efforts.

4. Strategic Objectives

- **Short-Term Objectives**: Set measurable goals for the near term, such as increasing brand awareness, launching a new product, or gaining market share.

- **Long-Term Goals**: Define broader objectives, such as establishing the brand as a leader in its therapeutic area or expanding into new markets.

Execution

Execution involves implementing the brand plan through various tactics and strategies to achieve the defined objectives.

1. Marketing and Communication Strategies

- **Integrated Marketing Campaigns**: Develop and execute integrated campaigns that leverage multiple channels (e.g., digital, print, events) to reach the target audience. Ensure consistency in messaging and branding across platforms.

- **Content Development**: Create engaging and informative content tailored to the needs of healthcare practitioners (HCPs), patients, and other stakeholders. This may include educational materials, case studies, and promotional content.

- **Digital Marketing**: Utilize digital channels such as social media, email marketing, and search engine optimization (SEO) to enhance brand visibility and engage with the audience. Implement targeted digital campaigns to reach specific segments.

2. Sales and Distribution

- **Sales Strategies**: Develop strategies to support the sales team, such as training programs, sales aids, and incentives. Ensure alignment between marketing and sales efforts to drive product adoption and market penetration.

- **Distribution Channels**: Identify and manage distribution channels to ensure product availability and accessibility. This includes working with wholesalers, distributors, and healthcare providers.

3. Monitoring and Evaluation

- **Performance Tracking**: Monitor key performance indicators (KPIs) to assess the effectiveness of marketing and sales activities. KPIs may include sales figures, market share, brand awareness, and engagement metrics.

- **Feedback Mechanisms**: Collect feedback from HCPs, patients, and other stakeholders to understand their perceptions of the brand and identify areas for improvement. Use this feedback to refine strategies and tactics.

- **Adjustments and Optimization**: Adjust marketing and sales strategies based on performance data and feedback. This could involve revising messaging, reallocating budgets, or exploring new opportunities.

4. Compliance and Ethics

- **Regulatory Adherence**: Ensure all marketing activities comply with regulatory requirements and industry guidelines. This includes accurate and balanced representations of the product's benefits and risks.

- **Ethical Considerations**: Adhere to ethical standards in marketing practices, including transparency, honesty, and respect for patient privacy.

Here is a case study that shows how brand planning and execution can help a company achieve its marketing objectives.

Case Study #31: Brand Planning and Execution

Company: Roche

Brand: Avastin (Bevacizumab)

Background:

Roche aimed to enhance the brand positioning of Avastin, a monoclonal antibody used in cancer treatment, to strengthen its market position and drive adoption.

Brand Planning:

- **Market Research**: Roche conducted comprehensive research to understand the needs of oncologists, patients, and payers. The research highlighted a need for more information on Avastin's benefits in combination with other therapies.

- **Brand Positioning**: Avastin was positioned as a groundbreaking therapy with a strong efficacy profile supported by clinical evidence. The unique value proposition (UVP) emphasized its role in extending patient survival and improving quality of life.

Execution:

- **Integrated Campaign**: Roche launched an integrated marketing campaign that included digital advertising, educational webinars for HCPs, and patient support materials. The campaign focused on showcasing clinical data and real-world evidence.

- **Sales Support**: The sales team received training and updated sales materials to effectively communicate Avastin's benefits and differentiate it from competitors.

- **Monitoring**: Roche tracked campaign performance through sales data, engagement metrics, and feedback from HCPs. This allowed for timely adjustments to optimize the campaign's impact.

Results:

- **Increased Market Share**: The campaign significantly increased Avastin's market share and sales volume.
- **Enhanced Brand Perception**: HCPs reported a better understanding of Avastin's benefits, leading to higher adoption rates.
- **Positive Patient Outcomes**: Patients experienced improved outcomes due to increased awareness and usage of therapy.

Key Takeaways:

- **Importance of Research**: Comprehensive market research and competitive analysis are crucial for effective brand positioning.
- **Integrated Approach**: Combining multiple marketing tactics ensures a cohesive message and maximizes reach.
- **Continuous Monitoring**: Ongoing evaluation and optimization based on performance data are essential for brand success.

Conclusion

Brand planning and execution are critical components of pharmaceutical marketing. Pharmaceutical companies can enhance their brand's market presence and achieve business goals by carefully developing its identity, positioning, and strategic objectives and effectively executing marketing and sales strategies. Continuous monitoring and adaptation ensure that strategies remain relevant and effective in a dynamic market environment.

22.1. The Elements of a Successful Brand Plan

A successful brand plan is a roadmap for creating and sustaining a strong brand presence in the market. It involves several key elements that define the brand's direction, strategy, and tactics. Here is a detailed description of each element:

1. **Brand Vision and Mission**
   - **Brand Vision**: Defines the long-term aspirational goals of the brand. It articulates what the brand aims to achieve and its ideal future state. For pharmaceutical brands, this often involves improving patient outcomes and advancing healthcare.
   - **Brand Mission**: This describes the brand's purpose and primary objectives. It explains why the brand exists and what it seeks to accomplish in the present. The mission should align with the company's mission and address specific needs in the healthcare market.

2. **Brand Identity**
   - **Brand Name and Logo**: The brand's name and visual identity (logo, colors, typography) should be distinctive and memorable and reflect its values and positioning. They play a crucial role in creating recognition and recall.
   - **Brand Personality** defines the human characteristics attributed to the brand. These could include attributes like innovation, trustworthiness, or compassion. The personality should resonate with the target audience and guide the brand's tone and communication style.
   - **Brand Value**: Core principles that guide the brand's actions and decisions. These values should be consistently reflected in all brand activities and communications.

3. **Market Analysis**
   - **Market Research**: Involves gathering and analyzing data about the market, including patient needs, healthcare provider preferences, and competitive dynamics. This helps in identifying opportunities and threats.

- **Segmentation**: Dividing the market into segments based on demographics, disease states, and treatment needs. This allows for targeted marketing strategies that address the specific needs of each segment.

- **Competitive Analysis** involves assessing competitors' strengths, weaknesses, and strategies. Understanding the competitive landscape helps identify areas for differentiation and develop strategies for gaining a competitive advantage.

4. Brand Positioning

- **Unique Value Proposition (UVP)**: Clearly articulates what makes the brand unique and why it is valuable to the target audience. The UVP should highlight the brand's key benefits and differentiators.

- **Positioning Statement**: A concise statement that communicates the brand's unique value and target audience. It serves as a guide for all marketing and communication efforts, ensuring consistency in messaging.

5. Strategic Objectives

- **Short-Term Objectives**: Specific, measurable goals to be achieved in the near term. These could include increasing brand awareness, launching a new product, or expanding into a new market.

- **Long-Term Goals**: Broader objectives aimed at achieving sustainable growth and establishing the brand as a leader in its field. These might involve building a strong market presence, enhancing brand reputation, or achieving significant market share.

6. Marketing and Communication Strategies

- **Integrated Marketing Campaigns**: Developing and executing campaigns that use a mix of channels (e.g., digital, print, events) to reach the target audience. Ensures a consistent message and maximizes impact.

- **Content Strategy**: Creating and distributing relevant and valuable content to engage with healthcare professionals

(HCPs), patients, and other stakeholders. This may include educational materials, case studies, and promotional content.

- **Digital Marketing** involves utilizing digital channels such as social media, email marketing, and SEO to enhance brand visibility and engagement. Targeted digital campaigns help reach specific audience segments.

7. Sales and Distribution

- **Sales Strategies**: Developing strategies to support the sales team, including training programs, sales aids, and incentives. Aligning marketing and sales efforts to drive product adoption and market penetration.

- **Distribution Channels**: Identifying and managing distribution channels to ensure product availability and accessibility. This involves working with wholesalers, distributors, and healthcare providers.

8. Monitoring and Evaluation

- **Performance Tracking**: Measuring key performance indicators (KPIs) to assess the effectiveness of marketing and sales activities. KPIs may include sales figures, market share, brand awareness, and engagement metrics.

- **Feedback Mechanisms**: We collect feedback from HCPs, patients, and other stakeholders to understand their perceptions of the brand and identify areas for improvement.

- **Adjustments and Optimization**: Making necessary adjustments to marketing and sales strategies based on performance data and feedback. This could involve revising messaging, reallocating budgets, or exploring new opportunities.

9. Compliance and Ethics

- **Regulatory Adherence**: Ensuring all marketing activities comply with regulatory requirements and industry guidelines. An accurate and balanced representation of the product's benefits and risks is essential.

- **Ethical Considerations**: Adhering to ethical standards in marketing practices, including transparency, honesty, and respect for patient privacy.

Example of a Successful Brand Plan

The following example demonstrates how each element of a brand plan is integrated to achieve a successful market presence and drive brand growth.

Case Study #32: Integrating All the Elements of the Brand Plan Helps Achieve A Significant Market Presence!

Company: Novartis

Brand: Cosentyx (Secukinumab)

Brand Vision and Mission:

- **Vision**: To transform patients' lives with autoimmune disease through innovative therapies.
- **Mission**: To deliver a breakthrough treatment for conditions like psoriasis and ankylosing spondylitis patients.

Brand Identity:

- **Name and Logo**: Cosentyx's branding includes a distinct logo and color scheme that conveys trust and innovation.
- **Personality**: Positioned as innovative, effective, and patient-focused.

Market Analysis:

- **Research**: Conducted extensive research on unmet needs in autoimmune disease treatment and competitor offerings.
- **Segmentation**: Targeted different segments based on disease type and severity.
- **Competitive Analysis**: Differentiated Cosentyx from other treatments, emphasizing its efficacy and safety profile.

Brand Positioning:

- **UVP**: A novel IL-17A inhibitor providing superior efficacy and safety compared to existing treatments.
- **Positioning Statement**: Cosentyx is a next-generation therapy that provides effective and sustained relief for patients with challenging autoimmune conditions.

Strategic Objectives:

- **Short-Term**: Increase market penetration in key therapeutic areas.
- **Long-Term**: Establish Cosentyx as a leading treatment for multiple autoimmune conditions.

Marketing and Communication Strategies:

- **Campaigns**: Integrated campaigns across digital, print, and professional channels.
- **Content**: Educational materials and case studies highlighting the clinical benefits of Cosentyx.
- **Digital Marketing**: Targeted digital campaigns to engage healthcare professionals and patients.

Sales and Distribution:

- **Sales Support**: Comprehensive training and resources for sales representatives.
- **Distribution**: Managed through global and regional distribution networks to ensure availability.

Monitoring and Evaluation:

- **Performance Tracking**: Monitored sales performance, market share, and brand perception.
- **Feedback**: Gathered feedback from HCPs and patients to refine marketing strategies.
- **Adjustments**: Made strategic adjustments based on market response and feedback.

Compliance and Ethics:

- **Regulatory Adherence**: Ensured all marketing materials were compliant with regulatory standards.
- **Ethical Practices**: Maintained transparency and integrity in all communications.

22.2. Execution Strategies for Brand Growth

Executing a brand growth strategy involves translating strategic plans into actionable steps that drive brand recognition, market share, and profitability. Here are key execution strategies for achieving brand growth:

1. Market Penetration

- **Objective**: increase the brand's market share within existing markets

- **Targeted Marketing Campaigns**: Launch campaigns focused on existing customer segments to increase brand loyalty and attract new customers. Utilize personalized marketing and data-driven approaches to enhance campaign effectiveness.

- **Promotional Activities**: Implement promotional strategies such as discounts, special offers, and loyalty programs to incentivize purchases and boost sales.

- **Distribution Expansion**: Increase product availability by expanding distribution channels, including partnerships with new distributors or entering new retail markets.

2. Product Innovation and Diversification

- **Objective**: Expand the brand's product portfolio and enter new markets.

- **New Product Development**: Invest in research and development to create innovative products that meet emerging customer needs or address gaps in the market. Ensure new products align with the brand's core values and target audience.

- **Product Line Extensions**: Introduce variations of existing products, such as different formulations, sizes, or flavors, to appeal to a broader audience.

- **Geographic Expansion**: Enter new geographic markets by adapting products and marketing strategies to fit local preferences and regulations.

3. Enhanced Customer Experience

- **Objective**: Improve customer satisfaction and loyalty through exceptional experiences.

- **Personalization**: Use data and analytics to personalize customer interactions, including tailored recommendations, content, and communications.

- **Customer Support**: Provide high-quality customer support through multiple channels, including online chat, phone, and email. Ensure quick issue resolution and provide helpful resources.

- **Feedback Mechanisms**: Implement systems to gather and act on customer feedback, including surveys, reviews, and focus groups. Use insights to make improvements and address customer needs.

4. Strategic Partnerships and Alliances

- **Objective**: Leverage partnerships to enhance brand visibility and access to new markets.

- **Collaborations**: Partner with other brands, healthcare organizations, or influencers to co-create content, sponsor events, or run joint campaigns. Choose partners whose values align with the brand and can provide complementary strengths.

- **Affiliate Programs**: Establish affiliate marketing programs where partners can earn commissions for driving sales or leads to the brand. This can expand reach and generate additional revenue. However, such programs should comply with the rules and regulatory guidelines governing the pharmaceutical industry.

5. Digital and Social Media Engagement

- **Objective**: Increase brand visibility and engagement through digital channels.

- **Content Marketing**: Develop and distribute valuable, relevant content that educates, informs, and engages the target audience. Content can include blog posts, videos, infographics, and webinars.

- **Social Media Strategy**: Utilize social media platforms to build brand awareness, engage with customers, and drive traffic to the brand's website. Tailor content for each platform and use social listening tools to monitor and respond to conversations.

- **Search Engine Optimization (SEO)**: Optimize the brand's website and content for search engines to improve organic search rankings and attract relevant traffic.

6. Data-Driven Decision Making

- **Objective**: Use data and analytics to inform strategic decisions and optimize performance.

- **Market Analysis**: Analyze market trends, customer behavior, and competitive landscape to identify opportunities and threats.

- **Performance Metrics**: Track key performance indicators (KPIs) such as sales growth, customer acquisition costs, and return on investment (ROI). Use these metrics to evaluate the effectiveness of marketing strategies and make data-driven adjustments.

- **A/B Testing**: Conduct A/B testing to compare different marketing tactics such as ad creatives or email subject lines, to determine what works better and optimize campaigns accordingly.

7. Brand Positioning and Messaging

- **Objective**: Strengthen the brand's market position through clear and consistent messaging.

- **Brand Messaging**: Develop and maintain consistent messaging that communicates the brand's unique value proposition and resonates with the target audience. Ensure all marketing materials and communication reflect this message.

- **Positioning Strategies**: Regularly review and adjust the brand's positioning to stay relevant in a changing market. Highlight the brand's unique strengths and differentiators in all promotional activities.

8. Regulatory Compliance and Risk Management

- **Objective**: Ensure all marketing activities comply with regulations and manage risks effectively.

- **Compliance Checks**: Regularly review marketing materials and strategies to ensure they adhere to regulatory guidelines and industry standards. Work with legal and compliance teams to address any potential issues.

- **Risk Mitigation**: Identify potential risks associated with brand growth, such as market entry barriers or regulatory changes, or develop contingency plans to address them.

Example of Execution Strategies for Brand Growth

Here is a case study highlighting how a global pharmaceutical major effectively managed and grew its oncology brand through targeted efforts across various areas.

Case Study #33: Pfizer's Impeccable Execution of Ibrance's Marketing Strategies!

Company: Pfizer

Brand: Ibrance (Palbociclib)

Market Penetration:

- **Targeted Campaigns**: Launched campaigns targeting oncologists and healthcare providers with personalized content about Ibrance's benefits in breast cancer treatment.

- **Promotions**: Offered educational webinars and free samples to support physicians in understanding and prescribing Ibrance.

Product Innovation and Diversification:

- **New Formulations**: Explored additional formulations and combinations with other cancer treatments to enhance efficacy and patient adherence.

- **Geographic Expansion**: Entered new international markets by adapting regulatory submissions and marketing strategies to local environments.

Enhanced Customer Experience:

- **Personalization**: Developed a patient support program with personalized treatment information and adherence resources.

- **Feedback**: Collected feedback from healthcare providers and patients to refine educational materials and support services.

Strategic Partnerships and Alliances:

- **Collaborations**: Partnered with cancer research organizations to support clinical trials and raise awareness about Ibrance.

- **Affiliate Programs**: Collaborated with advocacy groups to expand outreach and education efforts.

Digital and Social Media Engagement:

- **Content Marketing**: Created a series of educational videos and articles about breast cancer and Ibrance's role in treatment.
- **Social Media**: Engaged with patients and healthcare professionals through social media platforms, sharing success stories and treatment updates.

Data-Driven Decision Making:

- **Market Analysis**: Monitored market trends and competitor activities to adapt marketing strategies and maintain Ibrance's competitive edge.
- **Performance Metrics**: Tracked prescription rates and market share KPIs to measure campaign effectiveness.

Brand Positioning and Messaging:

- **Brand Messaging**: Emphasized Ibrance's innovative approach to treating hormone receptor-positive breast cancer, positioning it as a leading option for patients.
- **Positioning Strategies**: Regularly updated positioning to reflect new research findings and clinical data.

Regulatory Compliance and Risk Management:

- **Compliance Checks**: Ensured all promotional materials and activities complied with FDA regulations and industry guidelines.
- **Risk Mitigation**: Developed contingency plans for potential regulatory changes and market challenges.

22.3 Monitoring and Adjusting Brand Performance

Monitoring and adjusting brand performance is essential for maintaining and enhancing a brand's market position. This involves continuously tracking various metrics to assess the effectiveness of brand strategies and making necessary adjustments to optimize outcomes. Here is how to approach this process:

1. **Define Key Performance Indicators (KPIs)**

   - **Objective**: Establish metrics to measure brand performance.

   - **Sales Metrics**: Track sales volume, revenue growth, market share, and profitability. These indicators reflect the brand's market success and financial health.

   - **Customer Metrics**: Measure customer acquisition, retention rates, customer lifetime value (CLV), and customer satisfaction scores. These metrics gauge how well the brand meets customer needs and expectations.

   - **Marketing Metrics**: Analyze metrics such as return on investment (ROI), cost per acquisition (CPA), conversion, and engagement rates. These metrics help evaluate the effectiveness of marketing campaigns and strategies.

2. **Collect and Analyze Data**

   - **Objective**: Gather data to assess brand performance.

   - **Sales Data**: Monitor sales data from various channels, including online, retail, and direct sales. Use tools like CRM systems and sales dashboards to aggregate and analyze this data.

   - **Customer Feedback**: Collect feedback through surveys, reviews, and social media interactions. Use sentiment analysis tools to gauge overall customer sentiment and identify recurring issues.

   - **Marketing Analytics**: Utilize analytics platforms (Google Analytics, social media analytics) to track the performance of digital marketing efforts. Analyze traffic, engagement, and conversion data.

3. **Evaluate Brand Performance**

- **Objective**: Assess how well the brand is performing against objectives.

- **Benchmarking**: Compare brand performance against industry benchmarks and competitors. Identify areas where the brand excels or lags.

- **Trend Analysis**: Examine performance trends over time to understand how brand metrics evolve. Look for patterns that indicate growth opportunities or potential issues.

- **Performance Reviews**: Regular performance reviews assess progress against goals and objectives. Use these reviews to make informed decisions about strategy adjustments.

4. **Identify Issues and Opportunities**

- **Objective**: Detect areas needing improvement or growth opportunities.

- **Performance Gaps**: Identify discrepancies between expected and actual performance. Analyze the causes of these gaps to determine whether they stem from internal factors (marketing strategies) or external factors (market changes).

- **Opportunities**: Look for new opportunities based on market trends, customer needs, and competitive analysis. Explore ways to capitalize on these opportunities to enhance brand performance.

5. **Adjust Strategies and Tactics**

- **Objective**: Make adjustments to optimize brand performance.

- **Strategic Adjustments**: Based on performance insights, revise overarching brand strategies. This may involve repositioning the brand, updating the value proposition, or shifting target markets.

- **Tactical Changes**: Modify specific marketing tactics and campaigns to address identified issues or leverage new opportunities. This could include changing advertising channels, updating creative content, or revising promotional offers.

- **Resource Allocation**: Reallocate resources (e.g., budget, personnel) to areas showing the most promise or requiring additional support. Optimize spending based on what is yielding the best results.

6. Implement Feedback Loops

- **Objective**: Ensure continuous improvement through ongoing feedback.

- **Customer Feedback**: Regularly solicit customer feedback to understand their evolving needs and performance. Use this feedback to inform product or service improvements.

- **Employee Insights**: Gather input from brand management, marketing, and customer service employees. Their insights can provide valuable perspectives on performance and areas for improvement.

7. Communicate Changes and Monitor Results

- **Objective**: Ensure effective implementation of adjustments and track outcomes.

- **Communication**: Communicate any changes in strategy or tactics to relevant stakeholders, including internal teams and external partners. Ensure everyone is aligned and understands their roles in executing the revised plan.

- **Result Tracking**: Continue to monitor brand performance metrics to assess the impact of adjustments. Evaluate whether the changes have led to improvements and refine as needed.

Example of Monitoring and Adjusting Brand Performance

The following case study highlights how a pharmaceutical company can monitor and adjust its brand performance to grow continuously.

Case Study #34: How a Pharma Major Can Monitor and Adjust Brand Performance!

Company: AstraZeneca

Brand: Tagrisso (Osimertinib)

Define KPIs:

- **Sales Metrics**: Monitor sales volume, revenue growth, and market share in the EGFR-mutated non-small cell lung cancer (NSCLC) market.
- **Customer Needs**: Track patient adherence rates, satisfaction scores, and oncologist feedback.
- **Marketing Metrics**: Analyze ROI, cost per acquisition, and conversion rates for digital and field-based marketing efforts.

Collect and Analyze Data:

- **Sales Data**: Use sales dashboards to track performance across regions and channels.
- **Customer Feedback**: Collect feedback from patients and healthcare providers through surveys and focus groups.
- **Marketing Analytics**: Utilize analytics tools to assess the effectiveness of digital campaigns, including website traffic and engagement metrics.

Evaluate Brand Performance

- **Benchmarking**: Compare Tagrisso's performance with competitors like Gefitinib and Erlotinib.
- **Trend Analysis**: Review sales and customer feedback trends to identify areas of growth or concern.
- **Performance Reviews**: Conduct quarterly reviews to evaluate progress against sales targets and market penetration goals.

Identify Issues and Opportunities

- **Performance Gaps**: Identify regions where Tagrisso is underperforming and analyze potential causes, such as competitive actions or market conditions.

- **Opportunities**: Explore opportunities for expanding market share by targeting new patient segments or improving access through partnerships with oncology clinics.

Adjust Strategies and Tactics:

- **Strategic Adjustments**: Revise positioning strategies to differentiate Tagrisso from competitors based on new clinical data or patient needs.
- **Tactical Changes**: Update marketing materials and campaigns to address feedback from healthcare providers and enhance communication about Tagrisso's benefits.
- **Resource Allocation**: Allocate additional resources to regions with high growth potential or require more support.

Implement Feedback Loops

- **Customer Feedback**: Continuously use patient and provider feedback to improve support programs and patient education materials.
- **Employee Insights**: Gather insights from sales representatives and medical science liaisons to refine field strategies and enhance engagement with healthcare providers.

Communicate Changes and Monitor Results:

- **Communication**: Inform internal teams and external partners about strategy updates and new initiatives. Ensure alignment across all touchpoints.
- **Result Tracking**: Continue to track performance metrics to gauge the impact of adjustments and make further refinements as needed.

By following these steps, AstraZeneca can effectively monitor and adjust Tagrisso's performance to maximize its market success and address emerging challenges.

22.4 Example of Brand Plan for a Hypothetical Diabetes Medication Brand Glucotrol

Glucotrol Brand Plan

Brand Overview:

- **Brand Name**: Glucotrol
- **Therapeutic Area**: Diabetes Management
- **Product Attributes**:
  - Superior blood sugar control
  - Minimal side effects
  - Convenient once-daily dosage

Executive Summary

Glucotrol aims to establish itself as a leader in the diabetes medication market by highlighting its superior efficacy, safety profile, and convenience. This brand plan outlines the strategic initiatives to drive market penetration, increase physician adoption, and enhance patient adherence.

1. Market Analysis

1.1. Market Overview

- **Global Diabetes Market Size**: $XX billion (2023)
- **Growth Rate**: CAGR of X% (2023-2028)
- **Key Trends**: Increasing prevalence of diabetes, shift towards patient-centric care, demand for medications with better efficacy and fewer side effects.

1.2. Competitive Landscape

- **Key Competitors**:
  - Brand A: Twice-daily dosage, higher side effect profile.
  - Brand B: Once-daily dosge, moderate side effect profile, moderate efficacy.
  - Brand C: Twice-daily dosage, minimal side effects, lower efficacy.

1.3. SWOT Analysis

- **Strengths**:
  - Superior blood sugar control
  - Minimal side effects
  - Convenient once-daily dosage
- **Weaknesses**:
  - New entrant, lower brand awareness
  - Potentially higher cost
- **Opportunities**:
  - Growing diabetes population
  - Increasing demand for effective and convenient treatments
  - Potential for expansion into emerging markets
- **Threats**:
  - Strong competition from established brands
  - Regulatory changes
  - Generic medications

2. Target Audience

2.1. Primary Target Audience

- **Healthcare Professionals (HCPs)**:
  - Endocrinologists
  - General Practitioners
  - Diabetologists

2.2. Secondary Target Audience

- **Patients**:
  - Type 2 Diabetes Patients
  - Patients experiencing side effects from current medications
  - Patients seeking more convenient treatment options

3. Brand Positioning

3.1. Value Proposition

- "Glucotrol offers superior blood sugar control with minimal side effects in a convenient once-daily dosage, empowering patients to manage their diabetes effectively and easily."

3.2. Key Messages

- **Efficacy**: "Glucotrol provides superior blood sugar control compared to leading brands."
- **Safety**: "Experience minimal side effects with Glucotrol."
- **Convenience**: "Enjoy the simplicity of once-daily dosing with Glucotrol."

4. Marketing Strategy

4.1 Objectives:

- Achieve X% market share within the first year.
- Increase brand awareness among HCPs and patients.
- Drive patient adherence through education and support programs.

4.2. Market Mix (4 Ps)

- **Product**:
- **Formulation**: Once-daily oral tablet
- **Packaging**: Easy-to-open blister packs
- **Additional Features**: Patient information leaflet with detailed usage instructions
- **Price**:
  - **Pricing Strategy**: Competitive pricing with tiered discounts for bulk purchases by healthcare institutions.
  - **Reimbursement**: Work with insurance providers to ensure coverage.

Place:

- **Distribution Channels**: Pharmacies, hospitals, online platforms.

- **Geographic Focus**: Initial launch in major markets (US, Europe) and expansion into emerging markets.

Promotion:

- **HCP Engagement**:
  - Educational seminars and webinars
  - Detailing and sampling through sales representatives
  - Peer-to-peer interactions and endorsements
  - Patient Engagement
  - Digital marketing campaigns (social media, email marketing)
  - Patient support programs (hotlines, mobile apps)
  - Educational content (blogs, videos, brochures)

5. Implementation Plan

5.1. Timeline

Q1:

- Finalize branding and packaging
- Train sales and marketing teams
- Launch HCP educational campaigns

Q2:

- Begin patient-centric marketing initiatives
- Start distribution in key markets
- Launch the Glucotrol website and mobile app

Q3:

- Conduct market research and gather feedback
- Adjust strategies based on feedback
- Expand distribution to additional markets

Q4:

- Evaluate year-end performance
- Plan for next year's initiative
- Introduce new patient support features

5.2. Budget

- **Total Budget**: $X million

Allocation:

- Marketing and Promotions: 40%
- HCP Engagement: 25%
- Patient Support Programs: 20%
- Market Research: 10%
- Miscellaneous: 5%

6. Monitoring and Evaluation

6.1. Key Performance Indicators (KPIs)

- **Sales Metrics**: Market share, sales volume, revenue
- **Awareness Metrics**: Brand recognition, website traffic, social media engagement
- **Engagement Metrics**: HCP participation in educational programs, patient enrollment in support programs
- **Adherence Metrics**: Patient feedback, prescription refill rates

6.2 Feedback and Adjustments

- Regularly review KPIs and adjust strategies as needed
- Conduct surveys and focus groups to gather feedback from HCPs and patients
- Implement improvements based on feedback and performance data

Conclusion

Glucotrol is poised to become a leading brand in the diabetes medication market through its superior efficacy, minimal side effects, and convenient once-daily dosage. By executing this comprehensive brand plan, Glucotrol aims to improve the lives of diabetes patients and achieve significant market success.

Part VI. Leadership and Communication

In today's complex and fast-paced pharmaceutical industry, effective leadership of cross-functional teams is critical to the success of pharmaceutical product management. Leading these teams requires a deep understanding of the various functions and the ability to foster collaboration, align goals, and drive collective performance. Here are the details regarding what it takes to lead cross-functional teams:

1. **Definition and Importance**:
   - **Definition**: Cross-functional teams are groups composed of members from different departments or areas of expertise working together towards a common goal. In pharmaceutical product management, these teams typically include representatives from R&D, marketing, regulatory affairs, sales, and other relevant functions.
   - **Importance**: Effective cross-functional leadership ensures that diverse perspectives are integrated, projects are executed efficiently, and all aspects of the product lifecycle are managed cohesively. It is crucial for aligning team efforts, optimizing resources, and achieving strategic objectives.

2. **Building and Assembling the Team**:
   - **Identify Key Functions**: Based on the project's needs, determine which functions and expertise are required. This may include roles in clinical development, regulatory affairs, market access, commercial strategy, and more.
   - **Select Team Members**: Choose individuals with the necessary skills, experience, and knowledge. Ensure that they are motivated and aligned with the project goals.
   - **Define Roles and Responsibilities**: Clearly outline each team member's role and responsibilities to avoid confusion and ensure accountability.

3. **Facilitating Communication and Collaboration**:
   - **Establish Communication Channels**: Set up regular meetings, updates, and communication channels to keep everyone informed and engaged. Use collaboration tools and platforms to facilitate information sharing.

- **Foster Open Dialogue**: Encourage team members to share ideas, ask questions, and provide feedback. Create an environment where diverse opinions are valued and considered.
- **Address Conflicts**: Mediate and resolve conflicts that arise within the team. Ensure that disagreements are addressed constructively and that solutions are reached collaboratively.

4. **Setting Goals and Aligning Objectives**:
   - **Define Clear Objectives**: Set specific, measurable, achievable, relevant, and time-bound (SMART) goals for the team. Ensure that these objectives align with the overall product strategy and business goals.
   - **Communicate Vision**: Share the product vision and strategic objectives with the team to ensure everyone understands the big picture and their role in achieving it.

5. **Managing Project Execution**:
   - **Develop a Project Plan**: Create a detailed plan outlining task timelines, milestones, and deliverables. Assign responsibilities and set deadlines.
   - **Monitor Progress**: Track the project's progress against the plan. Use key performance indicators (KPIs) and regular status updates to assess performance and adjust as needed.
   - **Resource Allocation**: Ensure that resources (budget, personnel, time) are allocated effectively and efficiently. Adjust resource allocation based on project needs and priorities.

6. **Motivating and Leading the Team**:
   - **Inspire and Empower**: Motivate team members by recognizing their contributions, providing support, and empowering them to take ownership of their work.
   - **Provide Feedback**: Offer constructive feedback and guidance to help team members improve and grow. Celebrate successes and learn from setbacks.
   - **Promote Professional Development**: Encourage ongoing learning and development opportunities for team members to enhance their skills and knowledge.

7. **Ensuring Cross-Functional Integration**:

   - **Coordinate Activities**: Ensure that activities and outputs from different functions are integrated and aligned. Facilitate coordination between departments to avoid duplication and ensure consistency.

   - **Manage Dependencies**: Identify and manage dependencies between different functions to ensure that tasks are completed in the correct sequence and that delays are minimized.

8. **Evaluating and Improving Team Performance**:

   - **Conduct Reviews**: Regularly review team performance and project outcomes. Assess what worked well and what could be improved.

   - **Implement Improvements**: Use review insights to improve team processes, communication, and project management practices.

Implications for Pharma Product Managers

1. **Enhanced Efficiency**: Effective leadership and coordination of cross-functional teams lead to more efficient project execution and reduced time-to-market.

2. **Holistic Approach**: Integrating diverse expertise ensures a comprehensive approach to product development and management that addresses all aspects of the product lifecycle.

3. **Increased Innovation**: Collaboration across functions fosters creativity and innovation, leading to more effective solutions and strategies.

4. **Better Decision-Making**: Diverse perspectives and expertise contribute to more informed and well-rounded decision-making.

5. **Improved Team Morale**: Clear roles, effective communication, and recognition of contributions enhance team morale and engagement.

In summary, leading cross-functional teams involves assembling the right people, facilitating effective communication, setting and aligning goals, managing project execution, and continuously improving team performance. This leadership is essential for successful pharmaceutical product management and achieving strategic objectives.

Leading Cross-Functional Teams

In today's complex and fast-paced pharmaceutical industry, effective leadership of cross-functional teams is critical to the success of pharmaceutical product management. Leading these teams requires a deep understanding of the various functions and the ability to foster collaboration, align goals, and drive collective performance. Here are the details regarding what it takes to lead cross-functional teams:

1. **Definition and Importance**:
   - **Definition**: Cross-functional teams are groups composed of members from different departments or areas of expertise working together towards a common goal. In pharmaceutical product management, these teams typically include representatives from R&D, marketing, regulatory affairs, sales, and other relevant functions.
   - **Importance**: Effective cross-functional leadership ensures that diverse perspectives are integrated, projects are executed efficiently, and all aspects of the product lifecycle are managed cohesively. It is crucial for aligning team efforts, optimizing resources, and achieving strategic objectives.

2. **Building and Assembling the Team**:
   - **Identify Key Functions**: Based on the project's needs, determine which functions and expertise are required. This may include roles in clinical development, regulatory affairs, market access, commercial strategy, and more.

- **Select Team Members**: Choose individuals with the necessary skills, experience, and knowledge. Ensure that they are motivated and aligned with the project goals.

- **Define Roles and Responsibilities**: Clearly outline each team member's role and responsibilities to avoid confusion and ensure accountability.

3. **Facilitating Communication and Collaboration**:

- **Establish Communication Channels**: Set up regular meetings, updates, and communication channels to keep everyone informed and engaged. Use collaboration tools and platforms to facilitate information sharing.

- **Foster Open Dialogue**: Encourage team members to share ideas, ask questions, and provide feedback. Create an environment where diverse opinions are valued and considered.

- **Address Conflicts**: Mediate and resolve conflicts that arise within the team. Ensure that disagreements are addressed constructively and that solutions are reached collaboratively.

4. **Setting Goals and Aligning Objectives**:

- **Define Clear Objectives**: Set specific, measurable, achievable, relevant, and time-bound (SMART) goals for the team. Ensure that these objectives align with the overall product strategy and business goals.

- **Communicate Vision**: Share the product vision and strategic objectives with the team to ensure everyone understands the big picture and their role in achieving it.

5. **Managing Project Execution**:

- **Develop a Project Plan**: Create a detailed plan outlining task timelines, milestones, and deliverables. Assign responsibilities and set deadlines.

- **Monitor Progress**: Track the project's progress against the plan. Use key performance indicators (KPIs) and regular status updates to assess performance and adjust as needed.

- **Resource Allocation**: Ensure that resources (budget, personnel, time) are allocated effectively and efficiently. Adjust resource allocation based on project needs and priorities.

6. **Motivating and Leading the Team**:

- **Inspire and Empower**: Motivate team members by recognizing their contributions, providing support, and empowering them to take ownership of their work.

- **Provide Feedback**: Offer constructive feedback and guidance to help team members improve and grow. Celebrate successes and learn from setbacks.

- **Promote Professional Development**: Encourage ongoing learning and development opportunities for team members to enhance their skills and knowledge.

7. **Ensuring Cross-Functional Integration**:

- **Coordinate Activities**: Ensure that activities and outputs from different functions are integrated and aligned. Facilitate coordination between departments to avoid duplication and ensure consistency.

- **Manage Dependencies**: Identify and manage dependencies between different functions to ensure that tasks are completed in the correct sequence and that delays are minimized.

8. **Evaluating and Improving Team Performance**:

- **Conduct Reviews**: Regularly review team performance and project outcomes. Assess what worked well and what could be improved.

- **Implement Improvements**: Use review insights to improve team processes, communication, and project management practices.

Implications for Pharma Product Managers

1. **Enhanced Efficiency**: Effective leadership and coordination of cross-functional teams lead to more efficient project execution and reduced time-to-market.

2. **Holistic Approach**: Integrating diverse expertise ensures a comprehensive approach to product development and management that addresses all aspects of the product lifecycle.

3. **Increased Innovation**: Collaboration across functions fosters creativity and innovation, leading to more effective solutions and strategies.

4. **Better Decision-Making**: Diverse perspectives and expertise contribute to more informed and well-rounded decision-making.

5. **Improved Team Morale**: Clear roles, effective communication, and recognition of contributions enhance team morale and engagement.

In summary, leading cross-functional teams involves assembling the right people, facilitating effective communication, setting and aligning goals, managing project execution, and continuously improving team performance. This leadership is essential for successful pharmaceutical product management and achieving strategic objectives.

23.1 The Role of Leadership in Product Management

Leadership plays a crucial role in product management, influencing the product's success, the team's effectiveness, and the achievement of business objectives. Here is a detailed description of the role of leadership in product management:

Role of Leadership in Product Management

1. Vision and Strategy Development

- **Defining Vision**: Product management leaders are responsible for creating and articulating a clear and compelling vision for the product. This vision offers the product's purpose, goals, and strategic direction.

- **Strategic Planning**: Leaders develop and execute strategic plans that align with the company's objectives. This involves setting long-term goals, identifying key opportunities, and creating a roadmap for product development and market success.

2. Team Building and Motivation

- **Assembling the Right Team**: Effective product managers build and lead cross-functional teams that include members from various departments such as R&D, marketing, sales, and regulatory affairs.

- **Motivating Team Members**: Leaders inspire and motivate their teams by fostering a positive work environment, recognizing achievements, and encouraging professional growth. Motivation leads to higher engagement, productivity, and job satisfaction.

3. Decision-Making and Problem Solving

- **Making Informed Decisions**: Product managers make critical decisions about product development, marketing strategies, and resource allocation. Effective leadership involves gathering and analyzing data, considering diverse perspectives, and making well-informed decisions.

- **Addressing Challenges**: Leaders are responsible for identifying and addressing challenges during the product lifecycle. They implement problem-solving strategies, adjust plans to overcome obstacles, and keep the project on track.

4. Communication and Collaboration

- **Facilitating Communication**: Leaders ensure clear and effective communication within the team and with other stakeholders. This includes meeting regularly, providing updates, and ensuring everyone is aligned with the product goals.

- **Encouraging Collaboration**: Effective leaders foster a collaborative environment where team members work together towards common objectives. They facilitate collaboration between departments to ensure that all aspects of the product are integrated and aligned.

5. Stakeholder Management

- **Engaging Stakeholders**: Product managers must engage with various stakeholders, including executives, customers, healthcare professionals, and regulatory bodies. Leadership involves managing these relationships, addressing stakeholder needs, and meeting their expectations.

- **Managing Expectations**: Leaders are responsible for setting realistic expectations and managing stakeholder concerns. They communicate progress, address feedback, and negotiate compromises when necessary.

6. Project Management and Execution

- **Overseeing Execution**: Leaders oversee the execution of product plans, ensuring that projects are completed on time, within budget, and to the desired quality standards. They manage timelines, resources, and deliverables effectively.

- **Monitoring Performance**: Effective leaders track the performance of the product and the team, using key performance indicators (KPIs) and metrics to assess progress and make adjustments as needed.

7. Risk Management

- **Identifying Risks**: Leaders are responsible for identifying potential risks impacting the product's success. This includes assessing market risks, regulatory challenges, and operational issues.

- **Developing Mitigation Strategies**: Leaders develop and implement strategies to mitigate identified risks. This involves creating contingency plans, monitoring risk factors, and adapting strategies as necessary.

8. Driving Innovation

- **Encouraging Innovation**: Effective leaders promote a culture of innovation by encouraging creative thinking, supporting new ideas, and exploring emerging technologies. Innovation is essential for staying competitive and meeting evolving market needs.

- **Implementing New Approaches**: Leaders are responsible for adopting and implementing new approaches and technologies that enhance product development, marketing, and overall success.

9. Evaluating and Improving Processes

- **Conducting Reviews**: Leaders regularly evaluate the product's performance and the processes' effectiveness. They conduct post-launch reviews, gather feedback, and identify areas for improvement.

- **Implementing Improvements**: Based on evaluations, leaders implement process improvements to enhance efficiency, effectiveness, and overall product performance.

Implications for Product Managers

1. **Enhanced Product Success**: Strong leadership leads to better strategic planning, execution, and overall product success.

2. **Effective Team Dynamics**: Leadership ensures teams are motivated, collaborative, and aligned with the product's goals.

3. **Informed Decision-making**: Leadership ensures that teams are motivated, collaborative, and aligned with the product's goals.

4. **Stakeholder Satisfaction**: Leadership in managing stakeholder relationships ensures that expectations are met and concerns are addressed.

5. **Innovation and Growth**: Encouraging innovation and adopting new approaches help stay competitive and meet market demands.

In summary, leadership in product management involves defining vision and strategy, building and motivating teams, making informed decisions, facilitating communication, managing stakeholders, overseeing execution, managing risks, driving innovation, and continuously improving processes. Strong leadership is essential for achieving product success and meeting business objectives.

23.2 Building and Leading High-Performing Teams

Building and leading high-performance teams is a fundamental aspect of successful product management. High-performing teams are essential for achieving strategic goals, driving innovation, and executing complex projects effectively. Here is a detailed description of the key elements involved in building and leading high-performing teams:

1. **Assembling the Right Team**
   - **Identifying Needs**: Assess the skills and expertise required for the project or product. Determine the roles and responsibilities needed to achieve the team's goals.
   - **Recruiting Talent**: Select team members with the necessary skills, experience, and cultural fit. This may involve hiring new employees, promoting from within, or forming cross-functional teams from existing staff.
   - **Diverse Skill Sets**: Build a team with diverse skills and perspectives. Diversity in skills, backgrounds, and experiences can lead to more creative solutions and better decision-making.

2. **Setting Clear Goals and Expectations**
   - **Defining Objectives**: Clearly define the team's goals and objectives. Ensure these goals align with the overall product strategy and business objectives.
   - **Communicating Expectations**: Clearly communicate roles, responsibilities, and performance expectations to each team member. This helps set the right expectations and reduces ambiguity.
   - **Establishing KPIs**: Set measurable key performance Indicators (KPIs) to track progress and evaluate the team's performance. Regularly review these KPIs to ensure alignment with the project's goals.

3. Fostering a Collaborative Environment

- **Encouraging Open Communication**: Promote open and honest communication within the team. Encourage team members to share their ideas, feedback, and concerns.

- **Building Trust**: Be transparent, supportive, and respectful to foster trust among team members. Trust is essential for effective collaboration and teamwork.

- **Facilitating Collaboration**: Create opportunities for team members to collaborate and work together, especially in remote or hybrid work environments.

4. Providing Support and Resources

- **Offering Training and Development**: Provide opportunities for team members to develop their skills and knowledge. This may include training programs, workshops, or mentoring.

- **Allocating Resources**: Ensure the team has the resources, tools, and technology to perform their tasks effectively. This includes providing access to relevant software, data, and equipment.

- **Removing Obstacles**: Identify and address any obstacles or challenges hindering the team's performance. This could involve resolving conflicts, streamlining processes, or removing bureaucratic barriers.

5. Motivating and Engaging Team Members

- **Recognizing Achievements**: Acknowledge and celebrate the team's achievements and individual contributions. Recognition can boost morale and motivation.

- **Providing Feedback**: Offer regular, constructive feedback to team members. Focus on both strengths and areas for improvement and provide actionable advice.

- **Creating a Positive Work Environment**: Foster a positive and inclusive work environment where team members feel valued and supported. Encourage work-life balance and promote well-being.

6. Leading by Example

- **Demonstrating Commitment**: Show commitment to the team's goals and values. Lead by example and be a role model for the behaviors and attitudes you expect from the team.

- **Being Accessible**: Be approachable and available to team members. Provide support, guidance, and mentorship as needed.

- **Adapting Leadership Style**: Adjust your leadership style to fit the team's needs and the situation. Different team members and scenarios may require different approaches.

7. Managing Performance and Development

- **Conducting Performance Reviews**: Regularly evaluate team members' performance. Use performance reviews to assess progress, address issues, and set development goals.

- **Providing Opportunities for Growth**: Support team members' career growth and development. Identify opportunities for advancement and provide the necessary resources and support.

- **Addressing Underperformance**: Address performance issues promptly and constructively. Work with underperforming team members to identify the root causes and develop improvement plans.

8. Navigating Challenges and Conflicts

- **Resolving Conflicts**: Address conflicts and disagreements within the team promptly and fairly. Use conflict resolution techniques to mediate and find mutually acceptable solutions.

- **Messaging Change**: Effectively lead the team through changes and transitions. Communicate the reasons for change, involve team members in the process, and provide support during the transition.

- **Maintaining Focus**: Keep the team focused on the goals and objectives despite challenges. Use motivation and problem-solving skills to overcome obstacles and stay on track.

9. Evaluating Team Performance

- **Assessing Effectiveness**: Regularly assess the effectiveness of the team. Evaluate how well the team is achieving its goals and identify areas of improvement.
- **Soliciting Feedback**: Gather feedback from team members on team dynamics, leadership, and processes. Use this feedback to make improvements and enhance team performance.
- **Adjusting Strategies**: Adjust team strategies, processes, and structures as needed. Continuously refine and optimize the approach to improve results.

Implications for Product Managers

1. **Enhanced Team Performance**: Effective leadership and team-building lead to higher productivity, better quality outcomes, and successful project execution.

2. **Improved Collaboration**: A collaborative environment fosters innovation, creativity, and effective problem-solving.

3. **Increased Employee Satisfaction**: Supportive and engaging leadership contributes to higher job satisfaction, lower turnover, and a positive work culture.

4. **Successful Project Execution**: High-performing teams are more likely to meet deadlines, stay within budget, and achieve project goals.

In summary, building and leading high-performing teams involves assembling the right talent, setting clear goals, fostering collaboration, providing support, motivating team members, leading by example, managing performance, addressing challenges, and evaluating effectiveness. Strong leadership in these areas is crucial for achieving successful product outcomes and measuring business objectives.

23.3 Managing Conflicts and Driving Collaboration

Managing conflicts and driving collaboration is crucial for pharmaceutical product managers to ensure a productive, harmonious, and effective team environment. Here is a detailed description of these aspects:

Managing Conflicts

Conflicts can arise in any team due to differences in opinions, work styles, or goals. Effectively managing conflicts is essential to maintaining team cohesion and ensuring that projects stay on track.

1. **Recognition of Conflicts Early**

   - **Observation**: Pay attention to signs of conflict, such as changes in team dynamics, decreased communication, or visible tension among team members.

   - **Feedback**: Solicit feedback from team members to identify potential issues before they escalate.

2. **Understanding the Root Cause**

   - **Active Listening**: Listen carefully to all parties involved to understand their perspectives and concerns.

   - **Clarification**: Ask questions to get to the heart of the issue. Understanding the underlying causes of the conflict is the key to resolving it effectively.

3. **Addressing Conflicts Constructively**

   - **Open Dialogue**: Create a safe space for open and honest discussion. Encourage team members to express their views and feelings without fear of retaliation.

   - **Empathy**: Show empathy towards the individuals involved. Acknowledge their feelings and validate their concerns.

   - **Problem-Solving**: Work collaboratively to find solutions that address the root cause of the conflict. Focus on common goals and interests rather than personal grievances.

4. Mediation Techniques

- **Neutral Mediation**: As a product manager, act as a neutral mediator if necessary. Facilitate discussions without taking sides.

- **Conflict Resolution Strategies**: Use conflict resolution techniques such as negotiation, compromise, or arbitration to resolve disagreements.

5. Implementing Solutions

- **Action Plan**: Develop an action plan with clear steps and responsibilities to address the conflict. Ensure that all parties are committed to the solution.

- **Follow-Up**: Monitor the situation to ensure the conflict is resolved and relationships are restored. Provide additional support if needed.

6. Learning from Conflicts

- **Reflection**: Reflect on the conflict and the resolution process to learn from the experience.

- **Continuous Improvement**: Use conflict insights to improve team dynamics and prevent future issues.

Driving Collaboration

Collaboration is essential for achieving team goals, fostering innovation, and ensuring effective project execution. Here is how to drive collaboration within a team:

1. Building a Collaborative Culture

- **Encourage Teamwork**: Promote a culture of teamwork by encouraging collaboration and sharing ideas. Emphasize the importance of working together towards common goals.

- **Inclusivity**: Ensure that all team members feel included and valued. Create an environment where diverse perspectives are welcomed and respected.

2. Facilitating Communication

- **Open Channels**: Establish clear and open communication channels. Use email, instant messaging, and collaborative platforms to facilitate information sharing.

- **Regular Meetings**: Hold team meetings regularly to discuss progress, address issues, and brainstorm solutions. Ensure that meetings are structured and productive.

3. Defining Roles and Responsibilities

- **Clarity**: Clearly define roles and responsibilities to avoid confusion and overlap. Ensure that each team member understands their role and how it contributes to the team's objectives.

- **Accountability**: Hold team members accountable for their responsibilities while encouraging collaboration and support.

4. Encourage Idea Sharing

- **Brainstorming Sessions**: Organize brainstorming sessions to generate and explore new ideas. Encourage all team members to contribute their thoughts and suggestions.

- **Feedback Mechanism**: Provide constructive feedback on ideas and proposals. Encourage a culture of continuous improvement and learning.

5. Leveraging Strengths and Expertise

- **Utilizing Expertise**: Leverage team members' strengths and expertise to achieve the best results. Assign tasks based on individual functional areas to bring diverse perspectives and solutions.

- **Cross-Functional Collaboration**: Foster collaboration between different functional areas to bring diverse perspectives and solutions to the table.

6. Building Team Cohesion

- **Team Building Activities**: Organize team-building activities to strengthen relationships and build trust among team members. Activities can range from informal social events to structured team exercises.

- **Celebrating Success**: Recognize and celebrate team achievement and milestones. Acknowledge contributions and foster a sense of accomplishment.

7. Managing Remote Collaboration

- **Effective Tools**: Use collaboration tools and technologies to facilitate remote teamwork. Ensure that all team members have access to the necessary tools and resources.

- **Virtual Meetings**: Conduct virtual meetings to maintain communication and collaboration among remote or distributed team members.

Implications for Product Managers

- **Conflict Management**: Effective conflict management ensures a positive work environment, minimizes disruptions, and maintains team morale.

- **Collaboration**: Strong collaboration skills lead to more innovative solutions, efficient problem-solving, and successful project outcomes.

In summary, managing conflicts involves recognizing issues early, understanding their root causes, addressing them constructively, using mediation techniques, implementing solutions, and learning from the experience. Driving collaboration involves building a collaborative culture, facilitating communication, defining roles, encouraging idea-sharing, leveraging strengths, building team cohesion, and managing remote collaboration. Both skills are essential for ensuring a productive and harmonious team environment, contributing to the success of pharmaceutical product management.

Effective Communication Strategies

Effective communication is a cornerstone of successful pharmaceutical product management. It ensures clarity, fosters collaboration, and aligns team efforts with strategic goals. Here is a detailed description of effective communication strategies for pharmaceutical product managers:

1. Clear and Concise Messaging

A. Articulate Objectives

- **Define Goals**: Clearly articulate the goals and objectives of the communication. Whether it is a project update, strategic briefing, or stakeholder engagement, ensure the purpose is evident.

- **Simplify Complex Information**: Break down complex information into understandable segments. Use clear language and avoid jargon wherever possible.

B. Structured Communication

- **Organize Information**: Structure your communication logically. Use headings, bullet points, and summaries to make key points stand out.

- **Be Direct**: Get to the point quickly. Avoid unnecessary details and focus on what is most relevant to the audience.

2. Tailoring Communication to the Audience

A. Understand Your Audience

- **Audience Analysis**: Know your audience's needs, preferences, and levels of understanding. Tailor your communication style and content accordingly.
- **Personalization**: Customize messages for stakeholders, such as internal team members, external partners, healthcare professionals, or regulatory bodies.

B. Adapt Communication Channels

- **Select Appropriate Channels**: Choose the most effective communication channels for your audience. This might include emails, meetings, presentations, reports, or digital platforms.
- **Channel Consistency**: Maintain consistency in the messages delivered across different channels to avoid confusion.

3. Active Listening and Feedback

A. Engage in Active Listening

- **Listen Attentively**: Pay close attention to what others are saying without interrupting. Show that you value their input through non-verbal cues and acknowledgments.
- **Clarify and Confirm**: Ask clarifying questions if needed and confirm your understanding of the message to ensure accuracy.

B. Solicit Feedback

- **Encourage Input**: Invite feedback from your audience to gauge your understanding and address any concerns.
- **Respond to Feedback**: Act on feedback to improve communication and address issues or misunderstandings.

4. Regular and Timely Updates

A. Maintain Transparency

- **Frequent Updates**: Regularly update project progress, changes, or developments. Keeping stakeholders informed helps build trust and manage expectations.

- **Timelines**: Ensure that updates are timely and relevant. Avoid delays in communication, especially when addressing critical issues.

B. Use Structured Formats

- **Status Reports**: Utilize structured formats for regular status reports, such as progress summaries, dashboards, or key performance indicators (KPIs).

- **Meeting Agendas**: Share agendas in advance for meetings and distribute minutes afterward to ensure clarity and accountability.

5. Effective Presentation Skills

A. Engage Your Audience

- **Visual Aids**: Use visual aids such as slides, charts, and graphs to enhance understanding and keep the audience engaged.

- **Storytelling**: Incorporate storytelling techniques to make the presentation more relatable and memorable.

B. Practice Delivery

- **Rehearse**: Practice your presentation to ensure smooth delivery and to build confidence.

- **Manage Time**: Be mindful of time constraints and ensure you convey all key points within the allowed time.

6. Conflict Resolution and Negotiation

A. Address Conflicts Promptly

- **Identify Issues**: Address conflicts or disagreements promptly to prevent escalation. Use effective communication to resolve issues amicably.

- **Seek Solutions**: Focus on finding solutions rather than placing blame. Aim for a resolution that satisfies all parties involved.

B. Negotiate Effectively

- **Prepare for Negotiations** by understanding the other party's interests and goals. Prepare your arguments and be ready to propose mutually beneficial solutions.
- **Communicate Clearly**: During negotiations, communicate your position clearly and listen to the other party's concerns.

7. Building Relationships and Trust

A. Foster Positive Relationships

- **Engage Personally**: Build relationships by engaging personally with team members and stakeholders. Show genuine interest in their perspectives and contributions.
- **Show Appreciation**: Recognize and appreciate the efforts and achievements of others. Positive reinforcement helps build trust and morale.

B. Maintain Integrity

- **Be Honest**: Communicate honestly and transparently. Avoid exaggerations or withholding important information.
- **Follow-Through**: Honor commitments and follow through on promises. Reliability is key to maintaining trust and credibility.

8. Utilizing Technology for Communication

A. Leverage Digital Tools

- **Collaboration Platforms**: Use digital collaboration platforms (e.g., Slack, Microsoft Teams) for real-time communication and collaboration.
- **Project Management Tools**: Implement project management tools (e.g., Asana, Trello) to track progress and facilitate communication.

B. Enhance Accessibility

- **Virtual Meetings**: Utilize virtual meeting tools (e.g., Zoom, WebEx) to facilitate remote communication and ensure accessibility for all team members.

- **Document Sharing**: Use cloud-based document-sharing platforms (e.g., Google Drive, SharePoint) to access important files and information easily.

Implications for Product Managers

- **Enhanced Clarity**: Clear and concise communication helps prevent misunderstandings and ensures everyone is on the same page.
- **Improved Collaboration**: Tailoring communication and actively listening fosters better collaboration and teamwork.
- **Efficient Problem-Solving**: Effective conflict resolution and negotiation skills help address issues promptly and keep projects on track.
- **Stronger Relationships**: Building positive relationships and trust enhances stakeholder engagement and support.

In summary, effective communication strategies involve clear and concise messaging, tailoring communication to the audience, active listening and feedback, regular and timely updates, effective presentation skills, conflict resolution and negotiation, building relationships and trust, and utilizing technology. These strategies are essential for ensuring successful collaboration, managing projects, and achieving business objectives in pharmaceutical product management.

24.1 Crafting Clear and Compelling Messages

Clear and compelling messages are fundamental for effective communication, particularly in pharmaceutical product management. It involves presenting information understandably and engagingly to the intended audience. Here is a detailed description of how to craft clear and compelling messages:

1. Define the Objective

A. Clarify the Purpose

- **Identify Goals**: Determine the primary goal of your message. Are you informing, persuading, or requesting action?
- **Set Objectives**: Define what you want to achieve with the message. This could be increasing awareness, gaining approval, or motivating action.

B. Understand the Audience

- **Audience Analysis**: Know your audience and their needs, interests, and level of understanding. Tailor your message to address their specific concerns and preferences.
- **Segment the Audience**: If necessary, segment your audience into groups with similar characteristics or needs to deliver more targeted messages.

2. Structure the Message

A. Start with a Strong Opening

- **Hook**: Begin with a compelling hook or statement that grabs attention. This could be a surprising fact, a provocative question, or a relevant quote.
- **Set the Stage**: Provide context to help the audience understand the message's relevance.

B. Present Key Points Clearly

- **Main Ideas**: Focus on the core message and key points you want to convey. Avoid overwhelming the audience with too much information.

- **Organize Logically**: Arrange your points in a logical sequence. Use headings, bullet points, or numbered lists to make the information easier to follow.

C. Provide Evidence and Support

- **Data and Facts**: Use relevant data, statistics, and evidence to support your key points. This adds credibility and helps persuade the audience.

- **Examples and Anecdotes**: Include real-world examples or anecdotes to illustrate your points and make the message more reliable.

3. Use Clear and Concise Language

A. Avoid Jargon

- **Simple Language**: Use straightforward language and avoid technical jargon unless necessary for the audience. If jargon is used, provide explanations.

- **Clarify**: Ensure that each sentence is clear and to the point. Avoid complex sentences that could confuse the audience.

B. Be Specific

- **Precise Information**: Provide specific details rather than vague statements. This helps in making the message more actionable and understandable.

- **Direct Language**: Use active voice and direct language to make your message more engaging and impactful.

4. Craft an Engaging Message

A. Appeal to Emotions

- **Emotional Connection**: Connect with the audience emotionally by addressing their needs, concerns, or aspirations.

- **Storytelling**: Incorporate storytelling techniques to make the message more compelling and memorable.

B. Use Visuals

- **Visual Aids**: Enhance your message with visuals such as charts, infographics, or images. Visuals can clarify complex information and keep the audience engaged.

- **Design**: Ensure that visuals are well-designed and relevant to the message. Avoid cluttered or distracting elements.

5. Provide a Clear Call to Action

A. Specify Next Steps

- **Actionable Instructions**: Clearly outline what you want the audience to do next. Provide specific instructions or steps for taking action.

- **Motivation**: Explain why taking action is important and what benefits or outcomes will result.

B. Make it Easy

- **Accessibility**: Ensure that the call to action is easy to follow. Provide necessary resources, links, or contact information to facilitate action.

- **Follow-Up**: If needed, offer follow-up options or support to assist the audience in completing the action.

6. Review and Revise

A. Proofread

- **Check for Errors**: Review the message for grammatical, spelling, or factual errors. Ensure accuracy and professionalism.

- **Consistency**: Check for consistency in tone, style, and terminology.

B. Seek Feedback

- **Peer Review**: Have colleagues or stakeholders review the message and provide feedback. This can help identify areas for improvement.

- **Refine**: Revise the message based on feedback to enhance clarity and effectiveness.

Implications for Product Managers

- **Effective Communication**: Crafting clear and compelling messages helps ensure that stakeholders understand and support your product initiatives.

- **Engagement**: Engaging messages can build interest and motivate action, which is crucial for successful product launches and ongoing marketing efforts.

- **Alignment**: Well-crafted messages align with business objectives and resonate with the target audience, driving strategic goals forward.

In summary, crafting clear and compelling messages involves defining objectives, understanding the audience, structuring the message, using clear and concise language, creating engaging content, providing a clear call to action, and reviewing and revising the message. These practices are essential for effective communication and success in pharmaceutical product management.

24.2 Communication Channels and Tools in Pharma

In pharmaceutical product management, communication channels and tools are crucial for effectively reaching and engaging with various stakeholders, including healthcare professionals (HCPs), patients, payers, and regulatory bodies. Here is an overview of key communication channels and tools used in the pharmaceutical industry:

1. Communication Channels

A. Traditional Channels

Sales Representatives

- **Role**: Sales reps engage directly with HCPs to provide information about products, answer questions, and gather feedback.
- **Approach**: Face-to-face meetings, lunch-and-learns, and samples.
- **Medical Conferences and Events**
- **Role**: Provide opportunities to showcase new products, share research findings, and network with industry professionals.
- **Approach**: Presentations, booths, sponsorships.
- **Print Media**:
- **Role**: Reach a broad audience with targeted messaging through industry journals, magazines, and brochures.

B. Digital Channels

Email Marketing

- **Role**: Communicate with HCPs, patients, and other stakeholders with targeted information and updates.
- **Approach**: Newsletters, promotional emails, educational content.

Social Media

- **Role**: Engage with a wide audience, share content, and interact with stakeholders in real time.

- **Approach**: Platforms like LinkedIn, Twitter, Facebook, and Instagram for brand promotion, updates, and community engagement.

Websites

- **Role**: Serve as a central hub for product information, company news, and resources.
- **Approach**: Corporate sites, product-specific landing pages, and patient portals.

Mobile Apps

- **Role**: Provide interactive tools and resources for patients and HCPs, such as medication reminders or disease management support.
- **Approach**: Apps for tracking health, accessing educational content, and managing prescriptions.

Webinars and Virtual Meetings:

- **Role**: Facilitate real-time communication and training with HCPs and other stakeholders.
- **Approach**: Online seminars, workshops, and video conferences.

Digital Advertising:

- **Role**: Target specific audiences with tailored messages through various online platforms.
- **Approach**: Display ads, search engine marketing (SEM), and programmatic advertising.

2. Communication Tools

A. CRM (Customer Relationship Management) Systems

- **Role**: Manage stakeholder interactions, track engagement, and optimize communication strategies.
- **Features**: Contact management, sales tracking, email automation, and analytics.

B. Marketing Automation Platforms

- **Role**: Automate and manage marketing campaigns across various channels.

- **Features**: Campaign management, lead generation, email marketing, analytics.

C. Data Analytics Tools

- **Role**: Analyze data from various sources to gain insights into stakeholder behavior and campaign effectiveness.
- **Features**: Dashboard creation, trend analysis, performance metrics.

D. Content Management Systems (CMS)

- **Role**: Create, manage, and publish content on websites and digital platforms.
- **Features**: Content creation, editing, publishing workflows, and user access control.

E. Social Media Management Tools

- **Role**: Manage and optimize social media presence and engagement.
- **Features**: Post scheduling, monitoring, analytics, and social listening.

F. EHR (Electronic Health Records) Integration

- **Role**: Integrate with healthcare systems to streamline communication and data sharing.
- **Features**: Data synchronization, patient records access and compliance tracking.

G. Virtual Collaboration Tools

- **Role**: Facilitate teamwork and communication within and across teams.
- **Features**: Video conferencing, file sharing, project management, and real-time collaboration.

3. Best Practices for Effective Communication

A. Targeted Messaging

- **Personalization**: Customize messages based on the audience segment to ensure relevance and effectiveness.

- **Clarity**: Ensure that messages are concise and aligned with the communication objectives.

B. Multi-Channel Strategy

- **Integration**: Use a combination of channels to reach stakeholders through their preferred methods.
- **Consistency**: Maintain consistent messaging across all channels to reinforce key points.

C. Feedback and Adaptation

- **Monitoring**: Track the effectiveness of communication strategies and gather feedback from stakeholders.
- **Adaptation**: Adjust strategies and tactics based on feedback and performance data.

D. Compliance and Ethics

- **Regulations**: Adhere to industry regulations and ethical standards in all communications.
- **Transparency**: Ensure transparency in messaging and avoid misleading or deceptive practices.

In summary, communication channels and tools in pharma are diverse and play a crucial role in engaging with stakeholders. Leveraging a mix of traditional and digital channels and utilizing various CRM, marketing automation, and data analytics tools can help pharma product managers effectively communicate and achieve their objectives.

24.3 Effective Presentation Skills for Pharma Product Managers

Presentation skills are essential for pharma product managers (PMs) as they often need to communicate complex information, such as product data, marketing strategies, and performance results, to various stakeholders. Mastering these skills allows PMs to convey their message clearly and persuasively, build trust, and drive decisions that align with business objectives.

Here's a breakdown of key aspects of presentation skills for pharma product managers:

1. Understanding the Audience

1.1 Tailoring Content:

- **Target Audience Analysis:** Whether presenting to healthcare professionals (HCPs), internal teams, regulatory authorities, or marketing colleagues, understanding the audience's knowledge level, interests, and expectations is crucial.

- **Customizing Content:** Adjust the level of detail and focus based on the audience. For example, scientific data may need to be simplified for marketing teams but presented in depth for R&D professionals.

1.2 Addressing Key Concerns:

- **Focus on Relevance:** Highlight the aspects of the product or strategy that are most relevant to the audience, such as patient outcomes for healthcare providers or ROI for executives.

- **Anticipating Questions:** Be prepared to answer potential concerns or questions from stakeholders by addressing them proactively within the presentation.

2. Structuring the Presentation

2.1 Clear and Logical Flow:

- **Introduction, Body, Conclusion:** Start with a clear introduction, outlining the presentation's purpose, followed by logically arranged sections, and end with a strong conclusion summarizing key points.

- **Use of Data and Evidence:** Data plays a significant role in pharmaceutical marketing. Present clinical trial results, market research data, and sales figures effectively to support your claims.

2.2 Highlighting Key Takeaways:

- **Action-Oriented Approach:** Ensure the presentation has clear takeaways or action steps, guiding stakeholders on what to focus on and how to proceed.

3. Visual Communication

3.1 Designing Effective Slides:

- **Simplified Visuals:** Use graphs, charts, and visuals to simplify complex data. Avoid overloading slides with text— each slide should highlight key points.

- **Consistent Branding:** Ensure the presentation aligns with the brand's visual identity, reinforcing professionalism and consistency.

3.2 Using Infographics:

- **Translating Data into Graphics:** Convert clinical outcomes, sales trends, and market data into easy-to-understand infographics, making the information digestible and engaging.

4. Delivering with Confidence

4.1 Engaging the Audience:

- **Clear, Confident Speech:** Speak clearly and at a steady pace. Avoid jargon unless the audience is familiar with it.

- **Body Language and Eye Contact:** Use positive body language, maintain eye contact, and engage the audience by encouraging questions and interaction.

4.2 Handling Q&A Sessions:

- **Preparation for Queries:** Prepare thoroughly to answer questions, especially about sensitive topics like drug safety, efficacy, and regulatory compliance.

- **Clarifying Uncertainty:** If unsure of an answer, acknowledge it transparently and commit to following up with accurate information.

5. Storytelling Techniques

5.1 Narrative Approach:

- **Humanizing the Data:** Instead of just presenting data points, weave a story around the product's development, the patients it serves, or the impact it will have on healthcare outcomes.

- **Real-World Examples:** Include case studies, patient testimonials, or success stories to make the presentation more relatable and emotionally engaging.

5.2 Aligning with Brand Values:

- **Reinforcing Brand Messaging:** Ensure the storytelling aligns with the brand's core values and long-term vision, building a cohesive narrative around the product's benefits.

6. Technical Competence

6.1 Mastering the Tools:

- **Presentation Software Expertise:** Effectively use tools like PowerPoint, Keynote, or Google Slides. Be familiar with advanced features like animations, embedding videos, or live data links.

- **Use of Virtual Platforms:** In today's digital-first environment, pharma product managers must also be proficient in presenting via virtual platforms (Zoom, Microsoft Teams) and ensuring seamless engagement.

6.2 Addressing Technical Issues:

- **Backup Plans:** Always have a backup plan in case of technical difficulties (e.g., printed slides, alternative devices) to ensure the presentation runs smoothly.

7. Scientific Accuracy and Regulatory Considerations

7.1 Adherence to Compliance:

- **Regulatory Guidelines:** Pharma presentations often communicate clinical data and promotional information. The content must comply with regulatory standards like FDA or EMA guidelines to avoid legal issues.

- **Balanced Messaging:** Ensure the presentation does not exaggerate product benefits and provides a fair balance by disclosing potential risks and adverse effects, as required in the pharma industry.

7.2 Clinical and Scientific Rigor:

- **Data Integrity:** Always present accurate and up-to-date clinical data. Be transparent about the limitations of studies or data sources to build trust with the audience.

8. Practice and Feedback

8.1 Rehearsing the Presentation:

- **Continuous Improvement:** Practice the presentation multiple times to refine delivery, timing, and clarity. Simulate real-world conditions by rehearsing in front of colleagues or mentors.

- **Time Management:** Ensure the presentation fits within the allotted time without rushing or omitting important points.

8.2 Gathering Feedback:

- **Post-Presentation Review:** Seek feedback from peers or mentors to understand what worked well and areas for improvement, particularly regarding clarity, engagement, and the handling of technical information.

Conclusion

Pharma product managers must possess robust presentation skills to communicate complex product information, engage stakeholders, and drive successful marketing strategies. By combining scientific accuracy, storytelling, visual communication, and audience-focused delivery, product managers can ensure their presentations are impactful and effective in achieving business goals.

24.4 Effective Negotiation Skills for Pharma Product Managers

Effective negotiation is critical for product managers, who frequently interact with stakeholders, including internal teams, external partners, suppliers, and regulatory authorities. Successful negotiation allows product managers to achieve favorable outcomes while maintaining strong relationships with all involved parties. Here's an in-depth look at the essential negotiation skills for product managers:

1. Understanding Interests and Objectives

1.1 Clarifying Your Own Goals:

- **Define Objectives:** Before entering a negotiation, product managers must clearly understand their objectives, whether securing resources, defining project timelines, or resolving conflicts.

- **Prioritize Needs vs. Wants:** Identify the non-negotiable aspects and areas where you can be flexible.

1.2 Identifying Stakeholders' Interests:

- **Assess the Counterpart's Goals:** Understand what the other party aims to achieve in the negotiation. This helps identify common ground and potential trade-offs.

- **Build Empathy:** By considering the needs and challenges of others, you can foster cooperation and find mutually beneficial solutions.

2. Active Listening and Communication

2.1 Listening to Understand:

- **Gather Information:** Actively listen to understand the perspectives and concerns of other parties. This not only helps you gather valuable insights but also builds trust.

- **Ask Clarifying Questions:** If something is unclear, ask open-ended questions to gather more information or clarify the other party's position.

2.2 Clear and Concise Communication:

- **Articulate Proposals:** Present your ideas and proposals in a clear, structured, and concise manner. This ensures your objectives are well understood by all parties.

- **Maintain Open Dialogue:** Encourage a two-way conversation and ensure all voices are heard in the negotiation process.

3. Preparation and Research

3.1 Gathering Data:

- **Research Market and Competitor Insights:** Data-driven insights can strengthen your position in pricing, partnerships, or product development negotiations.

- **Know Your Numbers:** When negotiating with suppliers or vendors, consider cost breakdowns, timelines, and expected outcomes.

3.2 Anticipating Challenges:

- **Prepare for Objections:** Consider possible challenges or objections the other party might raise and prepare responses to address them.

- **Plan Alternatives:** If your initial offer is declined, preparing backup options and compromises allows for smoother negotiation.

4. Collaborative Problem-Solving

4.1 Seeking Win-Win Solutions:

- **Explore Options Together:** Rather than focusing on winning at the other party's expense, look for ways both sides can benefit from the agreement. This fosters long-term collaboration and partnership.

- **Trade-offs and Concessions:** Be willing to make concessions in less critical areas if doing so helps you achieve more important outcomes in the negotiation.

4.2   Conflict Resolution:

- **Addressing Disagreements:** If disagreements arise, focus on addressing the issue, not the person. Keep the conversation objective and solution-focused.

- **Reframe Challenges:** Use challenges as opportunities to brainstorm new solutions that could benefit both parties.

5. Emotional Intelligence and Managing Tension

5.1   Staying Calm Under Pressure:

- **Maintain Composure:** Product managers often face high-pressure negotiations. Managing your emotions and staying calm allows you to think more clearly and make rational decisions.

- **Defuse Tension:** If the negotiation becomes heated, focus on de-escalating the situation by addressing concerns calmly and steering the conversation back to shared objectives.

5.2   Reading Non-Verbal Cues:

- **Body Language Awareness:** Understanding non-verbal signals such as facial expressions, posture, and tone of voice can provide insights into the other party's feelings or concerns.

- **Responding Appropriately:** Based on these cues, adapt your negotiation strategy, adjusting your approach if you sense hesitation or resistance.

6. Flexibility and Adaptability

6.1   Adjusting Strategies:

- **Be Open to Change:** Product managers must be flexible during negotiations. If new information emerges or the other party makes a compelling point, be willing to adapt your approach.

- **Continuous Feedback:** As the negotiation progresses, adjust based on real-time feedback and the evolving dynamics of the conversation.

6.2   Finding Creative Solutions:

- **Think Outside the Box:** When faced with a stalemate, think creatively about alternative solutions that could satisfy both parties, whether through different payment structures, timelines, or resources.

- **Incorporate Compromise:** A creative compromise can sometimes lead to a better long-term outcome than the original goal.

7. Closing and Follow-Up

7.1   Securing Agreement:

- **Summarize Key Points:** Before concluding the negotiation, summarize the agreed-upon points to ensure mutual understanding and clarity.

- **Formalize the Agreement:** Document the terms, whether through an official contract, a memorandum of understanding, or a verbal agreement.

7.2   Building Long-Term Relationships:

- **Follow Up:** Follow up after the negotiation to ensure commitments are met. This helps maintain trust and keeps the relationship strong.

- **Feedback Loop:** Seek feedback from the other party on how the negotiation went. This helps refine your negotiation skills for future engagements.

Conclusion

For product managers, negotiation is a vital skill that affects not only immediate project outcomes but also long-term relationships and success in product management. Effective negotiation requires a balance of preparation, communication, empathy, and flexibility to achieve desired outcomes while fostering collaboration.

Building Stakeholder Relationships

Building stakeholder relationships is a critical aspect of successful pharmaceutical product management. Strong stakeholder relationships facilitate collaboration, enhance trust, and contribute to the overall success of a product. Here are the key steps of how to build effective stakeholder relationships:

1. Identifying and Mapping Stakeholders

A. Identification

- **Primary Stakeholders** include healthcare professionals (HCPs), patients, payers, and regulatory bodies.
- **Secondary Stakeholders** Include internal teams, investors, advocacy groups, and the general public.

B. Mapping and Analysis

- **Stakeholder Map**: Create a visual map identifying each stakeholder's influence, interests, and impact on the product.
- **Prioritization**: Assess stakeholders based on their level of influence and importance to prioritize engagement efforts.

2. Understanding Stakeholder Needs and Expectations

A. Needs Assessment

- **Interviews and Surveys**: Conduct interviews and surveys to gather insights into stakeholders' needs and expectations.

- **Focus Groups**: Organize focus groups to explore deeper issues and concerns related to the product or market.

B. Expectations Management

- **Clear Communication**: Ensure stakeholders have clear expectations about the product, its benefits, and its development.
- **Realistic Promises**: Make realistic promises and commitments based on what can be delivered.

3. Establishing Trust and Credibility

A. Transparent Communication

- **Honesty**: Provide honest and accurate information about the product, including its benefits and limitations.
- **Regular Updates**: Keep stakeholders informed about product developments, clinical trials, and market changes.

B. Reliability

- **Consistent Interaction**: Maintain consistent and reliable communication to build trust over time.
- **Follow-Through**: Follow through on promises and commitments to reinforce credibility.

4. Engaging Stakeholders Effectively

A. Tailored Communication

- **Customized Messaging**: Adapt messages to address different stakeholder groups' interests and needs.
- **Personal Touch**: Use personalized communication methods like direct emails or one-on-one meetings.

B. Active Listening

- **Feedback Mechanisms**: Implement feedback mechanisms such as surveys or feedback forms to understand stakeholder perspectives.
- **Responsive Actions**: Act on feedback and concerns to demonstrate that stakeholder input is valued and considered.

5. Building Long-Term Relationships

A. Ongoing Engagement

- **Regular Interaction**: Maintain regular contact with stakeholders through newsletters, updates, and events.
- **Value Addition**: Provide ongoing value through educational resources, updates, and support services.

B. Relationship Management

- **CRM Tools**: Utilize Customer Relationship Management (CRM) systems to track interactions and manage relationships.
- **Relationship Metrics**: Monitor relationship metrics to assess engagement levels and satisfaction.

6. Collaborating with Healthcare Professionals (HCPs)

A. Educational Initiatives

- **Scientific Education**: Offer educational programs and materials that inform HCPs about new research, clinical data, and treatment guidelines.
- **Continuing Medical Education (CME)**: Provide opportunities for HCPs to earn CME credits through your educational initiatives.

B. Advisory Boards

- **Expert Panels**: Form advisory boards with key opinion leaders (KOLs) to gain insights and feedback on product development and marketing strategies.
- **Regular Meetings**: Schedule regular meetings to discuss product performance and market trends.

7. Engaging with Patients

A. Patient Education

- **Informative Content**: Develop and distribute educational materials that help patients understand their condition and treatment options.
- **Support Services**: Offer services such as patient helplines and digital tools for managing health.

B. Patient Advocacy

- **Partnerships**: Partner with patient advocacy groups to understand patient needs and improve product offerings.
- **Patient Programs**: Support programs that enhance patient care and access to treatment.

8. Engaging with Payers

A. Value Demonstration

- **Economic Data**: Provide data on the product's economic value, cost-effectiveness, and clinical outcomes.
- **Value Propositions**: Develop compelling value propositions that align with payer priorities and concerns.

B. Negotiations

- **Pricing Strategies**: Work with payers to negotiate pricing and reimbursement terms that reflect the product's value.
- **Formulary Placement**: Ensure the product is included in the formulary and preferred drug lists.

9. Engaging with Regulatory Bodies

A. Compliance and Communication

- **Regulatory Submissions**: Prepare and submit the required documentation for product approval and regulatory compliance.
- **Ongoing Updates**: Maintain open communication with regulatory agencies to address issues and provide product development updates.

B. Policy Influence

- **Industry Engagement**: Participate in industry forums and discussions to influence regulatory policies and practices.
- **Advocacy**: Advocate for policies that support the product's market access and development.

10. Engaging with Internal Teams

A. Cross-Functional Collaboration

- **Team Alignment**: Ensure alignment between marketing, sales, R&D, finance, and other relevant departments.

- **Information Sharing**: Facilitate sharing information and insights across teams to support coordinated efforts.

B. Training and Development

- **Internal Training**: Train internal teams on product knowledge, market dynamics, and stakeholder engagement strategies.

- **Feedback Integration**: Incorporate feedback from internal teams into product strategies and decision-making processes.

11. Measuring Relationship Success

A. Key Performance Indicators

- **Engagement Metrics**: Track metrics such as stakeholder engagement levels, feedback quality, and relationship strength.

- **Outcome Measures**: Evaluate the impact of stakeholder relationships on product success and market performance.

B. Continuous Improvement

- **Feedback Analysis**: Analyze stakeholder feedback and relationship data to identify strengths and areas for improvement.

- **Strategy Adjustment**: Adjust engagement strategies based on performance data and stakeholder needs.

Building stakeholder relationships involves understanding their needs, establishing trust, and maintaining ongoing engagement through tailored communication and active listening. By leveraging various strategies and tools, pharmaceutical product managers can effectively build and sustain strong relationships with key stakeholders, which is crucial for product success and market impact.

25.1 Identifying Key Stakeholders in Pharma

Identifying key stakeholders in the pharmaceutical industry is essential for effective product management, marketing, and overall success. Stakeholders are individuals or groups interested in or are affected by the pharmaceutical product. Here is a detailed approach to identifying key stakeholders in pharma:

1. Understanding Stakeholder Categories

A. Primary Stakeholders

Healthcare Professionals (HCPs):

- **Doctors and Specialists**: Physicians who prescribe medications and influence treatment decisions.
- **Nurses and Pharmacists**: Involved in administering and dispensing drugs and providing patient care.
- **Healthcare Institutions**: Hospitals, clinics, and medical centers where treatments are provided.

Patients:

- **Current Patients**: Individuals receiving treatment or those benefiting from the new product.
- **Patient Advocacy Groups**: Organizations representing patient interests and advocating for better treatments and healthcare policies.

Payers:

- **Insurance Companies**: Entities that reimburse for treatments and negotiate drug prices.
- **Government Health Agencies**: Agencies that manage public health insurance and health policy in the United States, such as Medicare or Medicaid.

Regulatory Bodies:

- **FDA (Food and Drug Administration)**: US agency responsible for approving drugs and ensuring their safety and efficacy.
- **EMA (European Medicines Agency)**: European agency that regulates pharmaceuticals in the European Union.

B. Secondary Stakeholders

Internal Teams:

- **Marketing and Sales Teams**: Develop and execute marketing strategies and sales plans.
- **Research and Development (R&D)**: Responsible for drug development and clinical trials.
- **Finance and Operations**: Manage budgets, financial planning, and logistical operations.

Investors and Shareholders:

- **Financial Analysts**: Assess the company's performance and make investment recommendations.
- **Institutional Investors**: Entities that hold significant stakes in the company.

Media and Public Relations:

- **Journalists and Media Outlets**: Report on pharmaceutical products, industry news, and clinical developments.
- **Public Relations Agencies**: Manage the company's image and handle communications with the public.

Regulatory Consultants:

- **Compliance Experts**: Guide regulatory requirements and help navigate regulatory approval processes.

2. Stakeholder Identification Process

A. Mapping and Prioritization

- **Stakeholder Map**: Create a visual representation of identified stakeholders, highlighting their relationships to the product and each other.
- **Prioritization**: Assess stakeholders based on their influence, interest, and impact on the product. Prioritize those who have the most significant effect on the product's success.

B. Research and Data Collection

- **Market Research**: Use market research reports, surveys, and industry data to identify key players and their roles.

- **Industry Contacts**: Leverage existing industry contacts and networks to identify influential stakeholders.
- **Public Databases**: Consult public databases and registries for information on regulatory bodies and healthcare institutions.

3. Analyzing Stakeholder Impact

A. Influence and Interest

- **Influence**: Determine the level of influence each stakeholder has over the product's development, approval, and market success.
- **Interest**: Assess the level of interest each stakeholder has in the product and their potential motivations.

B. Relationship Dynamics

- **Relationships**: Identify how stakeholders interact with each other and the product. Understand the dynamics that could affect the product's market positioning and acceptance.
- **Potential Conflicts**: Recognize any potential conflicts of interest or disagreements between stakeholders.

4. Engaging Stakeholders

A. Tailored Communication

- **Customized Messaging**: Develop communication strategies tailored to the needs and interests of each stakeholder group.
- **Engagement Plans**: Create detailed engagement plans outlining how to interact with each stakeholder, including meetings, updates, and consultations.

B. Continuous Monitoring

- **Feedback Mechanisms**: Implement mechanisms to gather ongoing feedback from stakeholders and adjust engagement strategies as needed.
- **Relationship Management**: Regularly review and manage stakeholder relationships to ensure continued alignment and address emerging issues.

Examples of Key Stakeholders

- **New Drug Launch**: During the launch of a new drug, key stakeholders might include HCPs who will prescribe the medication, patients who will use it, payers who will reimburse for it, and regulatory bodies that need to approve it. Each group requires a tailored approach to engagement and communication.

- **Regulatory Approval**: Primary stakeholders include regulatory agencies like the FDA or EMA for regulatory approval. Secondary stakeholders include internal regulatory affairs teams and compliance consultants.

- **Market Access**: When gaining market access, key stakeholders include payers and insurance companies, who will decide on reimbursement levels and formulary placement.

Conclusion

Identifying key stakeholders involves understanding their categories, conducting thorough research, analyzing their influence and interests, and developing targeted engagement strategies. Effective stakeholder management is crucial for navigating the complexities of the pharmaceutical industry and ensuring successful product outcomes.

25.2 Strategies for Building Trust and Collaboration

Building trust and collaboration is vital in the pharmaceutical industry, where successful product management, marketing, and development rely heavily on strong relationships with various stakeholders. Here is a detailed approach to strategies for building trust and fostering collaboration:

1. Transparent Communication

A. Open Dialogue

- **Regular Updates**: Provide consistent and clear updates about product development, progress, and changes. This helps manage expectations and builds credibility.

- **Honest Feedback**: Share both successes and challenges openly. Address potential issues proactively to avoid surprises and demonstrate integrity.

B. Clarity and Consistency

- **Clear Messaging**: Ensure that all communications are unambiguous. Avoid jargon or overly technical language that could confuse stakeholders.

- **Consistency**: Maintain consistency in messaging and information across different channels and stakeholders to avoid mixed signals.

2. Building Credibility

A. Demonstrate Expertise

- **Showcase Knowledge**: Leverage your expertise and experience to provide valuable insights and solutions. Share relevant data, research findings, and industry knowledge to build authority.

- **Deliver on Promises**: Meet deadlines and commitments consistently. Reliability enhances your reputation and builds trust with stakeholders.

B. High-Quality Work

- **Exceed Expectations**: Strive to exceed expectations in deliverables, whether in product development, marketing materials, or stakeholder engagement.

- **Continuous Improvement**: Regularly assess and improve processes and outcomes to maintain high standards and adapt to evolving needs.

3. Building Relationships

A. Personal Connections

- **Networking**: Engage with stakeholders through industry events, conferences, and informal networking opportunities. Building personal relationships can strengthen professional bonds.

- **Understand Needs**: Understand stakeholders' goals, challenges, and preferences. Tailor your approach to align with their needs and priorities.

B. Mutual Respect

- **Value Contributions**: Acknowledge and respect the contributions and perspectives of all stakeholders. Recognize their expertise and contributions to collaborative efforts.

- **Empathy**: Show empathy towards stakeholders' concerns and challenges. Demonstrating understanding and compassion helps build stronger relationships.

4. Collaborative Approach

A. Joint Planning

- **Involve Stakeholders**: Include key stakeholders in planning and decision-making processes. Collaborative planning helps align goals and expectations.

- **Co-Creation**: Engage in co-creation activities where stakeholders contribute ideas and feedback, fostering a sense of ownership and commitment.

B. Teamwork and Coordination

- **Cross-Functional Teams**: Build and lead cross-functional teams with diverse expertise and perspectives. Facilitate collaboration among team members from different departments and specialties.

- **Shared Goals**: Align team objectives with the broader goals of the organization. Ensure that everyone understands and works towards common goals.

5. Effective Conflict Resolution

A. Address Issues Early

- **Timely Intervention**: Address conflicts and issues as soon as they arise. Early resolution prevents escalation and maintains a positive working environment.

- **Open Discussions**: Facilitate open discussions to understand different viewpoints and find mutually acceptable solutions.

B. Fair and Objective

- **Impartial Mediation**: Approach conflicts with neutrality and fairness. Focus on resolving the issue rather than personal disagreements.

- **Constructive Feedback**: Provide constructive feedback and work collaboratively to resolve differences. Aim for solutions that benefit all parties involved.

6. Enhancing Collaboration Through Technology

A. Collaboration Tools

- **Digital Platforms**: Use collaboration tools and platforms (e.g., Microsoft Teams, Slack, Asana) to facilitate communication, project management, and information sharing.

- **Data Sharing**: Implement secure data-sharing solutions for seamless access to important information and insights.

B. Virtual Meetings

- **Regular Check-Ins**: Schedule virtual meetings to discuss progress, address concerns, and keep everyone informed.

- **Interactive Sessions**: Use interactive formats and tools during virtual meetings to engage stakeholders and encourage active participation.

7. **Measuring and Improving Trust and Improving Collaboration**

A. Feedback Mechanisms

- **Surveys and Assessments**: Use surveys and assessments to gather feedback from stakeholders on their experiences and perceptions of collaboration.

- **Regular Reviews**: Conduct regular reviews of collaboration practices and relationships to identify areas for improvement.

B. Adjust and Evolve

- **Continuous Improvement**: Continuously refine and adapt strategies based on feedback and changing dynamics. Stay flexible and responsive to stakeholder needs and expectations.

- **Learning and Development**: Invest in training and development to enhance communication, collaboration, and conflict-resolution skills.

Conclusion

Building trust and collaboration in the pharmaceutical industry requires transparent communication, credibility, relationship building, collaborative approaches, effective conflict resolution, and the use of technology. By implementing these strategies, pharmaceutical product managers can foster stronger stakeholder relationships, leading to successful product outcomes and a positive working environment.

Part VII. Thriving in the Digital Age

The digital age has transformed every aspect of business, including pharmaceutical product management. For pharmaceutical product managers, embracing digital advancements is essential for staying competitive, improving efficiency, and enhancing stakeholder engagement. This section will explore how pharmaceutical product managers can thrive in the digital era by leveraging digital tools and strategies to drive success.

1. Embracing Digital Transformation

A. Understanding Digital Transformation

- **Definition**: Digital transformation involves integrating digital technology into all business areas, fundamentally changing how the company operates, and delivering value to customers.

- **Impact on Pharma**: In the pharmaceutical industry, digital transformation includes innovations in drug discovery, development, marketing, and patient engagement.

B. Key Areas of Transformation

- **Data Analytics**: Utilizing big data and analytics to gain insights into market needs, patient needs, and product performance.

- **Digital Marketing**: Employing digital channels and tools for targeted marketing, including social media, email campaigns, and online advertising.

- **Automation**: Implementing automated systems for data collection, report generation, and patient management tasks.

2. Leveraging Digital Tools and Technologies

A. Digital Marketing Tools

- **Social Media Platforms**: Using platforms like Twitter, LinkedIn, and Facebook to engage with healthcare professionals (HCPs) and patients, sharing information, and building brand awareness.

- **Email Marketing**: Creating personalized email campaigns to communicate with HCPs and patients, provide updates, and promote products.

B. Data Analytics and AI

- **Predictive Analytics**: Utilizing predictive models to forecast market trends, patient behaviors, and drug efficacy.

- **Artificial Intelligence (AI)**: Applying AI for drug discovery, clinical trial optimization, and personalized patient experiences.

C. Customer Relationship Management (CRM)

- **CRM Systems**: Implementing CRM systems to manage interactions with HCPs and patients, track engagement, and personalize communication.

- **AI-Powered CRM**: Using AI to enhance CRM capabilities, such as predictive lead scoring and automated follow-ups.

D. Digital Therapeutics and Remote Monitoring

- **Digital Therapeutics**: Developing and integrating digital solutions that complement traditional treatments and improve patient outcomes.

- **Remote Patient Monitoring**: Employing technology to monitor patients remotely, collect real-time data, and provide timely interventions.

3. Adapting to the Digital Era

A. Digital Skill Development

- **Training and Education**: Ensuring product managers and teams are trained in digital tools, data analytics, and digital marketing strategies.

- **Continuous Learning**: Staying updated with the latest digital trends and technologies through courses, webinars, and industry conferences.

B. Digital Strategy Integration

- **Strategic Planning**: Incorporating digital strategies into the overall product management plan, aligning digital initiatives with business objectives.

- **Execution and Monitoring**: Implementing digital strategies effectively and monitoring their performance to ensure they meet desired goals.

4. Overcoming Challenges

A. Data Privacy and Security

- **Compliance**: Adhering to data protection regulations such as GDPR and HIPAA to ensure the privacy and security of patient information.

- **Best Practices**: Implementing robust security measures and regularly auditing data protection practices.

B. Digital Literacy and Adoption

- **Resistance to Change**: Addressing resistance to digital tools and technologies through effective change management strategies and clear communication.

- **Training Programs**: Providing comprehensive training to ensure the successful adoption and utilization of digital tools.

C. Integration with Traditional Methods

- **Balancing Act**: Finding the right balance between digital and traditional methods to achieve optimal results.

- **Integration Strategies**: Develop strategies to integrate digital tools with existing processes and workflows seamlessly.

Conclusion

Thriving in the digital age requires pharmaceutical product managers to embrace digital transformation, leverage advanced tools and technologies, and continuously adapt to evolving digital trends. By integrating digital strategies into their product management processes, product managers can enhance efficiency, improve stakeholder engagement, and drive business success. As the digital landscape continues to evolve, staying informed and agile will be key to navigating the future of pharmaceutical product management.

Digital Marketing in Pharma

Digital marketing has become a pivotal strategy in the pharmaceutical industry, allowing companies to reach and engage with healthcare professionals (HCPs), patients, and other stakeholders more effectively than traditional methods. This section will delve into the various aspects of digital marketing in pharma, employing key strategies, tools, and best practices:

1. Overview of Digital Marketing in Pharma

A. Definition and Scope

- **Definition**: Digital marketing involves using digital channels and tools to promote products, services, and brands. In pharma, it encompasses strategies to enhance brand visibility, educate stakeholders, and drive product adoption.

- **Scope**: Digital marketing in pharma includes content marketing, social media engagement, email campaigns, search engine optimization (SEO), and online advertising.

B. Importance

- **Reach and Accessibility**: Digital marketing enables pharma companies to reach a global audience, including HCPs, patients, and caregivers.

- **Cost-Effectiveness**: Digital channels often offer more cost-effective solutions than traditional marketing methods.

- **Data-Driven Insights**: Digital marketing provides valuable data and analytics, helping companies understand audience behavior and optimize their strategies.

2. Key Digital Marketing Strategies

A. Content Marketing

- **Purpose**: To educate and engage stakeholders through valuable and relevant content.
- **Types of Content**: Includes blog posts, white papers, infographics, case studies, and videos.
- **Strategy**: Develop content that addresses the specific needs and interests of HCPs and patients, ensuring compliance with regulatory requirements.

B. Social Media Marketing

- **Platforms**: Key platforms include LinkedIn, Twitter (X now), Facebook, and Instagram.
- **Purpose**: To build brand awareness, engage with stakeholders, and share educational content.
- **Strategy**: Create platform-specific content, engage with followers through comments and messages, and use targeted ads to reach specific audiences.

C. Email Marketing

- **Purpose**: To communicate directly with HCPs, patients, and other stakeholders.
- **Types**: Includes newsletters, promotional emails, and educational content.
- **Strategy**: Personalize emails based on recipient preferences, segment email lists for targeted messaging, and use analytics to measure engagement and effectiveness.

D. Search Engine Optimization (SEO)

- **Purpose**: To improve the visibility of digital content in search engine results.
- **Techniques**: Includes keyword research, on-page optimization, and link building.
- **Strategy**: Optimize website and content for relevant keywords, ensure high quality and valuable content, and improve website usability and performance.

E. Online Advertising

- **Types**: Includes pay-per-click (PPC) ads, display ads, and sponsored content.
- **Purpose**: To drive traffic to websites and promote specific products or services.
- **Strategy**: Use targeted advertising to reach specific demographics, track ad performance, and optimize campaigns based on results.

3. Compliance and Regulation

A. Regulatory Considerations

- **Compliance**: Adhere to regulations like the FDA's digital advertising and promotion guidelines and other relevant local regulations.
- **Approval Process**: Ensure regulatory and legal teams review and approve all digital marketing materials before publication.

B. Data Privacy

- **Protection**: Comply with data protection laws such as GDPR or HIPAA to safeguard personal and medical information.

4. Measuring and Optimizing Performance

A. Key Metrics

- **Engagement Metrics**: Includes likes, shares, comments, and click-through rates (CTR).
- **Conversion Metrics**: Includes lead generation, product trials, and sales.
- **Performance Analytics**: Use tools like Google Analytics and social media analytics to track and analyze campaign performance.

B. Optimization Strategies

- **A/B Testing**: Test different versions of content and ads to determine which performs better.

- **Data Analysis**: Analyze performance data to identify trends and insights and adjust strategies accordingly.
- **Continuous Improvement**: Review and refine digital marketing strategies based on performance data and feedback.

5. Future Trends and Digital Marketing

A. Emerging Technologies

- **Artificial Intelligence (AI)**: AI-powered tools for personalization, chatbots, and predictive analytics.
- **Virtual Reality (VR) and Augmented Reality (AR)**: Interactive experiences for training and patient education.

B. Evolving Regulations

- **New Guidelines**: Stay updated with evolving regulations on digital marketing and data privacy.

C. Personalization and Engagement

- **Advanced Personalization**: Use AI and data analytics to create highly personalized content and experiences.

26.1 Overview of Digital Marketing Channels

Digital marketing channels are essential for pharmaceutical companies to connect with healthcare professionals (HCPs), patients, and other stakeholders. Each channel has unique attributes and advantages; understanding them helps optimize marketing strategies and achieve business objectives. Here's an overview of key digital marketing channels used in the pharmaceutical industry:

1. Website Marketing

Overview:

Pharmaceutical companies use their websites as a central hub for product information, research, and company news. Websites often include sections dedicated to HCPs, patients, and investors.

Benefits:

- Provides comprehensive product information, including efficacy, safety, and patient resources.
- Acts as a platform for educational content and patient support tools.
- Facilitates direct communication with stakeholders through contact forms and live chat options.

Examples:

- Pfizer's website offers detailed product information, clinical trial data, and resources for HCPs and patients.
- Novartis's site features interactive content about its drug pipeline and research initiatives.

2. Email Marketing

Overview:

Email marketing involves sending targeted messages to HCPs, patients, and other stakeholders. It can be used for product announcements, educational content, and updates on clinical trials.

Benefits:

- Personalized communication that can be segmented based on recipient interests and behaviors.

- Provides a direct and measurable way to engage with HCPs and patients.
- Enables automation for routine updates and reminders.

Examples:

- Merck uses email campaigns to update HCPs on new research and treatments relevant to their specialty.
- Eli Lilly leverages automated emails to inform patients about medication adherence and support programs.

3. Social Media Marketing

Overview:

Social media platforms like LinkedIn, Twitter (X now), Facebook, and Instagram are used to engage with HCPs and patients. These platforms are useful for building brand awareness, sharing content, and facilitating discussions.

Benefits:

- Allows for real-time interaction and engagement with a broad audience.
- Facilitates community building and patient advocacy.
- Provides insights into public perception and feedback through social listening tools.

Examples:

- Johnson & Johnson uses social media to share updates about their products and corporate social responsibility initiatives.
- GSK runs campaigns to raise awareness about health conditions and promote patient education through social media.

4. Content Marketing

Overview:

Content marketing involves creating and distributing valuable content to attract and engage target audiences. This includes blog posts, white papers, webinars, and videos.

Benefits:

- Establishes thought leadership and educates stakeholders about relevant topics.

- Enhances brand credibility and authority.
- Supports SEO efforts by generating relevant content that drives organic traffic.

Examples:

- Roche's blog features articles and case studies about advances in cancer research and patient care.
- Sanofi produces white papers and webinars to educate HCPs about new treatments and clinical developments.

5. Search Engine Optimization (SEO)

Overview:

SEO is optimizing website content and structure to improve visibility in search engine results. It helps attract organic traffic to a company's website.

Benefits:

- Increases online visibility and attracts qualified leads.
- Enhances user experience by providing relevant and accessible information.
- Supports long-term traffic growth through content optimization and keyword strategies.

Examples:

- GSK employs SEO strategies to enhance the visibility of their respiratory treatments, targeting regions with high prevalence rates.
- Novartis optimizes content related to their clinical trials to improve search engine rankings and attract researchers and HCPs.

6. Pay-Per-Click (PPC) Advertising

Overview:

PPC advertising involves placing ads on search engines or social media platforms where advertisers pay each time their ad is clicked. It's used for targeted advertising and lead generation.

Benefits:

- Provides immediate visibility and targeted reach based on user searches and interests.

- Allows for precise control over budget and ad performance.
- Facilitates retargeting of users who have previously interacted with the brand.

Examples:

- Pfizer uses PPC ads to promote specific medications and drive traffic to dedicated landing pages.
- AbbVie runs targeted PPC campaigns to attract HCPs to webinars and product information pages.

7. **Digital Therapeutics and Mobile Apps**

Overview:

Digital therapeutics and mobile apps offer interactive solutions for managing health conditions, supporting adherence, and providing educational resources.

Benefits:

- Enhances patient engagement and self-management of health conditions.
- Provides real-time data and insights for healthcare providers.
- Supports personalized care through tailored content and interventions.

Examples:

- Propeller Health's app helps asthma and COPD patients manage their conditions by tracking medication use and symptoms.
- Pear Therapeutics offers digital therapeutics for mental health conditions, providing evidence-based interventions through mobile platforms.

Conclusion

Each digital marketing channel offers unique advantages and can be strategically used to enhance pharmaceutical marketing efforts. Understanding how to leverage these channels effectively helps pharma product managers and marketers engage with their target audiences, drive brand success, and achieve their business objectives.

26.2 Email Marketing

Email marketing is a powerful tool for pharmaceutical product managers to engage with healthcare professionals (HCPs), patients, and other stakeholders. It allows for personalized communication, targeted messaging, and measurable results, making it a crucial component of a comprehensive digital marketing strategy. This section will cover the fundamentals of email marketing, best practices, and how to use email campaigns effectively in the pharmaceutical industry.

1. **Fundamentals of Email Marketing in Pharma**
   - **What is Email Marketing?**

     Email marketing involves sending commercial messages to a group of people using email. In the pharmaceutical industry, these messages can range from product updates and educational content to promotional offers and event invitations.

   - **Benefits of Email Marketing**
     - **Direct Communication**: Allows direct interaction with HCPs, patients, and other stakeholders.
     - **Personalization**: Enables tailored content based on recipient preferences and behaviors.
     - **Cost-Effective**: Generally costs less compared to other marketing channels.
     - **Measurable**: Provides detailed analytics on open rates, click-through rates, conversions, and more.

   - **Types of Email Campaigns**
     - **Newsletter**: Regular updates featuring a mix of content, such as articles, news, and product information.
     - **Promotional Emails**: Focused on specific promotions, discounts, or offers.
     - **Educational Emails**: Provide valuable information on medical conditions, treatment options, or research findings.

- **Event Invitations**: Announcements and invitations to webinars, conferences, or other events.
- **Follow-Up Emails**: Post-event or post-purchase follow-ups to maintain engagement.

2. **Best Practices for Email Marketing in Pharma**
   - **Building an Email List**
     - **Permission-Based Marketing**: Ensure that all email recipients have opted-in to receive communications. This is crucial for compliance with regulations like GDPR and CAN-SPAM.
     - **Segmentation**: Segment the email list based on various criteria such as profession, specialty, engagement level, and geographic location to tailor messaging and improve relevance.
   - **Crafting Effective Emails**
     - **Compelling Subject Lines**: Create subject lines that are clear and concise and encourage the recipient to open the email.
     - **Personalization**: Use the recipient's name and other personal details to make the email more relevant and engaging.
     - **Clear and Concise Content**: Keep the email content focused and to the point. Use bullet points, headings, and images to make it easily scannable.
     - **Strong Call-to-Action (CTA)**: Include a clear CTA that directs the recipient to take a specific action, such as clicking a link, signing up for an event, or downloading a resource.
   - **Design and Layout**
     - **Responsive Design**: Ensure emails are mobile-friendly and look good on all devices.
     - **Visual Appeal**: Use high-quality images and a clean layout to make the email visually appealing.
     - **Consistency**: Maintain consistent branding, including logos, colors, and fonts, across all emails.

- **Timing and Frequency:**
  - **Optimal Send Times**: Test different send times to determine when your audience will most likely engage with emails.
  - **Frequency**: Balance the frequency of emails to avoid overwhelming recipients while maintaining regular engagement.

3. **Measuring Email Marketing Success**
  - **Key Metrics**
    - **Open Rate**: The percentage of recipients who open the email. Indicates the effectiveness of the subject lines and sender name.
    - **Click-Through Rate (CTR)**: The percentage of recipients clicking links within the email. Measures the engagement level and relevance of the content.
    - **Conversion Rate**: The percentage of recipients who complete a desired action after clicking a link. Indicates the effectiveness of the email in driving specific outcomes.
    - **Bounce Rate**: The percentage of emails that were not successfully delivered. A high bounce rate can indicate issues with the email list or server.
    - **Unsubscribe Rate**: The percentage of recipients who opt out of receiving future emails. Helps identify issues with email frequency, content, or relevance.
  - **Analyzing and Optimizing Campaigns**
    - **A/B Testing**: Test different elements of the email (e.g., subject lines, CTAs, images) to see what resonates best with the audience.
    - **Performance Tracking**: Use email marketing software to track and analyze key metrics, identify trends, and optimize future campaigns.
    - **Feedback Loops**: Gather feedback from recipients to understand their preferences and improve email content and strategy.

4. Compliance and Ethical Considerations

- **Regulatory Compliance:**
  - **GDPR and CAN-SPAM**: Ensure all email marketing activities comply with relevant regulations, including obtaining explicit consent from recipients and providing easy opt-out options.
  - **Pharma-Specific Guidelines**: Follow industry-specific guidelines for promoting pharmaceutical products, including accurate representation of product information and balanced communication of benefits and risks.

- **Ethical Marketing**
  - **Transparency**: Be transparent about the email's purpose and the content's nature.
  - **Patient Privacy**: Protect patient information and ensure email communications comply with privacy laws and regulations.

Case Studies and Examples

The following two case studies demonstrate how email marketing can help achieve the campaign objectives.

Case Study #35: Educational Campaign for a New Treatment

Objective

The primary objective of this campaign was to raise awareness about new diabetes treatments among endocrinologists. The secondary objective was to educate them on the clinical benefits and efficacy of the new treatment to encourage its adoption in their practice.

Strategy

1. **Target Audience Identification**:

   - **Segmentation**: The email list included only endocrinologists and diabetes specialists. Further segmentation was done based on geographic regions to tailor the content to regional guidelines and practices.

2. **Content Creation**:

   - **Email Series**: Five emails were planned, each focusing on a different aspect of the new treatment.

     1. **Introduction Email**: Announced the new treatment, highlighting its innovative aspects and providing a brief overview.

     2. **Clinical Data Email**: Detailed the results from clinical trials, emphasizing efficacy, safety, and patient outcomes.

     3. **Patient Case Studies Email**: Shared real-world case studies showcasing the treatment's benefits and patient experiences.

     4. **Expert Opinions Email**: Included testimonials and insights from leading endocrinologists and researchers.

     5. **Follow-Up Email**: Summarized key points from previous emails and provided links to detailed resources and contact information for further inquiries.

3. Personalization and Design:

- **Personalization**: Each email was personalized with the recipient's name and relevant professional details (e.g., "Dear Dr. Smith, as an expert in diabetes care....).

- **Visual Design**: Emails were designed to be visually appealing, with high-quality images, graphs, and infographics illustrating key points. The layout was clean and easy to navigate, with clear headings and bullet points.

4. Call-to-Action (CTA):

- **Primary CTA**: Encouraged recipients to visit a dedicated landing page for detailed information and resources.

- **Secondary CTAs**: Included options to download clinical trial reports, watch video testimonials, and sign up for a follow-up webinar.

Results

- **Open Rates**: The series achieved an average open rate of 40 percent, significantly higher than the industry average of around 20 percent.

- **Click-Through Rates (CTR)**: The emails saw a higher CTR of 25 percent, indicating strong engagement with the content.

Conversions:

- **Webinar Sign-Ups**: Many recipients signed up for the follow-up webinar, providing an opportunity for further engagement.

- **Information Requests**: Many endocrinologists requested additional information and resources, showing interest in adopting the new treatment.

Analysis

- **Success Factors**:
  - **Segmentation and Personalization**: Targeting the right audience and personalizing the content were key to achieving high open and click-through rates.

- **Educational Content**: Providing valuable, well-researched information built trust and effectively demonstrated the treatment's benefits.
- **Engaging Design**: The emails' visual appeal and clear layout improved engagement.

Areas for Improvement:

- **Feedback Mechanisms**: Incorporating surveys or feedback forms within the emails could provide insights for further refinement.
- **Continuous Engagement**: Planning a follow-up series post-webinar to maintain momentum and address any remaining questions.

Case Study #36: Event Promotion for a Webinar

Objective

The main goal was to promote an upcoming webinar on advancements in oncology treatment, aiming to achieve high registration and attendance rates among oncologists and related healthcare professionals (HCPs).

Strategy

1. **Target Audience Identification**:
   - **Segmentation**: The email list was segmented to target oncologists, oncology nurses, and other HCPs involved in cancer treatment. Further segmentation was based on prior engagement levels, such as past webinar attendees and active newsletter subscribers.

2. **Content Creation**:
   - **Email Sequence**: Three emails were planned to build interest and ensure high attendance:
     1. **Announcement Email**: Introduced the webinar topic, date, and time, highlighting the key speakers and agenda.
     2. **Reminder Emails**: Sent two reminders — one week before and one day before the webinar — emphasizing the value of attending and providing easy registration links.
     3. **Follow-Up Email**: This email was sent after the webinar to thank attendees, share key takeaways, and provide access to the recorded sessions and additional resources.

3. **Personalization and Design**:
   - **Personalization**: Each email addressed the recipient by name and referred to their professional interests or past engagements (e.g., "As someone interested in oncology advancements...).

- **Visual Design**: Emails included professional graphics, speaker photos, and clear, concise text. The layout was designed to highlight key information and make registration straightforward.

4. **Call-to-Action (CTA)**:

   - **Primary CTA**: Encouraged recipients to register for the webinar with a prominent button.
   - **Secondary CTAs**: Included options to add the event to their calendar and share the webinar information with colleagues.

Results

- **Registration Rates**: Achieved a registration of 30 percent of email recipients, indicating strong interest in the webinar topic.
- **Attendance**: There was high attendance on the day of the webinar, with a significant proportion of registrants attending live.

Post-Webinar Engagement:

- **Resource Downloads**: Many attendees accessed the recorded session and additional resources provided in the follow-up email.
- **Feedback**: Positive feedback from participants, with many expressing interest in future webinars and related content.

Analysis

- **Success Factors**:
  - **Timely Reminders**: The reminder emails effectively prompted registrations and ensured high attendance.
  - **Speaker Credibility**: Highlighting well-known speakers and their expertise helped attract more registrants.
  - **Easy Registration Process**: A straightforward and user-friendly registration process minimizes barriers to signing up.

- **Areas for Improvement:**

- **Engagement During Webinar**: Incorporating more interactive elements during the webinar, such as live polls or Q&A sessions, could enhance engagement.

- **Long-Term Follow-Up**: Developing a series of follow-up emails to nurture relationships and provide ongoing value could strengthen long-term engagement.

By implementing these strategies and continually optimizing based on results and feedback, pharmaceutical product managers can effectively leverage email marketing to educate, engage, and convert their target audiences.

26.3 Designing Effective Email Campaigns

Email marketing remains one of the most powerful tools for engaging with healthcare professionals (HCPs), patients, and other stakeholders in the pharmaceutical industry. An effective email campaign requires careful planning, content creation, segmentation, and performance analysis. Here is a step-by-step guide to designing effective email campaigns:

1. Define Your Goals

Defining clear, measurable goals is crucial before creating an email campaign. Common objectives for pharmaceutical email campaigns include:

- Increasing brand awareness
- Promoting new product launches
- Educating HCPs about clinical data and guidelines
- Driving traffic to a website or webinar
- Encouraging patient adherence and engagement
- Gathering feedback or data through surveys

2. Understand Your Audience

Segment your audience to ensure the content is relevant and targeted. Common segments in pharma email campaigns include:

- HCPs (e.g., primary care physicians, specialists)
- Patients and caregivers
- Pharmacists
- Payers and insurers
- Regulatory bodies and stakeholders

3. Craft Compelling Content

Content is the core of any email campaign. It should be:

- **Relevant**: Tailored to the needs and interests of each segment.
- **Informative**: Providing valuable insights, updates, or educational content.

- **Engaging**: Using a conversational tone, compelling visuals, and clear calls to action (CTAs).

Components of a Great Email:

- **Subject Line**: Short, intriguing, and informative to encourage opens (e.g., "New Treatment Guidelines for Hypertension — What You Need to Know").

- **Preheader Text**: A summary that complements the subject line.

- **Body Content**: Keep it clear and concise, using a mix of text, images, and links. Use headings and bullet points for easy reading.

- **Personalization**: Address recipients by name and tailor content based on their interests and previous interactions.

- **Call to Action (CTA)**: A clear, compelling prompt encouraging the desired action (e.g., Download the Latest Clinical Study").

4. Design and Layout

A well-designed email enhances readability and engagement:

- **Responsive Design**: Ensure the email is optimized for desktop and mobile devices.

- **Visual Hierarchy**: Use headings, subheadings, and images to guide readers through the content.

- **Brand Consistency**: Align the design with your brand's visual identity, including logos, colors, and fonts.

- **Images and Multimedia**: Use high-quality images, infographics, and videos to support the content and break up text.

5. Compliance and Ethical Considerations

- Ensure all email content complies with relevant regulations and guidelines:

- **GDPR**: Obtain explicit consent for data collection and email communication from recipients in the European Union.

- **CAN-SPAM Act**: include a clear opt-out mechanism and avoid misleading subject lines.

- **Pharma-Specific Regulations**: Adhere to industry-specific guidelines, such as those from the FDA or EMA, regarding the promotion of pharmaceuticals.

6. Testing and Optimization

Before sending out the campaign, conduct thorough testing:

- **A/B Testing**: Test different subject lines, content, CTAs, and designs to identify what resonates best with your audience.
- **Technical Testing**: Ensure emails render correctly across various devices and email clients. Check for broken links or images.

7. Monitor and Analyze Performance

Track key performance metrics to evaluate your campaign's success and identify improvement areas.

- **Open Rate** is the percentage of recipients who open the email. This indicates the effectiveness of the subject line and the sender's reputation.
- **Click-Through Rate**: The percentage of recipients who clicked on the link within the email. This rate reflects the relevance and engagement of the content.
- **Conversion Rate**: The percentage of recipients who complete the desired action (e.g., signing up for a webinar or downloading a resource).
- **Bounce Rate** is the percentage of emails that could not be delivered. A high bounce rate may indicate issues with your mailing list.
- **Unsubscribe Rate**: Percentage of recipients who opted out of future emails. Monitor to ensure content remains relevant and valuable.

Here is a case study that demonstrates how email marketing can help drive a successful new product launch:

Case Study #37: Launching a New Medication — Glucotrol

Objective

Promote the launch of Glucotrol, a new diabetes medication with superior blood sugar control and a convenient once-daily dosage.

Pre-Launch Phase

1. **Audience Research and Segmentation**
   - **Objective**: Identify and segment the target audience to tailor the campaign effectively.

   Segments:
   - **Endocrinologists and Diabetes Specialists**: These are the primary prescribers who need to understand the clinical benefits of Glucotrol.
   - **Primary Care Physicians**: They manage many diabetes patients and can be key prescribers.
   - **Pharmacists**: They can influence patient adherence and provide recommendations.
   - **Patients and Caregivers**: The end users who will benefit from the medication.

2. **Teaser Campaign**
   - **Email 1: Subject Line** — "Coming Soon: A New Era in Diabetes Management."
   - **Content**: This brief teaser about the upcoming medication highlights the innovation and anticipated benefits without giving too many details.
   - **CTA**: "Save the Date for the Exclusive Webinar."
   - **Goal**: Create curiosity and anticipation.
   - **Email 2: Subject Line** – "Revolutionary Diabetes Treatment is Almost Here."
   - **Content**: More details on the launch date, emphasizing the unique benefits of Glucotrol, such as superior blood sugar control and once-daily dosage.

- **CTA**: "Register for Early Access Information."
- **Goal**: Increase pre-launch registrations for the webinar.

Launch Phase

3. Launch Announcement

- **Email 1:** Subject Line — "Introducing Glucotrol: Superior Blood Sugar Control with Once-Daily Convenience."
- **Content**: Detailed introduction to Glucotrol, including its mechanism of action, clinical trial results, and patient testimonials. Incorporate visuals such as infographics and videos.
- **CTA**: "Learn More" (link to a dedicated landing page with comprehensive information).
- **Goal**: Provide a compelling reason for HCPs to consider prescribing Glucotrol.
- **Email 2:** Subject Line — "Join us for the Glucotrol Launch Webinar."
- **Content**: Invite HCPs to a detailed webinar featuring KOL, which will discuss the clinical benefits and patient outcomes associated with Glucotrol.
- **CTA**: "Register Now."
- **Goal**: Drive registrations for the webinar to provide an interactive platform for discussing Glucotrol.

Educational Series

4. Ongoing Education

- **Email Series:** Subject Line — "Deep Dive into Glucotrol Mechanism and Benefits."
- **Email 1**: Focus on the pharmacodynamics and pharmacokinetics of Glucotrol.
- **Email 2**: Highlight clinical trial data, comparing Glucotrol to existing treatments.
- **Email 3**: Share patient case studies demonstrating improved outcomes with Glucotrol.

- **CTA**: "Download the Full Clinical Study" / "Read Patient Stories."
- **Goal**: Equip HCPs with in-depth knowledge to make informed prescribing decisions.

5. Patient and Caregiver Education

- **Email 1**: Subject Line – "Managing Diabetes with Glucotrol: What You Need to Know."
- **Content**: Simplified information about Glucotrol's benefits, how to use it, and what to expect. Include Q&A and myth-busting sections.
- **CTA**: "Learn More." / "Download Patient Guide."
- **Goal**: Educate patients and caregivers to ensure proper use and adherence.

Feedback and Optimization

6. Feedback Collection

- **Email**: Subject Line — "Your Experience with Glucotrol Matters."
- **Content**: Request feedback from HCPs and patients who have started using Glucotrol. Include a link to a detailed survey.
- **CTA**: "Provide Feedback."
- **Goal**: Gather insights to refine ongoing and future marketing efforts.

7. Continuous Engagement and Support

- **Email**: Subject Line — "Stay Updated on Glucotrol."
- **Content**: Share updates on new studies, patient support programs, and additional resources. Include links to new webinars or Q&A sessions with KOLs.
- **CTA**: "Access New Resources" / "Join Upcoming Webinar"
- **Goal**: Maintain engagement and support for HCPs and patients.

Performance Monitoring

- Metrics and Analysis
  - **Open Rates**: Monitored to evaluate the effectiveness of subject lines' and Preheader texts.

- **Click-Through Rates (CTR)**: Analyzed to determine engagement levels and the effectiveness of CTAs.
- **Conversion Rates**: These are tracked to measure the percentage of recipients who completed desired actions, such as registering for webinars or downloading resources.
- **Feedback and Survey Results**: Analyzed to understand Glucotrol's reception and identify improvement areas.

Real-World Impact

1. Increased Prescriptions

Post-launch, there was a noticeable increase in Glucotrol prescriptions among endocrinologists and primary care physicians, driven by the educational content and compelling clinical data in the emails.

2. Improved Patient Outcomes

Follow-up surveys indicated that patients experienced better blood sugar control with Glucotrol, validating the clinical claims made during the campaign.

3. Enhanced Brand Loyalty

Continuous engagement through educational series and updates helped build strong relationships with HCPs and patients, leading to higher brand loyalty and advocacy.

Conclusion

The strategic use of segmented email campaigns allowed for a successful launch of Glucotrol, highlighting the importance of tailored content, continuous education, and feedback collection in pharmaceutical marketing. By following these steps and leveraging real-world data, pharmaceutical product managers can create effective email campaigns that drive engagement, educate stakeholders, and achieve marketing objectives.

26.4 Measuring Email Marketing ROI

Measuring the return on investment (ROI) of email marketing campaigns is crucial for pharmaceutical product managers to understand the effectiveness of their efforts and make data-driven decisions. This process involves tracking key metrics, analyzing results, and calculating the financial impact of email marketing activities.

Key Metrics for Email Marketing

1. **Open Rate**:
   - **Definition**: The percentage of recipients who open the email.
   - **Importance**: Indicates the effectiveness of the subject line and the sender's credibility. A higher open rate suggests that the recipients perceive the email content as valuable and relevant.

2. **Click-Through Rate (CTR)**:
   - **Definition**: The percentage of recipients who click on one or more links within the email.
   - **Importance**: Measures the level of engagement and interest in the email content. A higher CTR indicates that the content is compelling and prompts recipients to take action.

3. **Conversion Rate**:
   - **Definition**: The percentage of recipients who complete a desired action (e.g., signing up for a webinar or downloading a resource) after clicking a link in the email.
   - **Importance**: Directly reflects the effectiveness of the email in achieving campaign goals. Higher conversion rates suggest successful content and CTA strategies.

4. **Bounce Rate**:
   - **Definition**: The percentage of emails that could not be delivered to the recipient's inbox.
   - **Importance**: Helps identify issues with the email list quality or delivery infrastructure. Lower bounce rates are preferable.

5. Unsubscribe Rate:

- **Definition**: The percentage of recipients who opt out of receiving future emails.

- **Importance**: This indicates recipient satisfaction and content relevance. A high unsubscribe rate may signal email frequency, relevance, or quality issues.

6. List Growth Rate:

- **Definition**: The rate at which your email list is growing, taking into account new subscribers and unsubscribes.

- **Importance**: Reflects overall health and expansion of your email audience.

Calculating ROI

1. Defining ROI:

- **Formula**: ROI = (Net Profit / Investment) x 100

- **Net Profit**: The financial gain from the email marketing campaign minus its associated costs.

- **Investment**: The total cost of running the email marketing campaign, including software, design, content creation, and labor.

2. Steps to Calculate ROI:

Step 1: Track Costs

- **Email Marketing Software**: Subscription fees for email marketing platforms (e.g., Mailchimp, Constant Contact).

- **Content Creation**: Costs for designing and writing the email content.

- **Labor**: Wages for team members involved in planning, executing, and analyzing the campaign.

- **Other Costs**: Any additional expenses, such as purchasing email lists or hiring external consultants.

Step 2: Measure Financial Gains

- **Direct Revenue**: Sales directly attributed to the email campaign. This can be tracked through unique promo codes, links, or landing pages.

- **Indirect Revenue**: Long-term sales influenced by the campaign, such as increased brand awareness leading to future purchases.
- **Cost Savings**: Savings from using email marketing instead of more expensive channels, like direct mail.

Step 3: Calculate Net Profit

Formula: Net Profit = Total Revenue - Total Costs

Step 4: Calculate ROI

Formula: ROI = (Net Profit / Total Costs) x 100

3. **Examples of Measuring ROI in Pharma**:

Example 1: Webinar Promotion Campaigns

Costs:

Email Marketing Software: $500

Content Creation: $1,000

Labor: $2,000

Total Costs: $3,500

Revenue:

Direct Revenue from Webinar Sign-Ups: $8,000

Indirect Revenue from Follow-Up Sales: $4,000

Total Revenue: $12,000

Net Profit:

Net Profit = $12,000 - $3,500 = $8,500

ROI:

ROI = ($8,500 / $3,500) x 100 = 243%

Example 2: New Product Launch Campaign

Costs:

Email Marketing Software: $1,000

Content Creation: $2,500

Labor: $4,000

Total Costs: $7,500

Revenue:

Direct Revenue from Pre-Orders: $15,000

Indirect Revenue from Increased Brand Awareness: $10,000

Total Revenue: $25,000

Net Profit:

Net Profit = $25,000 - $7,500 = $17,500

ROI:

ROI = ($17,500 / $7,500) x 100 = 233%

4. **Best Practices for Maximizing ROI**

   A. **Optimize Email Content and Design:**

   Personalization: Tailor content to individual recipient preferences and behaviors.

   Clear CTAs: Use strong and clear calls to action to drive conversions.

   A/B Testing: Test different elements (e.g., subject lines, images, CTAs) to identify what works best.

   B. **Improve List Quality**:

   Regular Cleaning: Remove inactive or invalid email addresses to improve deliverability.

   Segmentation: Segment your list to send more relevant and targeted content.

   C. **Enhance Engagement**:

   Value-Driven Content: Provide valuable information that addresses the needs and interests of your audience.

   Timing and Frequency: Send emails at optimal times and maintain a consistent but not overwhelming frequency.

   D. **Leverage Analytics**:

   Performance Tracking: Continuously monitor key metrics to understand what's working and what's not.

   Data-Driven Decisions: Use insights from analytics to refine and improve future campaigns.

By diligently tracking these metrics and calculating ROI, pharmaceutical product managers can ensure their email marketing efforts are effective and provide a significant return on investment. This approach helps optimize campaigns, improve engagement, and drive better business outcomes.

26.5 Strategies for Effective Digital Campaigns in Pharma

Effective digital campaigns in the pharmaceutical industry require careful planning, execution, and optimization. The strategies outlined below are essential for maximizing the impact of digital campaigns and achieving desired outcomes.

1. Define Clear Objectives

Overview:

Establishing clear, measurable objectives is the foundation of a successful digital campaign. Objectives should align with overall business goals and address specific needs, whether raising awareness, driving engagement, or increasing product adoption.

Strategies:

- **Set SMART Goals:** Specific, Measurable, Achievable, Relevant, and Time-bound goals help track progress and measure success.

- **Align with Business Goals:** Ensure campaign objectives support broader business objectives, such as increasing market share or launching a new product.

- **Identify Key Performance Indicators (KPIs):** To evaluate campaign performance and determine click-through rates (CTR), conversion rates, and engagement levels.

2. Understand the Target Audience

Overview:

Understanding the target audience is crucial for crafting relevant and engaging content. This includes knowing their demographics, preferences, pain points, and behavior.

Strategies:

- **Segment the Audience:** Use demographic, behavioral, and psychographic data to create audience segments for more personalized messaging.

- **Conduct Market Research:** Gather insights through surveys, focus groups, and data analysis to understand audience needs and preferences.

- **Develop Personas:** Create detailed personas representing different segments of the target audience to guide content creation and campaign strategies.

3. Develop Compelling Content

Overview:

Content is a critical component of digital campaigns. It should be relevant, engaging, and tailored to the target audience to drive interaction and conversions.

Strategies:

- **Create Value-Driven Content:** Focus on providing valuable information that addresses the needs and interests of the audience, such as educational articles, patient stories, or expert interviews.

- **Use a Multi-Format Approach:** To appeal to different preferences, employ various content formats, including blogs, videos, infographics, and webinars.

- **Ensure Compliance:** Adhere to regulatory guidelines and industry standards to ensure content is accurate, ethical, and compliant.

4. Leverage Digital Channels

Overview:

Utilizing the right digital channel mix helps reach and engage the target audience. Each channel offers unique advantages and should be chosen based on campaign goals and audience preferences.

Strategies:

- **Optimize Website and Landing Pages:** Ensure that websites and landing pages are user-friendly, informative, and optimized for conversions.

- **Utilize Social Media:** Engage with audiences on LinkedIn, Twitter (X now), Facebook, and Instagram through targeted posts, ads, and interactions.

- **Implement Email Marketing:** Use personalized email campaigns to communicate with HCPs, patients, and stakeholders about relevant updates and offers.

- **Leverage Paid Advertising:** Employ PPC (Pay-per-Click) and display ads to drive targeted traffic and increase visibility in search engines and social media platforms.

5. Personalize and Segment

Overview:

Personalization and segmentation enhance the relevance of messages and offers, improving engagement and conversion rates.

Strategies:

- **Segment Audiences Based on Behavior:** Use data to segment audiences by behavior, preferences, and engagement history for tailored messaging.

- **Personalize Content and Offers:** Create personalized content and offers based on user data, such as past interactions or preferences.

- **Use Dynamic Content:** Implement dynamic content that changes based on user profiles or behaviors to increase relevance and impact.

6. Optimize Campaign Performance

Overview:

Continuous optimization ensures that digital campaigns perform at their best and achieve the desired results.

Strategies:

- **Monitor Key Metrics:** Track KPIs such as CTR (Click-through rate), conversion rates, and engagement levels to assess campaign performance.

- **Conduct A/B Testing:** Test different content, messaging, and design variations to determine what works best and refine strategies accordingly.

- **Analyze Data and Adjust:** Use analytics tools to gain insights into campaign performance and make data-driven adjustments to improve outcomes.

7. Ensure Compliance and Ethics

Overview:

Compliance with regulatory guidelines and ethical standards is crucial in pharmaceutical marketing to avoid legal issues and maintain credibility.

Strategies:

- **Follow Regulatory Guidelines:** Adhere to industry regulations and guidelines set by authorities such as the FDA and EMA.

- **Ensure Transparency:** Be transparent about sponsorships, disclosures, and the nature of content.

- **Implement Ethical Practices:** Ensure all marketing practices are ethical and do not mislead or exploit the target audience.

8. Engage and Build Relationships

Overview:

Building and maintaining relationships with HCPs, patients, and other stakeholders is essential for long-term success.

Strategies:

- **Foster Engagement:** Encourage interaction through social media, forums, and community initiatives.

- **Provide Value:** Offer resources, support, and information to build trust and loyalty.

- **Respond to Feedback:** Actively listen to and address feedback from stakeholders to improve relationships and campaign effectiveness.

Conclusion

By implementing these strategies, pharmaceutical companies can design and execute effective digital campaigns that resonate with their target audience, drive engagement, and achieve business objectives. Continuous monitoring, optimization, and adherence to regulatory and ethical standards are key to success in the evolving digital landscape.

Customer Engagement Strategies

In today's pharmaceutical landscape, customer engagement is more than just a buzzword—it's a strategic imperative. With the growing complexity of healthcare systems, increased competition, and an empowered, digitally connected customer base, effectively engaging stakeholders is essential for success. For pharmaceutical companies, customer engagement encompasses interactions with various stakeholders, including healthcare professionals (HCPs), patients, payers, regulators, and caregivers.

This chapter delves into the art and science of customer engagement, exploring how pharmaceutical companies can create meaningful, trust-based relationships with their audiences. From leveraging digital platforms to crafting personalized marketing campaigns, pharma marketers need to understand the nuances of each customer segment and employ innovative strategies to engage them effectively. Pharmaceutical companies can use digital and traditional channels, integrate behavioral science, and apply real-world data to drive better health outcomes, ensure brand loyalty, and build long-term, sustainable relationships.

In this chapter, we will examine the core principles of customer engagement, explore proven strategies for creating value-driven interactions, and present real-world case studies showcasing successful engagement initiatives. Whether your goal is to improve patient adherence, build stronger connections with HCPs, or navigate regulatory requirements, a solid customer engagement strategy will be a cornerstone of your marketing success.

27.1   HCP Marketing

Healthcare Professional (HCP) marketing is a crucial component of pharmaceutical marketing. It engages and educates healthcare professionals about pharmaceutical products, treatments, and medical advancements. This section delves into the strategies, best practices, and ethical considerations involved in effectively marketing to HCPs.

1. **Importance of HCP Marketing**

   HCPs, including doctors, nurses, pharmacists, and other medical professionals, play a pivotal role in prescribing medications and recommending treatments. Effective HCP marketing can:

   - **Enhance Brand Awareness**: Ensure that HCPs are well-informed about new products and advancements, which will increase brand recognition.

   - **Drive Prescriptions**: HCPs participating in continuing medical education programs (CMEs) are more likely to prescribe new treatments they understand and trust.

   - **Improve Patient Outcomes**: Pharmaceutical companies can help ensure patients receive the best care by providing HCPs with accurate and comprehensive information.

2. **Strategies for Effective HCP Marketing**

   A. Educational Content:

   - **Scientific Literature and Research**: Providing HCPs access to peer-reviewed journals, clinical trial results, and case studies helps them make informed decisions.

   - **Continuing Medical Education (CME)**: Sponsoring CME programs and offering accredited educational resources can enhance HCPs' knowledge and skills.

B. Multi-Channel Engagement:

- **Digital Channels**: Utilizing websites, email campaigns, webinars, and social media to reach HCPs where they are most active.

- **Face-to-face interaction**: Engaging HCPs through conferences, medical symposiums, and sales representative visits.

C. Personalized Communication:

- **Segmentation**: Tailoring messages based on HCPs' specialties, preferences, and past interactions to increase relevance and engagement.

- **Customized Content**: Developing content that addresses different HCP segments' specific needs and concerns.

D. Thought Leadership and Advocacy:

- **Key Opinion Leaders (KOLs)**: Collaborating with respected experts in the medical field to endorse and share information about products.

- **Advisory Boards**: Establishing advisory boards with HCPs to provide insights and feedback on product development and marketing strategies.

3. **Ethical Considerations in HCP Marketing**

A. Compliance with Regulations:

- **Adhering to Guidelines**: Following guidelines set by regulatory bodies such as the FDA, EMA, and other national health authorities to ensure ethical marketing practices.

- **Transparency**: Disclosing any financial relationships or sponsorships in all communications and interactions.

B. Accurate and Balanced Information:

- **Honest Communication**: Providing accurate, up-to-date product information, including benefits and potential risks.

- **Avoiding Misleading Claims**: Ensuring that all promotional materials are based on solid scientific evidence and do not overstate efficacy or safety.

C. Respect for HCP Autonomy:

- **Non-Coercive Marketing**: Avoiding pressure tactics and respecting the professional judgment and decision-making processes of HCPs.

- **Supporting Informed Decisions**: Providing HCPs with comprehensive information to help them make the best choices for their patients.

3. Measuring the Effectiveness of HCP Marketing

A. Key Performance Indicators (KPIs):

- **Engagement Metrics**: Tracking metrics, such as open rates, click-through rates, and time spent on educational materials.

- **Behavioral Metrics**: Monitoring changes in prescription patterns, attendance at educational events, and participation in CME programs.

- **Feedback and Surveys**: Collecting feedback from HCPs to understand their needs, preferences, and perceptions of the marketing efforts.

B. ROI Analysis:

- **Direct ROI**: Calculating the financial return from increased prescriptions and sales directly attributed to marketing activities.

- **Indirect ROI**: Considering the long-term benefits of improved HCP relationships, enhanced brand reputation, and increased patient trust.

4. Future Trends in HCP Marketing

A. Digital Transformation:

- **Virtual Engagement**: Increasing use of virtual conferences, webinars, and tele-detailing to reach HCPs remotely.

- **AI and Data Analytics**: Leveraging artificial intelligence and advanced analytics to personalize communication and gain deeper insights into HCP behavior.

B. Integrated Marketing Approaches:

- **Omnichannel Strategies**: Creating seamless experiences across multiple channels to ensure consistent and effective communication with HCPs.

- **Collaborative Platforms**: Using platforms that facilitate collaboration and information sharing among HCPs, enhancing their ability to stay informed and engaged.

C. Patient-Centered Focus:

- **Patient Outcomes**: Shifting towards marketing strategies that emphasize the impact on patient outcomes, aligning with HCPs' primary concern for patient welfare.

- **Patient-HCP Collaboration**: Encouraging collaborative approaches where HCPs and patients are educated and empowered to make informed decisions.

By adopting these strategies and staying attuned to evolving trends, pharmaceutical product managers can effectively engage healthcare professionals, foster trust, and drive better patient outcomes through informed and ethical marketing practices.

27.2 Engaging Healthcare Professionals

Engaging healthcare professionals (HCPs) is a multifaceted process that requires a deep understanding of their needs, preferences, and the unique challenges they face in their daily practice. Effective engagement strategies can build trust, foster long-term relationships, and ultimately influence HCPs' prescribing behaviors and their perception of pharmaceutical products.

1. **Understanding HCP Needs and Preferences**

   A. Personalized Content:

   - **Relevance**: Tailor content to the specific interests, specialties, and patient populations of HCPs. Personalized communication increases the likelihood of engagement and trust.

   - **Timelines**: Provide timely information aligned with current medical trends, research updates, and treatment guidelines.

   B. Continuing Education

   - **Educational Resources**: Offer high-quality educational materials, including white papers, research articles, clinical guidelines, and case studies.

   - **Accredited Programs**: Develop and sponsor continuing medical education (CME) programs to help HCPs stay current with medical advancements and maintain their certifications.

2. **Multi-Channel Engagement Strategies**

   A. Digital Channels:

   - **Webinars and Online Conferences**: Host interactive webinars and virtual conferences featuring thought leaders and experts in the field.

   - **Email Campaigns**: Use segmented email campaigns to deliver relevant content directly to HCPs' inboxes. Track engagement metrics to refine and improve future campaigns.

- **Social Media**: Engage HCPs on professional social networks such as LinkedIn and specialized platforms like Sermo. Share valuable content, participate in discussions, and create a community around your brand.

B. Face-to-Face Interaction:

- **Sales Representatives**: Train sales representatives to provide valuable, science-based information during in-person visits. Focus on building relationships rather than just selling products.

- **Medical Conferences**: Attend and exhibit at medical conferences and symposiums. Use these opportunities to network, share knowledge, and gather insights from HCPs.

3. Building Relationships and Trust

A. Thought Leadership:

- **Key Opinion Leaders (KOLs)**: Collaborate with respected KOLs to endorse your products and share their expertise. Their influence can significantly impact the credibility and acceptance of your brand.

- **Advisory Boards**: Establish advisory boards consisting of HCPs to provide insights into product development, marketing strategies, and emerging trends.

B. Consistent Communication:

- **Regular Updates**: Keep HCPs informed about new research, product updates, and relevant medical news. Consistent communication helps maintain engagement and trust.

- **Transparent Practices**: Be transparent about product benefits, risks, and potential conflicts of interest. HCPs value honesty and integrity in their interactions with pharmaceutical companies.

4. Utilizing Technology and Data

A. Customer Relationship Management Systems (CRM) Systems:

- **Data Management**: Use CRM systems to track interactions with HCPs, monitor their preferences, and tailor communications accordingly.

- **Analytics**: Leverage data analytics to gain insights into HCP behaviors, preferences, and engagement patterns. Use these insights to refine marketing strategies and improve engagement.

B. Digital Tools:

- **Mobile Apps**: Develop apps that give HCPs easy access to product information, educational resources, and clinical tools.

- **Virtual Reality (VR) and Augmented Reality (AR)**: Use VR and AR to create immersive educational experiences, such as virtual lab tours or interactive 3D models of disease mechanisms.

5. Ethical Considerations in HCP Engagement

A. Compliance with Regulations:

- **Adherence to Guidelines**: Follow regulatory guidelines and industry codes of conduct to ensure ethical interactions with HCPs.

- **Documentation**: Maintain detailed records of all interactions and communications with HCPs to ensure transparency and accountability.

B. Respecting HCP Autonomy:

- **Non-Coercive Practices**: Avoid high-pressure sales tactics and respect the professional judgment of HCPs.

- **Informed Decisions**: Provide comprehensive and balanced information to support HCPs in making informed decisions about patient care.

6. Measuring Engagement Success

A. Key Performance Indicators (KPIs):

- **Engagement Metrics**: Track metrics, such as open rates, click-through rates, webinar attendance, and social media interactions to gauge engagement levels.

- **Behavioral Metrics**: Monitor changes in prescribing patterns, participation in educational programs, and feedback from HCPs.

B. Feedback and Continuous Improvement:

- **Surveys and Feedback**: Regularly collect feedback from HCPs to understand their needs, preferences, and perceptions of your marketing efforts.

- **Iterative Approach**: Use feedback to continuously refine and improve engagement strategies, ensuring they remain relevant and effective.

Case Studies in HCP Engagement

Here are two case studies that show how tailored approaches, leveraging both digital and traditional methods, can effectively engage HCPs and drive positive outcomes in pharmaceutical marketing.

Case Study #38: Digital CME Program

Challenge:

A pharmaceutical company introduced a new treatment for a chronic condition but faced challenges raising awareness and understanding among healthcare professionals (HCPs). To increase its adoption, the company needed an effective way to educate HCPs about the new treatment's benefits, usage guidelines, and clinical evidence.

Solution:

The company developed a series of digital Continuing Medical Education (CME) programs, leveraging the convenience and reach of online platforms. Key elements of the solution included:

Interactive Case Studies: The CME programs featured detailed case studies that allowed HCPs to explore real-world scenarios where the new treatment was applied. These case studies provided practical insights into the treatment's efficacy, patient selection criteria, and management of potential side effects.

Video Lectures: Leading experts in the relevant medical field were invited to deliver video lectures. These lectures covered the underlying science, clinical trial results, and comparative advantages of the new treatment. The video format allowed HCPs to engage with the content at their own pace.

Live Q&A Sessions: The company organized live Q&A sessions with expert speakers to foster interaction and address specific queries. HCPs could submit their questions in advance or during the sessions, ensuring their concerns were directly addressed.

Supplementary Materials: Participants were provided additional resources, such as downloadable slide decks, research articles, and treatment guidelines, for further reading and reference.

Outcome:

The digital CME programs achieved high participation rates, indicating strong interest and engagement from HCPs. Key outcomes included:

Increased Knowledge and Confidence: Post-program surveys revealed that many HCPs reported a significant increase in their understanding of the new treatment. They felt more confident prescribing the treatment and managing their patients' conditions.

Positive Feedback: Participants appreciated the digital format's convenience and flexibility. The interactive elements, such as case studies and live Q&A sessions, were well-received.

Impact on Prescriptions: Following the CME programs, the company observed a notable uptick in prescriptions for the new treatment. This increase was attributed to the enhanced education and confidence of HCPs in using the treatment.

Case Study #39: Multi-Channel Marketing Campaign

Challenge:

A pharmaceutical company noticed declining prescriptions for one of its established products. Many HCPs who had previously prescribed the product had not done so in the past year. The company needed to re-engage these HCPs and remind them of the product's benefits and clinical evidence.

Solution:

The company launched a comprehensive multi-channel marketing campaign to re-engage HCPs through various touchpoints. Key elements of the solution included:

Email Newsletters: A series of targeted email newsletters were sent to the HCPs. These newsletters updated recent clinical data, highlighted patient success stories, and shared practical tips for integrating the product into treatment protocols. Each email was personalized based on the HCP's specialty and prescribing history.

Targeted Social Media Ads: The company utilized social media platforms, particularly LinkedIn, to run targeted ads. These ads promoted key messages about the product, linked to detailed resources, and encouraged HCPs to participate in webinars and virtual events.

In-Person Visits by Sales Representatives: Recognizing the value of personal interaction, the company deployed its sales representatives to visit the targeted HCPs. These visits addressed concerns or misconceptions, provided updated information, and discussed the latest clinical evidence supporting the product's use.

Outcome:

The multi-channel campaign successfully re-engaged many HCPs, resulting in several positive outcomes.

Increased Engagement: Combining digital and in-person touchpoints ensured that the messages reached HCPs through their preferred channels. This multi-faceted approach enhanced overall engagement.

Positive Feedback: HCPs appreciated the personalized communication and the effort to provide relevant, updated information. Integrating patient success stories and practical tips made the content more relatable and actionable.

The Uptick in Prescriptions: Within six months of the campaign launch, the company observed a 20 percent increase in prescriptions for the product. Feedback from sales representatives and post-campaign surveys indicated that the re-engagement efforts directly impacted renewed interest and trust in the product.

27.3 Tools and Techniques for HCP Marketing

Effectively marketing to healthcare professionals (HCPs) involves using various tools and techniques designed to engage, educate, and influence them. This multi-channel approach combines digital and traditional methods to ensure comprehensive reach and impact. Here are some key tools and techniques used in HCP marketing:

1. **Digital Tools**

   A. Customer Relationship Management (CRM) Systems:
   - **Purpose**: Manage and analyze HCP interactions and data throughout the engagement lifecycle.
   - **Features**: Segmentation, personalized communication, tracking interactions, and generating insights.
   - **Examples**: Salesforce Health Cloud and Veeva CRM.

   B. Email Marketing Platforms
   - **Purpose**: Deliver targeted and personalized content to HCPs.
   - **Features**: Segmentation, automation, A/B testing, and detailed analytics.
   - **Examples**: Mailchimp, HubSpot, and Constant Contact.

   C. Webinars and Virtual Events:
   - **Purpose**: Provide educational content and interactive sessions with experts.
   - **Features**: Live streaming, Q&A sessions, on-demand viewing, and integration with CRM systems.
   - **Examples**: Zoom, WebEx, GoToWebinar.

   D. Social Media Platforms
   - **Purpose**: Engage with HCPs on professional and specialized networks.
   - **Features**: Content sharing, targeted advertising, and community building.
   - **Examples**: LinkedIn, Sermo, Doximity.

E. Content Management Systems (CMS)

- **Purpose**: Create, manage, and distribute educational and promotional content.
- **Features**: Content scheduling, SEO optimization, and analytics.
- **Examples**: WordPress, Drupal, Adobe Experience Manager

F. Mobile Apps:

- **Purpose**: Provide HCPs easy access to product information, clinical tools, and educational resources.
- **Features**: Interactive tools, push notifications and offline access.
- **Examples**: Epocrates, Medscape, UpToDate

G. Data Analytics and Business Intelligence (BI) Tools:

- **Purpose**: Analyze engagement data and generate actionable insights.
- **Features**: Data visualization, predictive analytics, and reporting.
- **Examples**: Tableau, PowerBI, SAS Analytics

2. **Traditional Tools**

A. Sales Representatives:

- **Purpose**: Build relationships, provide personalized information, and gather feedback.
- **Features**: In-person visits, detailed product presentations, and follow-up communications.
- **Examples**: Field sales teams with mobile CRM tools for real-time data access.

B. Print Materials:

- **Purpose**: Offer tangible resources that HCPs can reference and share.
- **Features**: Brochures, flyers, scientific reports, and product monographs.
- **Examples**: Customized print materials distributed during conferences and sales visits.

C. Medical Conferences and Symposia:

- **Purpose**: Network with HCPs, share the latest research, and showcase products.
- **Features**: Booths, presentations, workshops, and networking events.
- **Examples**: Annual meetings of medical associations and specialized symposia.

D. Advisory Boards and Focus Groups:

- **Purpose**: Gather insights from key opinion leaders (KOLs) and frontline HCPs.
- **Features**: Regular meetings, structured feedback sessions, and collaborative discussions.
- **Examples**: Advisory boards for product development and marketing strategy input.

3. **Integrated Techniques**

   A. Multi-Channel Campaigns:

- **Purpose**: Reach HCPs through a combination of digital and traditional channels.
- **Features**: Coordinated messaging across emails, social media, in-person visits, and events.
- **Examples**: A campaign that includes email newsletters, LinkedIn ads, and follow-up sales visits.

   B. Personalization and Segmentation:

- **Purpose**: Deliver tailored content based on HCPs' specialties, preferences, and behavior.
- **Features**: Dynamic content, personalized email subject lines, and targeted advertising.
- **Examples**: Segmenting HCPs by specialty and sending targeted educational content relevant to their field.

   C. Continuing Medical Education (CME) Programs:

- **Purpose**: Provide accredited educational opportunities to enhance HCP knowledge.
- **Features**: Online courses, live webinars, interactive case studies, and certification.

- **Examples**: CME programs hosted on platforms like Medscape or offered through partnerships with medical schools.

D. Virtual and Augmented Reality (VR/AR):

- **Purpose**: Create immersive educational experiences for HCPs.
- **Features**: Virtual lab tours, 3D models of disease mechanisms, and interactive training modules.
- **Examples**: VR simulations for surgical training and AR apps for understanding drug mechanisms.

E. Feedback and Surveys:

- **Purpose**: Gather direct input from HCPs to improve marketing efforts and product offerings.
- **Features**: Online surveys, post-event feedback forms, and in-person interviews.
- **Examples**: SurveyMonkey for online surveys and feedback forms distributed at conferences.

Tools and Techniques: Some Examples in Action

Example #1: CRM Systems in Action

A pharmaceutical company uses Salesforce Health Cloud to manage interactions with cardiologists. The system tracks each HCP's engagement with educational content, webinar attendance, and email campaign responses. This data helps the company tailor future communications and identify potential advocates for their new cardiovascular drug.

Example #2: Multi-Channel Campaign Success

To promote a new oncology drug, a company launches a campaign that includes targeted LinkedIn ads, personalized email newsletters, and in-person visits by oncology-trained sales representatives. The campaign results in increased webinar attendance, higher email engagement rates, and a significant rise in product inquiries from oncologists.

Example #3: VR for Educational Engagement

A pharmaceutical firm develops a VR app that allows HCPs to explore the mechanism of a new biological treatment for rheumatoid arthritis. The immersive experience helps HCPs understand how the drug interacts with immune cells, leading to better comprehension and higher prescribing confidence.

By effectively utilizing these tools and techniques, pharmaceutical companies can create robust and impactful marketing strategies that engage HCPs, foster trust, and ultimately improve patient outcomes.

27.4 Developing Patient Education Programs

Developing effective patient education programs is essential for empowering patients with the knowledge and tools they need to manage their health conditions. These programs aim to improve health literacy, enhance adherence to treatment plans, and ultimately lead to better health outcomes. Here's a detailed look at the steps involved in developing successful patient education programs:

1. **Assessing Patient Needs**

    A. Identifying Target Audience:

    - **Demographics**: Consider the age, gender, cultural background, and socioeconomic status of the target population.
    - **Health Literacy**: Assess the general health literacy levels to tailor educational materials appropriately.

    B. Understanding Patient Challenges:

    - **Common Barriers**: Identify barriers patients face in managing their conditions, such as lack of knowledge, misconceptions, and logistical issues.
    - **Feedback Collection**: Use surveys, focus groups, and interviews to gather insights directly from patients and healthcare providers.

2. **Defining Program Objectives**

    A. Specific Goals:

    - **Knowledge Improvement**: Enhance patient understanding of their condition and treatment options.
    - **Behavioral Changes**: Encourage healthier behaviors and better adherence to treatment plans.

    B. SMART Objectives:

    - **Specific, Measurable, Achievable, Relevant, and Time-bound:** Ensure each objective meets these criteria to facilitate evaluation and success.

3. **Designing Educational Content**

    A. Content Development:

    - **Accurate Information**: Ensure all content is evidence-based and reviewed by medical experts.

- **Patient-Friendly Language**: Use simple, non-technical language that is easy to understand.

- **Cultural Sensitivity**: Tailor content to be culturally relevant and respectful of diverse patient populations.

B. Formats and Channels

- **Printed Materials**: Brochures, leaflets, and booklets for distribution in clinics and hospitals.

- **Digital Tools**: Websites, mobile apps, and online portals for easy access to information.

- **Multimedia Resources**: Videos, infographics, and podcasts to cater to different learning preferences.

C. Interactive Elements

- **Quizzes and Surveys**: Engage patients with interactive quizzes to reinforce learning and gather feedback.

- **Tools and Calculators**: Provide tools like symptom checkers, medication reminders, and health trackers.

4. Implementing the Program

A. Training Healthcare Providers:

- **Education Techniques**: Train healthcare providers on how to use educational materials effectively and communicate with patients.

- **Support Resources**: Provide additional resources and guidelines for healthcare providers to support patient education.

B. Multi-Channel Distribution:

- **Clinics and Hospitals**: Distribute printed materials and display posters in waiting areas.

- **Digital Platforms**: Launch and promote websites and apps through social media and email campaigns.

- **Community Outreach**: Collaborate with community organizations to reach a broader audience.

5. Engaging Patients

A. Interactive Sessions:

- **Workshops and Seminars**: Organize workshops and seminars to educate patients more engagingly and interactively.
- **Webinars**: Conduct online webinars to reach patients who prefer digital learning.

B. Support Networks:

- **Peer Support Groups**: Facilitate the creation of peer support groups where patients can share experiences and support each other.
- **Patient Ambassadors**: Engage patient ambassadors who can share their success stories and motivate others.

6. Evaluating Effectiveness

A. Feedback Mechanisms:

- **Surveys and Questionnaires**: Collect feedback from patients and healthcare providers to assess the program's impact.
- **Focus Groups**: Conduct focus groups to gather in-depth insights into the program's strengths and areas for improvement.

B. Outcome Metrics:

- **Engagement Metrics**: Track website visits, app usage, and attendance at educational sessions.
- **Behavioral Metrics**: Monitor changes in patient behaviors, such as medicaiton adherence and lifestyle modifications.
- **Health Outcomes**: Evaluate improvements in health outcomes, such as symptom management and reduced hospitalizations.

7. Continuous Improvement

A. Iterative Revisions:

- **Content Revisions**: Regularly update educational materials to reflect the latest medical research and patient feedback.

- **Program Adjustments**: Make necessary adjustments to the program based on evaluation results and emerging patient needs.

B. Ongoing Support:

- **Follow-Up**: Provide ongoing support to patients through follow-up sessions and additional resources.
- **Community Building**: Foster a sense of community among patients to ensure sustained engagement and support.

By following these steps, pharmaceutical companies and healthcare providers can develop comprehensive patient education programs that empower patients, improve health literacy, and enhance overall health outcomes.

27.5 Leveraging Digital Tools for Patient Engagement

Digital tools have transformed how pharmaceutical companies and healthcare providers engage with patients. By utilizing these technologies, they can provide personalized, accessible, and interactive education and support, leading to better patient outcomes. Here's a detailed look at how digital tools can be leveraged for patient engagement:

1. Patient Portals

A. Centralized Access:

- **Medical Records**: Patients can access their medical records, lab results, and medicaiton history.
- **Appointment Scheduling**: Online booking systems for convenient appointment scheduling.

B. Communication:

- **Secure Messaging**: Direct communication with healthcare providers for queries and follow-ups.
- **Notifications**: Automated reminders for appointments, medicaiton refills, and health check-ups.

2. Mobile Apps

A. Medication Management:

- **Reminders**: Alerts for medication schedules, dosage, and refills.
- **Tracking**: Tools to track medication adherence and symptoms.

B. Health Monitoring:

- **Fitness and Diet Tracking**: Integration with wearable devices to monitor physical activity, diet, and sleep.
- **Chronic Condition Management**: Apps tailored to specific conditions (e.g., diabetes, hypertension) to track relevant health metrics.

C. Educational Content:

- **Videos and Tutorials**: Step-by-step guides on medication usage, lifestyle tips, and disease management.
- **Interactive Tools**: Quizzes, calculators, and symptom checkers.

3. Telehealth Services

A. Virtual Consultations:

- **Convenience**: Access to healthcare providers from the comfort of their homes.
- **Accessibility**: Reaching patients in remote or underserved areas.

B. Follow-Up Care:

- **Continuous Monitoring**: Regular virtual check-ins to monitor patient progress.
- **Patient Support**: Real-time support and adjustments to treatment plans.

4. Social Media and Online Communities

A. Information Sharing:

- **Educational Campaigns**: Social media platforms for awareness campaigns, health tips, and updates.
- **Live Sessions**: Q&A sessions, webinars, and live chats with healthcare professionals.

B. Peer Support:

- **Community Groups**: Online forums, and social media groups for patients with similar conditions to share experiences and support each other.
- **Patient Stories**: Sharing success stories and testimonials to inspire each other.

5. Wearable Devices

A. Real-Time Data Collection:

- **Health Metrics**: Continuous monitoring of vital signs, activity levels, and sleep patterns.
- **Alerts**: Notifications for abnormal readings or health events.

B. Integration with Health Apps:

- **Data Synchronization**: Seamless integration with mobile health apps for comprehensive health tracking.
- **Personalized Insights**: Data-driven insights and recommendations for lifestyle adjustments.

6. **Virtual Reality (VR) and Augmented Reality (AR)**

A. Immersive Education:

- **Simulations**: VR simulations to educate patients about surgical procedures, treatment plans, and disease progression.
- **Interactive Learning**: AR-enhanced educational materials for a more engaging learning experience.

B. Therapy and Rehabilitation:

- **VR Therapy**: Virtual environments for pain management, mental health therapy, and physical rehabilitation.
- **AR Guidance**: Real-time AR guidance for exercises and rehabilitation protocols.

7. **Chatbots and AI Assistants**

A. Immediate Support:

- **24/7 Assistance**: AI-powered chatbots provide round-the-clock support for common queries and concerns.
- **Symptom Checkers**: Interactive tools to assess and suggest possible actions.

B. Personalized Care:

- **Tailored Recommendations**: AI Algorithms analyze patient data to provide personalized health tips and medication reminders.
- **Behavioral Insights**: Understanding patient behavior and preferences to offer customized engagement strategies.

8. **E-Learning Platforms**

A. Structured Courses:

- **Health Literacy**: Online courses covering various health topics to improve patient knowledge.

- **Condition-Specific Education**: Detailed courses on managing specific chronic conditions.

B. Interactive Learning:

- **Quizzes and Assessments**: Tools to reinforce learning and assess patient understanding.

- **Certificates**: Completion certificates to motivate and recognize patient effort.

Implementing Digital Tools for Patient Engagement

1. **Needs Assessment:**

   - **Patient Surveys**: Conduct surveys to understand patient preferences and needs.

   - **Data Analysis**: Analyze patient data to identify gaps in current engagement strategies.

2. **Content Development:**

   - **User-Friendly Design**: Ensure digital tools are easy to navigate and use.

   - **Multilingual Support**: Provide content in multiple languages to cater to diverse patient populations.

   - **Regular Updates**: Keep content current and relevant with regular updates.

3. **Integration and Interoperability:**

   - **Seamless Integration**: Ensure digital tools can integrate with existing healthcare systems and electronic health records (EHRs).

   - **Data Security**: Implement robust security measures to protect patient data.

4. **Training and Support:**

   - **Provider Training**: Train healthcare providers on incorporating digital tools into patient care.

   - **Patient Education**: Offer tutorials and support for patients to maximize their use of digital tools.

5. **Evaluation and Feedback:**

   - **Usage Metrics**: Monitor engagement metrics such as app downloads, active users, and session durations.

- **Patient Feedback**: Regularly collect feedback from patients to improve digital tools and address issues.

By leveraging these digital tools, pharmaceutical companies and healthcare providers can significantly enhance patient engagement, improving health literacy, adherence to treatment plans, and overall better health outcomes.

Data-Driven Decision-Making

Data-driven decision-making involves using data analysis and insights to guide business decisions rather than relying solely on intuition or anecdotal evidence. In pharmaceutical product management, this approach helps optimize product development, marketing strategies, and overall business outcomes by leveraging quantitative and qualitative data.

Key Components:

1. **Data Collection:**

   - **Clinical Data:** Includes results from clinical trials, patient outcomes, and safety reports.

   - **Market Data:** Encompasses information about market trends, competitor performance, and customer preferences.

   - **Sales Data:** Involves sales figures, revenue metrics, and distribution data.

   - **Customer Data:** Includes feedback from healthcare professionals (HCPs), patients, and stakeholders.

2. **Data Analysis:**

   - **Descriptive Analytics:** Summarizes past data to understand what has happened. For example, analyzing sales trends to identify growth patterns.

   - **Predictive Analytics:** Uses historical data to forecast future outcomes. For instance, predicting market demand based on previous product launches.

- **Prescriptive Analytics:** Provides recommendations for actions based on data insights. For example, suggesting marketing strategies to improve product adoption.
- **Diagnostic Analytics:** Examines data to determine the reasons behind past outcomes. For instance, it analyzes the factors behind a decline in product sales.

3. **Data Integration:**
   - **Data Warehousing:** Consolidates data from various sources into a central repository for easier access and analysis.
   - **Data Mining:** Extracts useful patterns and correlations from large datasets.
   - **Business Intelligence Tools:** Utilizes software and tools to visualize data and generate reports (e.g., Tableau, Power BI).

4. **Decision-Making Process:**
   - **Identify Objectives:** Define the goals and questions to address (e.g., increasing market share, optimizing marketing spend).
   - **Gather Data:** Collect relevant data from various sources.
   - **Analyze Data:** Use statistical methods and analytical tools to interpret the data.
   - **Generate Insights:** Derive actionable insights from the analysis.
   - **Make Decisions:** Base decisions on the insights obtained rather than relying solely on intuition.
   - **Monitor and Evaluate:** Continuously track the outcomes of decisions and adjust strategies as needed based on new data.

Applications in Pharma Product Management:

1. **Product Development:**
   - **Clinical Trials:** Analyze trial data to determine the efficacy and safety of a new drug and adjust trial designs or formulations as needed.

- **Market Needs:** Use market research data to identify unmet needs and guide the development of new products.

2. **Marketing Strategies:**

- **Targeting:** Target specific segments more effectively using customer and market data.

- **Campaign Optimization:** Analyze performance metrics of marketing campaigns to optimize messaging and channels.

3. **Sales Performance:**

- **Sales Forecasting:** Predict future sales based on historical data and market trends.

- **Market Segmentation:** Identify and target high-value customer segments based on purchasing patterns and preferences.

4. **Operational Efficiency:**

- **Supply Chain Management:** Analyze data to optimize inventory levels, reduce costs, and improve distribution efficiency.

- **Cost Management:** Use financial data to identify cost-saving opportunities and optimize budget allocations.

5. **Regulatory Compliance:**

- **Adverse Event Reporting:** Monitor and analyze adverse event reports to ensure compliance and address safety concerns promptly.

Benefits:

- **Improved Accuracy:** Data-driven decisions are based on objective evidence, reducing the risk of errors.

- **Enhanced Efficiency:** Streamlines processes and optimizes resource allocation.

- **Better Outcomes:** Increases the likelihood of achieving desired results by aligning strategies with data insights.

- **Competitive Advantage:** Provides a deeper understanding of market dynamics and customer needs, leading to more effective strategies.

Challenges:

- **Data Quality:** Ensuring data accuracy and completeness can be challenging.

- **Complexity:** Analyzing large datasets requires specialized skills and tools.

- **Integration:** Integrating data from diverse sources can be complex.

28.1 The Importance of Data in Pharma Marketing

Data has become a foundation of effective marketing strategies in today's pharmaceutical industry. With the rise of digital technologies, pharma companies can access vast amounts of data generated from clinical trials, patient interactions, healthcare providers, and real-world outcomes. Collecting, analyzing, and leveraging this data enables pharma marketers to make informed decisions, personalize their outreach, and optimize campaigns for better results. In an environment where precision and evidence-based strategies are critical, understanding the importance of data in pharmaceutical marketing can unlock new levels of success, driving both patient outcomes and business growth.

1. **Understanding Customer Needs:**

   - **Patient Insights:** Data helps understand patient needs, preferences, and behaviors. By analyzing patient demographics, health conditions, and treatment preferences, pharma marketers can tailor their strategies to address specific needs.

   - **Healthcare Professional (HCP) Preferences:** Data on HCP practices, prescribing habits, and interactions with pharmaceutical companies can guide the development of targeted marketing strategies and educational materials.

2. **Optimizing Marketing Strategies:**

   - **Segmentation:** Data enables precise segmentation of target audiences based on various criteria, such as geographic location, disease prevalence, and patient demographics. This helps in crafting targeted marketing messages and campaigns.

   - **Personalization:** By leveraging data, pharma companies can create personalized marketing campaigns that resonate with individual HCPs and patients, enhancing engagement and effectiveness.

3. Enhancing Campaign Effectiveness:

- **Performance Tracking:** Data allows for real-time monitoring and analysis of marketing campaign performance. Key performance indicators (KPIs) such as engagement, conversion, and ROI can be assessed to determine campaigns' effectiveness.

- **A/B Testing:** Marketers can use data to conduct A/B testing, comparing different versions of marketing materials or strategies to identify the most effective approach.

4. Improving Decision-Making:

- **Informed Decisions:** Data-driven insights help pharma marketers make informed decisions about marketing strategies, resource allocation, and budget management. This reduces reliance on intuition and enhances strategic planning.

- **Predictive Analytics:** Advanced analytics can forecast future trends, patient behaviors, and market dynamics, allowing for proactive adjustments to marketing strategies.

5. Enhancing Competitive Advantage:

- **Market Analysis:** Data provides insights into market trends, competitive activities, and emerging opportunities. This helps pharma companies stay ahead of competitors and capitalize on new market opportunities.

- **Benchmarking:** Comparing performance metrics with industry benchmarks allows pharma marketers to identify areas for improvement and adopt best practices.

6. Ensuring Regulatory Compliance:

- **Adherence to Regulations:** Data helps ensure marketing practices comply with regulatory requirements by tracking and documenting interactions with HCPs, patient outreach efforts, and promotional activities.

- **Risk Management:** Analyzing data helps in identifying potential compliance risks and implementing corrective measures to avoid regulatory issues.

7. **Optimizing Resource Allocation:**

   - **Budget Efficiency:** Data-driven insights allow for more efficient allocation of marketing budgets by identifying high-impact channels and strategies. This ensures that resources are invested in areas that deliver the best results.

   - **Resource Planning:** Data helps forecast resource needs and plan marketing activities effectively, reducing wastage and improving overall efficiency.

8. **Enhancing Customer Engagement:**

   - **Tailored Content:** Pharma companies can create relevant and engaging content for their target audiences by analyzing customer interactions and preferences data.

   - **Multi-Channel Strategies:** Data enables the development of integrated multi-channel marketing strategies that reach customers through their preferred channels, enhancing engagement and interaction.

9. **Measuring and Demonstrating ROI:**

   - **Effectiveness Measurement:** Data provides a clear picture of the ROI of marketing efforts by linking marketing activities to sales outcomes and other key business metrics.

   - **Accountability:** Demonstrating the impact of marketing activities through data helps secure buy-in from stakeholders and justify marketing expenditures.

10. **Continuous Improvement:**

    - **Feedback Loop:** Data creates a feedback loop that allows for continuous improvement of marketing strategies. Pharma marketers can refine and enhance their approaches by analyzing performance data.

28.2 Tools and Techniques for Data Analysis

In today's data-driven pharmaceutical marketing, effectively analyzing and interpreting vast amounts of data is crucial for making informed marketing decisions. The tools and techniques for data analysis provide pharma product managers and marketers with the ability to uncover valuable insights from diverse data sources, including clinical studies, patient behavior, and market trends. By leveraging advanced statistical methods, data visualization tools, and machine learning algorithms, pharma professionals can optimize their marketing strategies, improve patient outcomes, and stay competitive in an ever-evolving healthcare environment. This section explores the essential tools and techniques that empower data-driven decision-making in pharmaceutical marketing.

1. **Data Analysis Tools**

    1.1. **Excel:**

    - **Functionality:** Widely used for data organization, basic statistical analysis, and creating charts and graphs.

    - **Use Cases:** Analyzing small to medium-sized datasets, generating summary statistics, and visualizing data trends.

    1.2. **SQL (Structured Query Language):**

    - **Functionality:** Used to manage and query relational databases.

    - **Use Cases:** Extracting and manipulating large datasets from databases, performing complex queries, and generating reports.

    1.3. **R (Programming Language):**

    - **Functionality:** A programming language and software environment for statistical computing and graphics.

    - **Use Cases:** Advanced statistical analysis, data visualization, and modeling. Useful for handling large datasets and performing complex analyses.

    1.4. **Python:**

    - **Functionality:** A programming language with extensive libraries for data analysis (e.g., pandas, NumPy, scikit-learn).

- **Use Cases:** Data manipulation, statistical analysis, machine learning, and visualization.

1.5. Tableau:

- **Functionality:** A data visualization tool that helps create interactive and shareable dashboards.

- **Use Cases:** Visualizing complex datasets, creating interactive reports, and facilitating data-driven decision-making.

1.6. Power BI:

- **Functionality:** A business analytics tool from Microsoft for creating interactive reports and dashboards.

- **Use Cases:** Integrating data from various sources, generating visualizations, and performing real-time data analysis.

1.7. SAS (Statistical Analysis System):

- **Functionality:** A software suite for advanced analytics, business intelligence, and data management.

- **Use Cases:** Complex statistical analysis, predictive modeling, and data management.

1.8. SPSS (Statistical Package for the Social Sciences):

- **Functionality:** Software for statistical analysis in social science.

- **Use Cases:** Conducting statistical tests, surveys, and trend analyses.

1.9. Google Analytics:

- **Functionality:** A web analytics service for tracking and reporting website traffic.

- **Use Cases:** Monitoring website performance, analyzing user behavior, and optimizing digital marketing efforts.

2. Data Analysis Techniques

2.1 Descriptive Statistics:

- **Functionality:** Summarizes and describes the main features of a dataset.

- **Techniques:** Mean, median, mode, standard deviation, and frequency distributions.
- **Use Cases:** Understanding basic trends and characteristics of data.

2.2. Inferential Statistics:

- **Functionality:** Makes inferences and predictions about a population based on a sample.
- **Techniques:** Hypothesis testing, confidence intervals, regression analysis.
- **Use Cases:** Concluding patient behaviors, treatment effectiveness, and market trends.

2.3. Predictive Analytics:

- **Functionality:** Uses historical data and statistical algorithms to forecast future outcomes.
- **Techniques:** Regression analysis, time series analysis, machine learning models (e.g., decision trees, neural networks).
- **Use Cases:** Forecasting drug sales, predicting patient outcomes, and identifying market opportunities.

2.4. Sentiment Analysis:

- **Functionality:** Analyzes text data to determine the sentiment or emotion expressed.
- **Techniques:** Natural language processing (NLP), text mining.
- **Use Cases:** Monitoring patient and HCP feedback and understanding public perception of pharmaceutical products.

2.5. Segmentation Analysis:

- **Functionality:** Divides a market into distinct groups based on shared characteristics.
- **Techniques:** Cluster analysis, factor analysis.
- **Use Cases:** Identifying target segments for marketing campaigns and personalizing outreach efforts.

2.6. Cohort Analysis:

- **Functionality:** Analyzes data from subjects who share a common characteristic over a specific period.
- **Techniques:** Tracking performance over time and analyzing retention rates.
- **Use Cases:** Evaluating the long-term effectiveness of treatments and understanding patient journeys.

2.7. A/B Testing:

- **Functionality:** Compares two variable versions to determine which performs better.
- **Techniques:** Randomized controlled trials, statistical analysis of test results.
- **Use Cases:** Testing different marketing messages and optimizing digital content.

2.8. Data Visualization:

- **Functionality:** Presents data in graphical formats to make interpretation easier.
- **Techniques:** Charts, graphs, heatmaps, dashboards.
- **Use Cases:** Communicating data insights and tracking performance metrics.

2.9. Machine Learning and AI:

- **Functionality:** Uses algorithms to analyze data and make predictions or decisions.
- **Techniques:** Supervised learning, unsupervised learning, reinforcement learning.
- **Use Cases:** Predicting patient responses, automating data analysis, enhancing personalized marketing.

3. Applications in Pharma Marketing

3.1. Patient Segmentation:

- Use data analysis to segment patients based on demographics, behaviors, and treatment needs for targeted marketing.

3.2. HCP Targeting:

- Analyze HCP data to identify key opinion leaders and tailor marketing strategies to their preferences and practices.

3.3. Campaign Optimization:

- Leverage A/B testing and predictive analytics to optimize marketing campaigns and improve ROI.

3.4. Market Forecasting:

- Use predictive models to forecast market trends and adjust strategies accordingly.

3.5. Competitive Analysis:

- Analyze competitor data to understand market positioning and identify opportunities for differentiation.

Pharma product managers can effectively utilize these tools and techniques to gain deeper insights, make informed decisions, and enhance their marketing strategies.

28.3 Leveraging Data for Strategic Decision-Making

In the pharmaceutical industry, leveraging data effectively is crucial for making informed, strategic decisions that can enhance marketing effectiveness, improve patient outcomes, and drive business growth. Here's how pharma product managers can harness data for strategic decision-making:

1. Data Collection and Integration

1.1. Collecting Relevant Data:

- **Types of Data:** Gather data from various sources, including clinical trials, market research, patient feedback, sales data, and digital interactions.

- **Methods:** Use surveys, focus groups, electronic health records (EHRs), and digital analytics tools to collect comprehensive and accurate data.

1.2. Integrating Data Sources:

- **Data Warehousing:** Centralize data from disparate sources into a single repository to facilitate comprehensive analysis.

- **Data Integration Tools:** Employ tools and platforms that enable seamless integration and data synchronization across systems.

2. Data Analysis and Interpretation

2.1. Descriptive Analytics:

- **Purpose:** Summarize historical data to understand past performance and identify trends.

- **Techniques:** Visualizing and interpreting historical data using statistical summaries, charts, and dashboards.

2.2. Predictive Analytics:

- **Purpose:** Forecast future trends and outcomes based on historical data.

- **Techniques:** Apply machine learning algorithms, regression models, and time-series analysis to predict market trends, patient behaviors, and sales performance.

2.3. Prescriptive Analytics:

- **Purpose:** Provide actionable recommendations based on data analysis.
- **Techniques:** Use optimization models and simulation tools to recommend strategies for product launches, marketing campaigns, and resource allocation.

2.4. Sentiment and Text Analysis:

- **Purpose:** Understand public perception and patient sentiment regarding products and treatments.
- **Techniques:** Utilize natural language processing (NLP) and sentiment analysis tools to analyze feedback from social media, forums, and patient reviews.

3. Data-Driven Strategic Decision-Making

3.1. Market Segmentation:

- **Approach:** Analyze patient and HCP data to identify distinct market segments based on needs, behaviors, and demographics.
- **Application:** Develop targeted marketing strategies and personalized outreach efforts for different segments.

3.2. Competitive Analysis:

- **Approach:** Gather and analyze data on competitors' products, pricing, marketing strategies, and market share.
- **Application:** Identify competitive advantages, market gaps, and opportunities for differentiation.

3.3. Product Development and Lifecycle Management:

- **Approach:** Use clinical and market data to guide product development, optimization, and lifecycle management.
- **Application:** Make informed decisions about product enhancements, line extensions, and phase-out strategies.

3.4. Campaign Effectiveness:

- **Approach:** Measure and analyze the performance of marketing campaigns using metrics such as engagement, conversion rates, and ROI.

- **Application:** Adjust strategies and tactics based on data-driven insights to optimize campaign effectiveness.

3.5. Resource Allocation:

- **Approach:** Analyze data on sales performance, market potential, and resource utilization.

- **Application:** Allocate resources efficiently to maximize impact and achieve strategic goals.

3.6. Risk Management:

- **Approach:** Use data to identify potential risks and vulnerabilities in marketing strategies and product performance.

- **Application:** Develop risk mitigation strategies and contingency plans based on data-driven insights.

4. Tools and Technologies

4.1. Business Intelligence (BI) Platforms:

- **Tools:** Tableau, Power BI, QlikView

- **Functionality:** Provide advanced data visualization, reporting, and analytics capabilities.

4.2. Advanced Analytics and Machine Learning:

- **Tools:** Python, R, SAS, IBM Watson

- **Functionality:** Support complex data analysis, predictive modeling, and AI-driven insights.

4.3. CRM Systems:

- **Tools:** Salesforce, HubSpot

- **Functionality:** Manage customer interactions, track sales activities, and analyze customer data.

4.4. Data Integration Platforms:

- **Tools:** Talend, Informatica, Microsoft Azure Data Factory

- **Functionality:** Facilitate data integration, transformation, and synchronization across systems.

5. Examples

5.1. Personalized Marketing Campaigns:

- **Example:** A pharmaceutical company uses predictive analytics to segment patients based on their likelihood to respond to specific marketing messages, leading to increased engagement and conversion rates.

5.2. Product Development Decisions:

- **Example:** Data from clinical trials and market research helps a company decide on the optimal formulation of a new drug, ensuring it meets the needs of the target patient population.

5.3. Competitive Positioning:

- **Example:** Competitor data analysis reveals a market gap that the company capitalizes on by launching a differentiated product, gaining a competitive edge.

Leveraging data for strategic decision-making enables pharmaceutical product managers to make informed, evidence-based decisions that drive product success and business growth. By integrating data analysis into their decision-making processes, they can better understand market dynamics, optimize strategies, and respond proactively to industry changes.

AI and Emerging Technologies in Pharma

Artificial Intelligence (AI) and emerging technologies are transforming the pharmaceutical industry by enhancing research and development, optimizing operations, and improving patient outcomes. Here's an overview of how these technologies are being utilized and their implications for the pharmaceutical sector:

1. Artificial Intelligence (AI)

1.1. Drug Discovery and Development:

- **AI for Target Identification:** AI algorithms analyze biological data to identify potential drug targets and biomarkers. Machine learning models can predict how different compounds interact with these targets, speeding up the discovery of new drugs.

- **Predictive Modeling:** AI predicts the efficacy and safety of drug candidates by analyzing historical data and modeling the interactions between drugs and biological systems.

1.2. Clinical Trials:

- **Patient Recruitment:** AI-driven tools streamline patient recruitment by analyzing electronic health records (EHRs) to identify suitable candidates for clinical trials based on specific criteria.

- **Trial Design:** AI assists in designing clinical trials by simulating various scenarios and optimizing protocols to enhance trial efficiency and accuracy.

1.3. Drug Repurposing:

- **AI for Repurposing:** AI algorithms identify existing drugs that could be repurposed for new indications by analyzing vast drug interactions and patient outcomes datasets.

1.4. Personalized Medicine:

- **Genomic Data Analysis:** AI analyzes genomic data to tailor treatments to individual patients, improving the efficacy of therapies and reducing adverse effects.

1.5. Pharmacovigilance:

- **Adverse Event Detection:** AI systems monitor and analyze patient reports to detect adverse drug reactions (ADRs) and predict potential safety issues, enhancing drug safety monitoring.

2. Machine Learning and Data Analytics

2.1. Predictive Analytics:

- **Market Forecasting:** Machine learning models predict market trends, patient needs, and competitive dynamics, helping pharma companies make informed business decisions.

- **Sales and Marketing Optimization:** AI analyzes sales data and customer interactions to optimize marketing strategies and tactics.

2.2. Real-World Evidence (RWE):

- **Data Integration:** AI integrates data from various sources, such as EHRs, insurance claims, and patient registries, to provide insights into drug effectiveness and patient outcomes in real-world settings.

2.3. Natural Language Processing (NLP):

- **Text Mining:** NLP techniques extract valuable insights from unstructured data sources, such as medical literature, clinical notes, and patient feedback.

3. Emerging Technologies

3.1. Blockchain:

- **Data Security:** Blockchain technology provides secure and transparent tracking of drug supply chains, ensuring the authenticity and safety of pharmaceutical products.

- **Clinical Trials:** Blockchain can enhance the integrity and transparency of clinical trial data by creating immutable records of trial protocols and results.

3.2. Internet of Things (IoT):

- **Remote Monitoring:** IoT devices monitor patient health metrics in real-time, providing data for personalized treatment adjustments and improving patient adherence to therapies.

- **Smart Drug Delivery Systems:** IoT-enabled drug delivery systems track and manage medication administration, ensuring accurate dosing and adherence.

3.3. Virtual Reality (VR) and Augmented Reality (AR):

- **Medical Training:** VR and AR are used for immersive medical training and simulations, enhancing the education and skills of healthcare professionals.

- **Patient Engagement:** AR applications improve patient education by visualizing complex medical concepts and treatment plans in an interactive format.

3.4. Digital Therapeutics (DTx):

- **Behavioral Interventions:** DTx platforms deliver evidence-based therapeutic interventions via digital platforms, helping manage chronic conditions and improve patient outcomes.

- **Personalized Programs:** AI-driven DTx solutions provide personalized treatment programs based on patient data and behavioral insights.

4. Implications for Pharma Product Managers

4.1. Strategic Planning:

- **Data-Driven Insights:** AI and emerging technologies provide valuable insights for strategic planning, enabling product managers to make informed decisions based on comprehensive data analysis.

4.2. Innovation and Competitive Advantage:

- **Adoption of New Technologies:** Embracing AI and emerging technologies positions pharma companies as innovators in the industry, offering competitive advantages through enhanced R&D and operational efficiency.

4.3. Patient-Centric Approaches:

- **Personalized Treatments:** AI-driven insights and digital therapeutics enable product managers to develop patient-centric solutions that address individual needs and improve treatment outcomes.

4.4. Operational Efficiency:

- **Automation and Optimization:** AI and emerging technologies automate routine tasks, optimize processes, and enhance operational efficiency, reducing costs and accelerating time-to-market.

4.5. Regulatory Considerations:

- **Compliance and Ethics:** Product managers must navigate regulatory considerations and ethical implications associated with AI and emerging technologies, ensuring compliance with industry standards and guidelines.

By leveraging AI and emerging technologies, pharmaceutical product managers can drive innovation, enhance operational efficiency, and deliver more effective and personalized solutions to meet the evolving needs of patients and the healthcare industry.

29.1 The Role of AI in Drug Development and Marketing

Artificial Intelligence (AI) is revolutionizing drug development and pharmaceutical marketing by enhancing efficiency, accuracy, and innovation across various stages. Here's an in-depth look at how AI is transforming these critical areas:

AI in Drug Development

1. Drug Discovery:

- **Target Identification:** AI algorithms analyze vast datasets, including genomic, proteomic, and metabolomic data, to identify potential biological targets for drug development. Machine learning models can predict which proteins or genes are associated with specific diseases, helping researchers focus on the most promising targets.

- **Compound Screening:** AI models use predictive analytics to evaluate chemical compounds' potential efficacy and safety. These models can simulate how compounds interact with biological targets, significantly speeding up the initial screening process.

2. Preclinical Research:

- **Predictive Modeling:** AI enhances predictive modeling by analyzing historical data to forecast how new drugs might behave in biological systems. This includes predicting potential side effects and efficacy and reducing the need for extensive animal testing.

- **Bioinformatics:** AI-driven tools analyze complex biological data to understand disease mechanisms and identify biomarkers that new drugs could target.

3. Clinical Trials:

- **Patient Recruitment:** AI helps identify suitable candidates for clinical trials by analyzing patient records, genetic information, and previous trial data. This targeted approach improves the likelihood of finding appropriate participants, accelerating recruitment.

- **Trial Design:** AI assists in designing clinical trials by optimizing protocols based on simulations and historical data. This includes determining the most effective dosage, treatment duration, and endpoints to achieve meaningful results.

- **Monitoring and Compliance:** AI tools monitor patient adherence and detect deviations from trial protocols in real-time, ensuring compliance and enhancing data integrity.

4. **Drug Repurposing:**

- **Existing Drug Analysis:** AI can analyze data on existing drugs to identify new therapeutic uses. By evaluating how well-known drugs interact with different biological targets or diseases, AI helps discover new indications for established medications.

5. **Personalized Medicine:**

- **Genomic Data Analysis:** AI analyzes genetic data to develop personalized treatment plans based on individual genetic profiles. This approach allows for more precise targeting of therapies, improving efficacy and reducing adverse effects.

AI in Pharmaceutical Marketing

1. **Customer Insights and Segmentation:**

- **Data Analysis:** AI analyzes customer data, including purchase history, behavior, and preferences, to identify trends and segments within the market. This enables more targeted marketing strategies and personalized messaging.

- **Predictive Analytics:** AI models predict customer needs and preferences, helping pharmaceutical companies anticipate market trends and tailor their marketing efforts accordingly.

2. **Content Creation and Optimization:**

- **Automated Content Generation:** AI-powered tools can create personalized marketing content, including emails, social media posts, and advertisements. These tools use natural language processing (NLP) to generate content that resonates with specific target audiences.

- **Content Optimization:** AI analyzes marketing content performance to identify the most effective messages and formats. This data-driven approach helps optimize content strategies for better engagement and conversion rates.

3. Campaign Management:

- **Targeted Advertising:** AI enables more precise targeting of online ads by analyzing user behavior and demographics. This ensures that marketing messages reach the right audience, increasing the likelihood of engagement and conversion.

- **Performance Tracking:** AI tools monitor and analyze the performance of marketing campaigns in real-time. This includes tracking key metrics such as click-through rates, conversion rates, and return on investment (ROI), allowing for timely adjustments and optimization.

4. Sentiment Analysis:

- **Online Monitoring:** AI-driven sentiment analysis tools monitor social media and online forums to gauge public opinion about pharmaceutical products and brands. This helps companies understand patient perceptions, identify potential issues, and adjust marketing strategies accordingly.

5. Customer Relationship Management (CRM):

- **Personalized Communication:** AI enhances CRM systems by providing insights into customer behavior and preferences. This enables pharmaceutical companies to deliver personalized communication and build stronger relationships with healthcare professionals (HCPs) and patients.

6. Market Access and Pricing:

- **Pricing Optimization:** AI analyzes market data, including competitor pricing and reimbursement information, to optimize drug pricing strategies. This ensures competitive pricing while maximizing revenue and market access.

Implications for Pharmaceutical Product Managers

1. **Strategic Planning:**
   - **Data-Driven Decisions:** AI provides actionable insights that inform strategic planning and decision-making, helping product managers make informed choices about drug development and marketing strategies.

2. **Innovation:**
   - **Enhanced R&D:** AI accelerates the drug development process, enabling product managers to bring innovative therapies to market more quickly and efficiently.

3. **Personalization:**
   - **Targeted Marketing:** AI-driven insights allow product managers to develop personalized marketing strategies that resonate with the specific patient and HCP segments.

4. **Efficiency:**
   - **Streamlined Processes:** AI automates routine tasks, optimizes processes, and improves operational efficiency, reducing costs and enhancing productivity.

5. **Compliance and Risk Management:**
   - **Regulatory Compliance:** AI tools assist in ensuring compliance with regulatory requirements and managing risks by providing real-time monitoring and analysis.

By leveraging AI in drug development and marketing, pharmaceutical product managers can drive innovation, enhance efficiency, and deliver more effective and personalized solutions, ultimately improving patient outcomes and achieving business success.

29.2   Emerging Technologies Shaping the Pharmaceutical Industry

Emerging technologies are rapidly transforming the pharmaceutical industry, driving innovation and improving various aspects of drug development, manufacturing, marketing, and patient care. Here's an overview of some key emerging technologies:

1. **Artificial Intelligence (AI) and Machine Learning (ML)**

   Applications:

   - **Drug Discovery:** AI algorithms analyze large datasets to identify potential drug candidates and predict their interactions with biological targets.

   - **Predictive Analytics:** ML models forecast patient responses to treatments, optimize clinical trial designs, and identify new drug indications.

   - **Personalized Medicine:** AI helps tailor treatments to individual patients based on their genetic and health data, improving efficacy and reducing adverse effects.

   Impact:

   - Accelerates drug discovery and development processes.

   - Enhances the precision of personalized medicine.

   - Improves patient outcomes through targeted therapies.

2. **Genomics and Precision Medicine**

   Applications:

   - **Genomic Sequencing:** Advances in genomic sequencing technologies provide detailed insights into genetic variations and their impact on disease and drug response.

   - **Biomarker Discovery:** Identifying biomarkers that predict disease risk or treatment response will enable more targeted and effective therapies.

   Impact:

   - Enables the development of personalized treatments based on genetic profiles.

   - Improves understanding of complex diseases and their underlying mechanisms.

3. Blockchain Technology

Applications:

- **Data Security:** Blockchain ensures secure and transparent clinical trial data and patient records management.
- **Supply Chain Management:** Provides an immutable ledger for tracking the provenance of pharmaceuticals, reducing counterfeiting, and ensuring drug integrity.

Impact:

- Enhances data security and transparency.
- Improves traceability and reduces fraud in the supply chain.

4. Digital Therapeutics (DTx)

Applications:

- **Software-Based Treatments:** Develops digital health interventions delivered through mobile apps and digital platforms to manage chronic conditions and improve patient adherence.
- **Behavioral Change:** Uses digital tools to promote lifestyle changes and support mental health.

Impact:

- Provides scalable and accessible treatment options.
- Enhances patient engagement and adherence to therapy.

5. Telemedicine and Remote Patient Monitoring (RPM)

Applications:

- **Telehealth Platforms:** Facilitates remote consultations between patients and healthcare professionals, improving access to care.
- **RPM Devices:** Monitors patient health metrics in real-time using wearable devices and connected technologies.

Impact:

- Expands access to healthcare services.
- Enables continuous monitoring and timely interventions.

6. Augmented Reality (AR) and Virtual Reality (VR)

Applications:

- **Training and Education:** Uses AR and VR to simulate drug development processes and train healthcare professionals.
- **Patient Engagement:** Provides immersive experiences for patient education and therapy adherence.

Impact:

- Enhances training and educational experiences.
- Improves patient understanding and engagement with treatment protocols.

7. 3D Printing

Applications:

- **Custom Drug Delivery Systems:** Creates personalized drug delivery devices and formulations.
- **Bio-printing:** Develops tissue and organ models for research and regenerative medicine.

Impact:

- Allows for personalized and on-demand production of pharmaceuticals.
- Advances research in tissue engineering and regenerative medicine.

8. Internet of Things (IoT)

Applications:

- **Smart Devices:** Integrates IoT technology into medical devices for real-time data collection and patient monitoring.
- **Data Integration:** Connects various health devices and systems to streamline data collection and analysis.

Impact:

- Provides continuous health monitoring and data collection.
- Improves coordination and efficiency in healthcare delivery.

9. Robotics and Automation

Applications:

- **Manufacturing:** Automates drug manufacturing processes, improving precision and efficiency.

- **Clinical Trials:** Uses robotic systems for sample handling and data collection in clinical research.

Impact:

- Enhances efficiency and accuracy in drug production.
- Streamlines clinical trial operations and data management.

10. **Advanced Analytics and Big Data**

Applications:

- **Data Integration:** This process combines data from multiple sources, including electronic health records (EHRs), clinical trials, and patient surveys.

- **Predictive Modeling:** Uses advanced analytics to predict patient outcomes, optimize treatment plans, and identify market trends.

Impact:

- Provides actionable insights for decision-making and strategy development.

- Enhances understanding of patient populations and treatment efficacy.

Implications for Pharmaceutical Product Managers

1. **Strategic Decision-Making:**

- Leverage emerging technologies to make informed decisions about drug development, marketing strategies, and patient engagement.

2. **Innovation and Competitive Advantage:**

- Stay ahead of competitors by adopting and integrating the latest technologies into product development and commercialization strategies.

3. **Efficiency and Cost Reduction:**

- Use automation and digital tools to streamline operations, reduce costs, and improve drug development and marketing efficiency.

4. **Personalization and Patient-Centricity:**

- Apply advancements in genomics, AI, and digital therapeutics to deliver personalized and patient-centered solutions.

5. Regulatory and Compliance Considerations:

- Stay informed about regulatory changes and ensure emerging technologies comply with industry standards and regulations.

By embracing these emerging technologies, pharmaceutical product managers can drive innovation, enhance operational efficiency, and deliver more effective and personalized solutions, ultimately improving patient outcomes and achieving business success.

29.3 Preparing for the Future: Adapting to Technological Changes

Adapting to technological changes is crucial for pharmaceutical product managers to stay competitive and drive innovation. Here's a comprehensive guide on how to prepare for the future and effectively adapt to technological advancements:

1. Embrace a Growth Mindset

Overview:

- **Adopt a Growth Mindset:** Cultivate an attitude of continuous learning and adaptability. Recognize that technological advancements are opportunities for growth rather than threats.

Actions:

- Encourage ongoing education and professional development.
- Stay informed about emerging technologies and industry trends.

2. Invest in Technology and Infrastructure

Overview:

- **Upgrade Systems:** Invest in modern technologies and infrastructure to support new digital tools and platforms.

Actions:

- Evaluate current systems and identify areas needing upgrades.
- Implement scalable solutions that can integrate with emerging technologies.

3. Foster a Culture of Innovation

Overview:

- **Promote Innovation:** Create an organizational culture that encourages experimentation and innovation.

Actions:

- Encourage cross-functional collaboration to generate new ideas.

- Implement innovation programs and incentives for employees.

4. Develop Digital Competencies

Overview:

- **Build Skills:** Equip your team with the skills to effectively leverage digital tools and technologies.

Actions:

- Provide training on digital marketing, data analytics, AI, and other relevant technologies.
- Hire or develop talent with expertise in emerging technologies.

5. Integrate Data-Driven Decision-Making

Overview:

- **Leverage Data:** Utilize data and analytics to make informed decisions and optimize strategies.

Actions:

- Invest in data analytics tools and platforms.
- Develop processes for collecting, analyzing, and applying data insights.

6. Stay Ahead with Research and Development

Overview:

- **Invest in R&D:** Focus on research and development to explore new technologies and their applications in pharma.

Actions:

- Collaborate with research institutions and technology partners.
- Support internal R&D initiatives focused on technology-driven solutions.

7. Monitor Regulatory Changes

Overview:

- **Understand Regulations:** Stay informed about regulatory requirements related to new technologies.

Actions:

- Monitor updates from regulatory bodies like the FDA, EMA, and other relevant agencies.
- Ensure compliance with regulations in the development and implementation of new technologies.

8. Enhance Cybersecurity Measures

Overview:

- **Protect Data:** Implement robust cybersecurity measures to safeguard sensitive data and technology systems.

Actions:

- Invest in cybersecurity solutions and regularly update security protocols.
- Train staff on cybersecurity best practices and threat awareness.

9. Foster Collaboration with Technology Partners

Overview:

- **Build Partnerships:** Collaborate with technology providers, startups, and industry experts to access cutting-edge solutions.

Actions:

- Establish strategic partnerships and alliances with technology firms.
- Participate in industry forums and conferences to stay connected with technology trends.

10. Focus on Patient-Centric Solutions

Overview:

- **Prioritize Patients:** Ensure that technological advancements enhance patient outcomes and experiences.

Actions:

- Develop technologies and solutions that address patient needs and preferences.
- Use patient feedback to guide technology development and implementation.

11. **Implement Agile Methodologies**
    Overview:
    - **Adopt Agility:** Use agile methodologies to respond quickly to technological changes and market demands.

    Actions:
    - Implement agile project management practices.
    - Encourage iterative development and rapid prototyping.

12. **Prepare for Technological Disruptions**
    Overview:
    - **Anticipate Changes:** Be proactive in preparing for potential disruptions caused by emerging technologies.

    Actions:
    - Conduct scenario planning to anticipate the impact of technological changes.
    - Develop contingency plans to address potential challenges and disruptions.

13. **Continuously Evaluate Technology Impact**
    Overview:
    - **Assess Effectiveness:** Regularly evaluate the impact of new technologies on your operations and strategy.

    Actions:
    - Implement metrics and KPIs to measure the success of technology implementations.
    - Conduct regular reviews to assess technology performance and make necessary adjustments.

Conclusion

Preparing for the future and adapting to technological changes involves a proactive approach, strategic investments, and a commitment to continuous learning and innovation. By embracing emerging technologies, fostering a culture of innovation, and leveraging data-driven decision-making, pharmaceutical product managers can position themselves and their organizations for long-term success in an increasingly digital and technology-driven landscape.

Trends in Digital Health Telemedicine

The healthcare landscape is undergoing a seismic shift, driven by technological advancements redefining how care is delivered and managed. Among the most transformative changes are the growing digital health and telemedicine trends. These innovations empower healthcare providers to deliver services more efficiently, patients to engage more actively in managing their health, and pharmaceutical companies to tailor their offerings to meet evolving needs.

Digital health encompasses many tools and solutions, from telemedicine and mobile health applications to wearable devices and digital therapeutics, all aimed at improving patient outcomes and streamlining healthcare delivery. Meanwhile, once seen as a niche option for rural or underserved populations, telemedicine has emerged as a mainstream approach, enabling virtual consultations, remote monitoring, and real-time data sharing. The COVID-19 pandemic accelerated this adoption, pushing telemedicine to the forefront of healthcare strategies across the globe.

In this chapter, we will explore the current and emerging trends shaping the future of digital health and telemedicine. From expanding virtual care to the rise of artificial intelligence (AI) in diagnostics and treatment, we will explore how these technologies transform the patient experience and the broader healthcare ecosystem. Furthermore, we will examine the implications of these trends for pharmaceutical companies and product managers, who are increasingly leveraging digital tools to drive innovation in drug development, marketing, and patient engagement.

As we navigate the digital age, understanding and adapting to these trends will be crucial for stakeholders across the healthcare sector. Integrating technology into healthcare delivery is no longer a distant vision but an immediate reality, reshaping how care is provided and consumed. This chapter serves as a guide to understanding these evolving dynamics and their profound impact on the future of healthcare.

30.1 Overview of Digital Health Trends

Digital health is a rapidly evolving field that leverages technology to improve healthcare delivery, enhance patient outcomes, and increase efficiency across the healthcare system. Here is an overview of the key trends shaping the digital health landscape:

1. Telemedicine Expansion

A. Widespread Adoption:

- **Virtual Consultations**: Telemedicine has become a mainstream consultation option, providing patients with convenient access to healthcare services from their homes.

- **Broadening Scope**: Services now include primary care, specialist consultations, mental health services, and chronic disease management.

B. Integration with Traditional Care

- **Hybrid Models**: Combining in-person and virtual care to provide more flexible and comprehensive patient care.

- **Continuity of Care**: Using telemedicine for follow-up visits, monitoring chronic conditions, and post-operative care.

2. Growth of Digital Therapeutics

A. Evidence-Based Solutions:

- **Clinical Validation**: Digital therapeutics (DTx) are gaining recognition for their effectiveness in managing diabetes, mental health disorders, and respiratory diseases.

- **Regulatory Approvals**: Increased support from regulatory bodies like the FDA facilitates the development and integration of DTx into standard care practices.

B. Integration with Conventional Therapies:

- **Combination Approaches**: Using DTx alongside traditional treatments to enhance patient outcomes and adherence.

3. Advancements in Remote Patient Monitoring (RPM)

A. Wearable Technology:

- **Health Tracking Devices**: The proliferation of wearables like smartwatches and fitness trackers that monitor heart rate, sleep, and activity levels.

- **Real-Time Data Collection**: Continuous data collection enables timely intervention and personalized care.

B. Data Integration:

- **Comprehensive Health Data**: Integrating data from various sources, including wearables, home devices, and electronic health records (EHRs), creates a holistic view of patient health.

- **Predictive Analytics**: Utilizing AI and machine learning to predict health issues and personalize treatment plans based on collected data.

4. **Artificial Intelligence and Machine Learning**

   A. Personalized Medicine:

   - **Tailored Treatment Plans**: AI algorithms develop personalized treatment plans based on genetic, environmental, and lifestyle data.

   - **Enhanced Decision Support**: AI tools assist healthcare providers with clinical decision-making, improving diagnostic accuracy and treatment recommendations.

   B. Diagnostic Innovations:

   - **Image Analysis**: AI-driven analysis of medical images to detect abnormalities and assist in diagnoses.

   - **Natural Language Processing**: Extracting insights from unstructured data in patient records and research articles to support clinical decisions.

5. **Data Privacy and Security**

   A. Compliance with Regulations:

   - **Data Protection**: Ensuring compliance with data protection regulations such as GDPR and HIPAA to safeguard patient information.

   - **Cybersecurity Measures**: Implementing robust cybersecurity practices to protect against data breaches and unauthorized access.

   B. Patient Consent

   - **Transparency**: Providing clear information on patient data use and obtaining informed consent to maintain trust and ensure ethical practices.

6. **Expansion of Health Platforms and Ecosystems**

   A. **Integrated Healthcare**:

   - **Unified Systems**: Development of platforms that integrate telemedicine, remote monitoring, EHRs, and other services into a single interface for improved care coordination.

   - **Interoperability**: Enhancing the ability of different health systems and platforms to work together and share data.

   B. **Patient Engagement Tools**:

   - **Mobile Health Apps**: Empowering patients to manage their health proactively through mobile apps that offer access to health information, appointment scheduling, and communication with providers.

   - **Virtual Health Communities**: Online forums and support groups that provide education and peer support for patients with chronic conditions.

7. **Future Outlook and Emerging Trends**

   A. **Personalized and Precision Medicine**:

   - **Advanced Personalization**: Continued advancements in personalized medicine, with treatments tailored to individual genetic profiles and health data.

   B. **Integration with Smart Technologies**:

   - **Smart Home Health Solutions**: Integrating digital health technologies with smart home devices for enhanced monitoring and management of health conditions.

   C. **Global Expansion**

   - **Interest Access**: Expansion of digital health solutions to underserved and remote areas, improving healthcare access and addressing health disparities.

   D. **Regulatory Evolution**:

   - **Adaptive Regulations**: Evolving regulatory frameworks to accommodate new digital health technologies and ensure patient safety.

These trends illustrate how digital health transforms healthcare by improving accessibility, efficiency, and patient outcomes. Integrating digital solutions into healthcare systems will likely deepen as technology advances, offering providers, patients, and stakeholders new opportunities and challenges.

30.2 The Role of Telemedicine in Patient Care

Telemedicine, the practice of delivering healthcare services through digital communication technologies, has become a transformative force in patient care. Its role encompasses a wide range of functions that enhance the accessibility, efficiency, and quality of healthcare. Here is a comprehensive look at how telemedicine impacts patient care:

1. **Increased Accessibility to Healthcare**

   A. Remote Access:
   - **Geographical Reach**: Patients in rural or underserved areas can access healthcare services without needing long-distance travel, overcoming geographical barriers.
   - **Convenience**: Provides access to healthcare services for patients who have mobility issues or who cannot take time off work to attend in-person appointments.

   B. Specialty Care:
   - **Expert Consultations**: Enables patients to consult with specialists who may not be available locally, improving access to expert care for complex conditions.

2. **Convenience and Efficiency**

   A. Flexible Scheduling:
   - **Appointment Flexibility**: Patients can schedule appointments outside traditional office hours, reducing wait times and making healthcare more accessible.
   - **Reduced Travel**: Eliminates the need for travel time and associated costs, making it easier for patients to attend appointments.

   B. Streamlined Processes:
   - **Quick Consultations**: Allows quick consultations for minor ailments or follow-up visits, which can be handled efficiently via video calls or messaging.
   - **Enhanced Coordination**: Facilitates better coordination between healthcare providers, improving the overall management of patient care.

3. Continuity of Care

A. Ongoing Monitoring:

- **Chronic Disease Management**: Provides a platform for continuous monitoring of chronic conditions, allowing for timely adjustments to treatment plans.
- **Follow-Up Care**: Ensures continuity of care through follow-up appointments, reducing the risk of missed or delayed care.

B. Integration with Health Records:

- **EHR Integration**: Telemedicine platforms often integrate with electronic health records (EHRs), providing a complete view of the patient's health history and ensuring informed decision-making.

4. Improved Patient Engagement

A. Patient Empowerment:

- **Self-Management Tools**: Offers tools and resources for patients to manage their health proactively, such as symptom checkers, educational materials, and self-monitoring apps.
- **Interactive Communication**: Interactive communication channels enhance patient engagement, allowing patients to ask questions and receive timely responses.

B. Personalized Care:

- **Tailored Advice**: Provides personalized advice and treatment recommendations based on real-time data and patient interactions.

5. Cost-Effectiveness

A. Reduced Costs:

- **Lower Healthcare Costs**: Reduces the need for in-person visits, decreasing overall healthcare costs for patients and providers.
- **Decreased Emergency Room Visits**: This tool helps manage minor issues remotely, potentially reducing unnecessary emergency room visits and hospital admissions.

B. Efficient Resource Utilization:

- **Optimized Use of Resources**: Addressing issues that do not require physical examination allows for more efficient use of healthcare resources.

6. **Enhanced Quality of Care**

A. **Evidence-Based Practice**:

- **Access to Evidence-Based Guidelines**: The latest clinical guidelines and evidence-based practices ensure high-quality care.

- **Second Opinions**: Facilitates the acquisition of second opinions and reviews from specialists, contributing to more accurate diagnoses and treatment plans.

B. **Early Intervention**:

- **Timely Intervention**: Enables early intervention for symptoms or conditions that might otherwise go untreated, improving patient outcomes.

7. **Challenges and Considerations**

A. **Technology Barriers**:

- **Digital Literacy**: Patients and providers must have the technology and digital literacy to use telemedicine platforms effectively.

- **Technology Access**: Ensuring that all patients, including those with limited access to technology, can benefit from telemedicine services.

B. **Privacy and Security**:

- **Data Protection**: Ensuring robust security measures to protect patient data and comply with privacy regulations.

C. **Regulatory Compliance**:

- **Licensing and Reimbursement**: Navigating varying state and national regulations regarding licensing, reimbursement, and the scope of practice for telemedicine services.

Conclusion

Telemedicine is crucial in modernizing patient care by improving accessibility, efficiency, and engagement. It offers a flexible and convenient solution for various healthcare needs, from routine consultations to chronic disease management. While these are challenging to address, the ongoing evolution and adoption of telemedicine are likely to continue reshaping the landscape of patient care in positive and impactful ways.

30.3 Implications for Pharma Product Managers

The rise of digital health, including telemedicine, digital therapeutics, remote patient monitoring, and technological advancements, presents several implications for pharmaceutical product managers. These implications shape how they approach product development, marketing, and lifecycle management. Consider the following implications:

1. **Product Development and Innovation**

   A. Integration of Digital Solutions

   - **Enhanced Product Offerings**: Pharma product managers must consider integrating digital health solutions with traditional pharmaceutical products, such as combining drugs with digital therapeutics or remote monitoring capabilities.

   - **Innovation in Drug Delivery**: The rise of digital technologies opens opportunities for innovative drug delivery systems, such as small pill bottles that track adherence.

   B. Collaboration with Tech Companies

   - **Partnerships**: Collaborations with technology companies to develop integrated health solutions and digital tools that complement pharmaceutical products.

   - **Cross-Disciplinary Teams**: Building teams that include experts in digital health to drive innovation and develop integrated solutions.

2. **Marketing and Communication Strategies**

   A. Leveraging Digital Channels

   - **Enhanced Engagement**: Utilizing digital channels for marketing and communication, including social media, mobile apps, and digital platforms, to engage with healthcare professionals (HCPs) and patients more effectively.

   - **Personalized Content**: Utilizing digital channels for marketing and communication, including social media, mobile apps, and digital platforms, to engage with healthcare professionals (HCPs) and patients more effectively.

- **Personalized Content**: Creating personalized content and experiences based on digital health data and insights.

B. Educating Stakeholders

- **Training and Support**: Providing education and support to HCPs and patients on using digital tools and technologies associated with pharmaceutical products.

- **Evidence-Based Marketing**: Leveraging data from digital health solutions to support marketing claims and demonstrate the product's value.

3. **Lifecycle Management**

   A. Ongoing Monitoring and Adaptation

   - **Real-Time Feedback**: Using data from digital health solutions to monitor product performance and patient outcomes in real-time.

   - **Adaptation Strategies**: Adjusting marketing strategies, product features, or support services based on insights gathered from digital health tools.

   B. Managing Digital Therapeutics

   - **Lifecycle Support**: Managing the lifecycle of digital therapeutics alongside traditional pharmaceuticals, including updates, regulatory compliance, and user support.

   - **Value Demonstration**: Demonstrating the value of integrated digital therapeutics and their impact on patient outcomes.

4. **Regulatory Compliance Considerations**

   A. Navigating Regulations

   - **Compliance**: Staying informed about evolving regulatory requirements for digital health solutions and ensuring that products meet all necessary standards.

   - **Data Privacy**: Ensuring compliance with data privacy regulations (e.g., GDPR, HIPAA) when dealing with patient data collected through digital health tools.

B. Adapting to Changes

- **Regulatory Updates**: Adapting to changes in regulatory frameworks that impact digital health technologies and integrating these changes into product development and marketing strategies.

5. **Competitive Landscape**

   A. Staying Ahead

   - **Market Trends**: Monitoring trends in digital health to stay competitive and identify opportunities for innovation and differentiation.

   - **Competitive Intelligence**: Gathering intelligence on competitors' digital health strategies and products to inform strategic decisions and maintain a competitive edge.

   B. Strategic Positioning

   - **Differentiation**: Leveraging digital health solutions to differentiate products in the market and offer added value to stakeholders.

6. **Customer Engagement and Support**

   A. Enhancing Patient Experience

   - **Digital Tools**: Providing digital tools and resources to support patients in managing their health and adhering to treatment plans.

   - **User Support**: Offering robust support services to help patients and HCPs use digital helath solutions effectively.

   B. Feedback Mechanisms

   - **Patient Insights**: Collecting and analyzing patient feedback from digital health tools to improve products and services.

7. **Future Trends and Adaptation**

   A. Embracing Emerging Technologies

   - **Adopting Innovations**: Staying informed about emerging technologies such as AI, machine learning, and blockchain that could impact the pharmaceutical industry.

   - **Strategic Planning**: Incorporating insights from emerging trends into strategic planning and future product development.

B. Preparing for Disruptions

- **Anticipating Changes**: Preparing for potential disruptions in the healthcare landscape due to technological advancements and adjusting strategies accordingly.

Pharma product managers must navigate a complex and rapidly evolving digital landscape. By leveraging digital health trends, they can enhance product development, marketing, and lifecycle management while addressing regulatory, competitive, and customer engagement challenges. Adapting to these changes effectively will be crucial for maintaining a competitive edge and delivering value in the digital health era.

Part VIII. Career Development and Advancement

In the dynamic and rapidly evolving pharmaceutical industry, product manager career development and advancement opportunities are more abundant than ever. As the role of the pharmaceutical product manager grows in complexity—encompassing strategic planning, cross-functional leadership, digital transformation, and a deep understanding of regulatory landscapes—it becomes increasingly important for professionals in this field to enhance their skills and competencies continuously.

Navigating the path to career growth involves a clear understanding of the key capabilities required to excel in product management, a commitment to lifelong learning, and the ability to adapt to emerging industry trends. Product managers can position themselves for advancement within their organizations or across the broader pharmaceutical industry by focusing on technical knowledge and leadership skills. This section explores the essential elements of career development in pharma product management, offering insights into pathways for growth, skills required for success, and strategies for achieving long-term professional goals.

Career Pathways and Advancement for Product Managers

The pharmaceutical industry offers a dynamic and rewarding career for those interested in product management. As the sector continues to grow, driven by innovation in drug development, digital health, and evolving market demands, the role of a pharmaceutical product manager has become more essential and multi-faceted. A product manager in this industry acts as the linchpin, coordinating cross-functional teams, navigating regulatory landscapes, and devising marketing strategies that ensure a product's success throughout its lifecycle.

This chapter will explore the various career pathways available to pharmaceutical product managers, highlighting the skills, experiences, and qualifications required to excel in the role. We will delve into the stages of career progression—from entry-level product management roles to senior leadership positions—and the strategic decisions that can propel a professional to higher responsibilities.

Additionally, this chapter will examine the core competencies crucial for career advancement, such as leadership, business acumen, and the ability to adapt to digital transformation in the industry.

Whether you're an aspiring product manager or a seasoned professional seeking new challenges, this chapter will provide a roadmap for leveraging opportunities, overcoming obstacles, and achieving long-term success in pharmaceutical product management.

31.1 Exploring Career Pathways in Pharma Product Management

Pharmaceutical product managers (PMs) play a critical role in driving the success of pharmaceutical products throughout their lifecycle. The career path for PMs offers numerous opportunities for growth and advancement as they gain experience and expertise in areas such as marketing, strategy, and cross-functional leadership. Below is an overview of potential career pathways and advancement opportunities for pharmaceutical product managers:

1. Entry-Level Roles

1.1 Assistant Product Manager / Associate Product Manager:

- **Responsibilities:** Entry-level positions involve supporting the product management team with market research, competitive analysis, and tactical marketing activities. These roles provide exposure to product launches, marketing campaigns, and cross-functional collaboration.

- **Skills Developed:** Market analysis, data interpretation, project coordination, and teamwork.

2. Mid-Level Roles

2.1 Product Manager (PM):

- **Responsibilities:** Product managers oversee the strategy, planning, and execution of product launches, manage the product lifecycle, and collaborate with cross-functional teams (marketing, sales, regulatory, R&D). They also work closely with stakeholders to ensure alignment on product goals.

- **Skills Developed:** Leadership, strategic planning, market analysis, stakeholder management, and brand management.

2.2 Senior Product Manager (SPM):

- **Responsibilities:** Senior PMs take on greater strategic responsibilities, managing a larger product portfolio or more complex products. They lead the development of marketing strategies, oversee product positioning, and ensure the achievement of business goals. SPMs also mentor junior product managers and help guide cross-functional teams.

- **Skills Developed:** Advanced strategic thinking, cross-functional leadership, budgeting, long-term planning, and project management.

3. Advanced Roles

3.1 Product Director / Group Product Manager:

- **Responsibilities:** In these roles, individuals manage a portfolio of products or lead multiple product teams. They are responsible for high-level strategic planning, portfolio management, product innovation, and ensuring that each product aligns with the company's broader business objectives. Directors work closely with senior leadership and play a key role in decision-making.

- **Skills Developed:** Strategic leadership, portfolio management, high-level decision-making, and business acumen.

3.2 Marketing Director / Head of Product:

- **Responsibilities:** This role involves leading the entire product marketing function. Marketing directors are responsible for setting marketing and product strategies, driving product innovation, and ensuring marketing initiatives support business growth. They work closely with executive teams to align marketing efforts with corporate strategy and lead large cross-functional teams.

- **Skills Developed:** Leadership of large teams, business strategy, external communication, and change management.

4. Senior Executive Roles

4.1 Vice President (VP) of Marketing / Product Management:

- **Responsibilities:** The VP oversees the entire marketing or product management department. This includes setting the strategic vision, managing large teams, and working with the C-suite to ensure the company's growth and success.

VPs drive product commercialization and oversee all aspects of product management, from R&D to post-launch marketing.

- **Skills Developed:** High-level business strategy, executive leadership, corporate governance, and cross-functional influence.

4.2 Chief Marketing Officer (CMO) / Chief Product Officer (CPO):

- **Responsibilities:** CMOs or CPOs are responsible for a company's overall marketing or product strategy at the executive level. They drive innovation, product portfolio growth, and alignment between marketing and business objectives. These roles are often involved in mergers, acquisitions, and major strategic decisions that impact the entire organization.

- **Skills Developed:** Leadership at the executive level, corporate governance, innovation, and business growth strategy.

5. Lateral Career Moves

5.1 Market Access / Commercial Strategy Roles:

- **Responsibilities:** In market access or commercial strategy roles, PMs focus on pricing, reimbursement, and ensuring that products are available and accessible in key markets. These roles require a strong understanding of health economics, payer strategies, and regulatory environments.

- **Skills Developed:** Health economics, payer relations, regulatory knowledge, and pricing strategy.

5.2 Business Development:

- **Responsibilities:** In business development, PMs explore new opportunities for partnerships, collaborations, and product licensing. This role involves assessing the commercial potential of new markets and identifying growth opportunities through external partnerships.

- **Skills Developed:** Strategic partnerships, market entry strategies, and negotiation.

6. Academic and Consulting Opportunities

6.1 Academia:

- Product managers can move into academic roles, leveraging their experience to teach future product managers or marketers. They can also contribute to academic research in pharmaceutical marketing, innovation, or regulatory affairs.

- **Skills Developed:** Teaching, research, industry-academic collaboration, and thought leadership.

6.2 Consulting:

- PMs can transition into consulting roles, offering strategic advice to pharmaceutical companies on product development, marketing strategies, market access, and portfolio management. Consultants play an advisory role in optimizing product performance across the industry.

- **Skills Developed:** Strategic advising, client management, industry expertise, and analytical thinking.

Key Skills for Career Advancement

1. **Cross-Functional Leadership:** Leading teams across functions such as R&D, regulatory, marketing, and sales is critical for success in senior roles.

2. **Strategic Thinking:** Developing and executing long-term strategies aligned with business goals is essential for career growth.

3. **Data-Driven Decision-Making:** Utilizing data insights for market analysis, customer segmentation, and performance measurement is key.

4. **Innovation:** Driving product innovation and adapting to new trends, such as digital health and AI, is vital for staying competitive.

5. **Communication and Influence:** Effectively communicating with internal and external stakeholders is crucial, particularly when advocating for product strategies and investments.

6. **Regulatory Knowledge:** Understanding the regulatory landscape is important in navigating product approvals, compliance, and market access strategies.

Conclusion

Pharmaceutical product management offers a dynamic and rewarding career path, with opportunities to grow into strategic leadership roles across marketing, business development, and executive management. As the industry evolves with new technologies and shifting healthcare landscapes, product managers who embrace continuous learning, strategic thinking, and cross-functional leadership will find ample opportunities for career advancement.

31.2 Skills and Experience Needed for Advancement

To advance in pharmaceutical product management, professionals should cultivate a blend of strategic, leadership, and technical skills. Focusing on continuous learning, gaining practical experience, and actively seeking growth opportunities will equip them for leadership roles and career progression.

Here's a comprehensive look at essential skills for success in pharmaceutical product management:

Essential Skills for Success in Pharma Product Management

1. **Strategic Thinking:**
   - **Overview:** Ability to develop long-term plans aligned with the company's vision and market needs.
   - **Importance:** Ensures product strategies are forward-looking and adaptable to market changes.
   - **Application:** Involves creating a roadmap for product development, market entry, and lifecycle management.

2. **Scientific Knowledge:**
   - **Overview:** Understanding of drug development, pharmacology, and clinical research.
   - **Importance:** Enables informed decisions about product features, benefits, and positioning.
   - **Application:** Helps evaluate clinical trial data, understand therapeutic areas, and communicate scientific information effectively.

3. **Business Acumen:**
   - **Overview:** Financial literacy, market analysis, and understanding of the pharmaceutical business model.
   - **Importance:** Facilitates effective budgeting, forecasting, and strategic business planning.
   - **Application:** Analyze financial statements, manage budgets, and understand market dynamics.

4. **Project Management:**

   - **Overview:** Skills to plan, execute, and oversee projects efficiently.

   - **Importance:** Ensures timely and successful product development and launch.

   - **Application:** Includes managing cross-functional teams, timelines, and resources.

5. **Regulatory Knowledge:**

   - **Overview:** Understanding of regulatory requirements and compliance processes.

   - **Importance:** Ensures that products meet legal and regulatory standards.

   - **Application:** Involves navigating the approval process and ensuring adherence to regulatory guidelines.

6. **Marketing and Sales Expertise:**

   - **Overview:** Knowledge of marketing strategies, sales tactics, and HCP engagement.

   - **Importance:** Drives successful product positioning and market penetration.

   - **Application:** Includes developing marketing campaigns, sales strategies, and customer engagement plans.

7. **Data Analysis and Interpretation:**

   - **Overview:** Ability to analyze market data, clinical data, and performance metrics.

   - **Importance:** Supports evidence-based decision-making and strategy adjustments.

   - **Application:** Involves using data analytics tools to evaluate market trends and product performance.

8. **Leadership and Team Management:**

   - **Overview:** Skills to lead and manage cross-functional teams effectively.

   - **Importance:** Ensures team alignment and productivity in achieving product goals.

- **Application:** Setting clear objectives, motivating team members, and resolving conflicts.

9. **Communication Skills:**
   - **Overview:** Ability to convey information clearly and persuasively.
   - **Importance:** Facilitates effective stakeholder engagement and information dissemination.
   - **Application:** Involves crafting compelling messages, presenting data, and engaging with various stakeholders.

10. **Adaptability and Innovation:**
    - **Overview:** Flexibility to adapt to changing market conditions and embrace new technologies.
    - **Importance:** Ensures that product strategies remain relevant and competitive.
    - **Application:** Includes staying updated on industry trends, technological advancements, and adjusting strategies as needed.

11. **Customer and Market Insight:**
    - **Overview:** Understanding customer needs and market dynamics.
    - **Importance:** Helps in creating products and strategies that meet market demands.
    - **Application:** Involves conducting market research, analyzing customer feedback, and identifying growth opportunities.

By developing and honing these skills, pharmaceutical product managers can effectively navigate the industry's complexities, drive successful product outcomes, and contribute to the overall success of their organizations.

Networking and Mentorship

In the fast-paced, ever-changing world of pharmaceutical marketing and product management, staying connected, informed, and adaptable is crucial. The ability to network effectively, find and nurture meaningful mentorship relationships, and commit to continuous learning can be key differentiators between a product manager who merely keeps up and one who excels.

Networking is not just about making contacts; it's about building relationships that provide mutual support, insights, and growth opportunities. Within the pharma industry, where cross-functional collaboration and regulatory complexities abound, a robust professional network can open doors to new ideas, best practices, and market intelligence.

Mentorship takes networking to the next level by offering personal guidance and wisdom from those who have successfully navigated the industry's challenges. The right mentor can accelerate learning, provide critical feedback, and offer perspectives that enhance personal and professional development.

Equally important is the commitment to continuous learning. In an industry shaped by scientific advancements, emerging technologies, and shifting regulatory landscapes, keeping skills and knowledge up to date is essential. A product manager's ability to adapt, innovate, and lead in the digital age depends on ongoing investment in education through formal training, industry events, or self-driven exploration of new tools and trends.

This chapter delves into strategically leveraging networking, mentorship, and continuous learning to thrive in pharmaceutical marketing and product management. By understanding how these elements interact and contribute to professional success, you can build a foundation for long-term growth, innovation, and leadership in the industry.

32.1 Building a Professional Network in Pharma

Building a strong professional network is essential for success in the pharmaceutical industry. Networking can open doors to new opportunities, provide access to valuable resources, and help advance a career. Here's how to effectively build and maintain a professional network in the pharma industry:

1. Understanding the Importance of Networking

1.1 Career Advancement:

- **Opportunities:** Networking helps uncover job opportunities, collaborations, and partnerships that may not be advertised publicly.
- **Visibility:** Being active in the industry increases visibility and can lead to career advancements.

1.2 Knowledge Sharing:

- **Industry Trends:** Connecting with professionals can help you stay informed about the latest industry trends, regulatory changes, and technological advancements.
- **Best Practices:** Sharing experiences and challenges with peers can lead to learning best practices and innovative solutions.

1.3 Building Relationships:

- **Mutual Support:** Building strong relationships with colleagues, mentors, and industry leaders can support and guide your career.
- **Collaboration:** Networking facilitates collaborations that can lead to new projects, research opportunities, and business ventures.

2. Identifying Key Stakeholders

2.1 Internal Networks:

- **Colleagues:** Build relationships with colleagues across different departments, such as R&D, regulatory affairs, marketing, and sales.

- **Management:** Cultivate relationships with supervisors and higher management to gain insights and guidance.

2.2 External Networks:

- **Industry Peers:** Connect with professionals from other pharmaceutical companies, academia, and research institutions.

- **Key Opinion Leaders (KOLs):** Engage with KOLs who influence the industry and can provide valuable insights and mentorship.

- **Regulatory Bodies:** Connect with professionals to stay updated on compliance and regulatory issues.

2.3 Professional Associations:

- **Industry Groups:** Join industry-specific organizations like the Pharmaceutical Research and Manufacturers of America (PhRMA), International Society for Pharmaceutical Engineering (ISPE), or Drug Information Association (DIA).

- **Networking Events:** Attend conferences, seminars, webinars, and workshops to meet new contacts and strengthen existing relationships.

3. Networking Strategies

3.1 Attend Industry Events:

- **Conferences and Trade Shows:** Participate in pharmaceutical industry conferences, trade shows, and exhibitions where you can meet professionals from different sectors.

- **Workshops and Seminars:** Attend workshops and seminars that offer opportunities to learn and network with peers.

3.2 Online Networking:

- **LinkedIn:** Use LinkedIn to connect with industry professionals, join relevant groups, and participate in discussions.

- **Professional Forums:** Engage in industry-specific forums and online communities where you can share knowledge and network.

3.3 Informational Interviews:

- **Reach out:** Request informational interviews with professionals whose careers you admire or who work in your area of interest.

- **Learn and Connect:** Use these opportunities to learn about their experiences and build a connection.

3.4 Mentorship:

- **Seek Mentors:** Identify mentors who can guide you in your career and help you navigate the complexities of the pharmaceutical industry.

- **Mentorship Programs:** Participate in formal mentorship programs offered by professional organizations or your company.

4. Building and Maintaining Relationships

4.1 Follow-Up:

- **After Events:** Follow up with new contacts after meeting them at events, sending a personalized message to keep the connection alive.

- **Regular Communication:** Contact your network through emails, phone calls, or social media.

4.2 Offer Value:

- **Share Knowledge:** Contribute to discussions, share articles, and offer insights that can benefit your network.

- **Collaborate:** Be open to collaboration and help others in your network when possible.

4.3 Be Authentic:

- **Genuine Interest:** Show genuine interest in others' work and achievements. Building trust is key to long-lasting professional relationships.

- **Integrity:** Maintain integrity and professionalism in all your interactions.

5. Leveraging Your Network

5.1 Career Development:

- **Job Opportunities:** Use your network to learn about job openings and get referrals.

- **Skill Development:** Seek advice and recommendations from your network to enhance your skills and knowledge.

5.2 Business Growth:

- **Partnerships:** Identify potential partners for business ventures, research projects, or marketing collaborations.

- **Market Insights:** Gain insights into market trends, customer needs, and competitive landscape from your network.

5.3 Thought Leadership:

- **Speaking Opportunities:** Leverage your network to secure speaking engagements at conferences and industry events.

- **Publishing:** Collaborate on articles, white papers, or research publications to establish yourself as a thought leader.

6. Adapting to the Digital Age

6.1 Virtual Networking:

- **Webinars and Online Conferences:** Participate in virtual events that offer networking opportunities through chat rooms, breakout sessions, and virtual booths.

- **Social Media:** Use platforms like Twitter (X, now) and LinkedIn to engage with industry leaders, participate in discussions, and share content.

6.2 Online Communities:

- **Join Groups:** Be active in online communities focused on the pharmaceutical industry to expand your network beyond geographical limitations.

- **Host Virtual Meetups:** Organize virtual meetups or discussion groups to connect with like-minded professionals.

7. Overcoming Networking Challenges

7.1 Time Management:

- **Prioritize:** Focus on high-impact networking activities that align with your career goals.
- **Scheduled Networking:** Set aside regular time for networking activities, both online and offline.

7.2 Overcoming Introversion:

- **Preparation:** Before networking events, prepare conversation starters and questions to ease anxiety.
- **One-on-One Interactions:** Focus on building deep, one-on-one connections rather than engaging with large groups.

7.3 Maintaining Long-Distance Relationships:

- **Regular Check-Ins:** Schedule regular check-ins with geographically distant contacts to maintain the relationship.
- **Use Technology:** Leverage video calls and social media to stay connected despite the distance.

Conclusion

Building a professional network in pharma requires a strategic approach, continuous effort, and genuine relationship-building. By actively participating in industry events, leveraging online platforms, and nurturing relationships, you can create a robust network that supports your career growth, provides valuable insights, and opens doors to new opportunities in the pharmaceutical industry.

32.2 Finding and Leveraging Mentorship Opportunities

Mentorship is crucial in professional development, particularly in specialized fields like pharmaceutical product management. Here's how to effectively find and leverage mentorship opportunities:

1. Identifying Potential Mentors

1.1 Internal Company Networks:

- **Formal Programs:** Many companies offer structured mentorship programs where experienced professionals are paired with newer employees. Participating in these programs can provide direct access to seasoned industry veterans.

- **Informal Connections:** Building relationships with senior colleagues who can offer guidance and insights based on their experiences.

1.2 Industry Associations and Conferences:

- **Professional Organizations:** Joining industry associations like the Pharmaceutical Marketing Research Group (PMRG) or Drug Information Association (DIA) can help you connect with experienced professionals.

- **Networking Events:** Attending industry conferences, seminars, and workshops provides opportunities to meet potential mentors who can offer career and personal development guidance.

1.3 Online Platforms:

- **LinkedIn:** Utilize LinkedIn to identify and connect with leaders in your field. Engage with their content, participate in discussions, and seek advice or guidance.

- **Mentorship Platforms:** MentorCruise, Ten Thousand Coffees, and MicroMentor offer dedicated spaces to find mentors aligned with your career goals.

2. Approaching a Potential Mentor

2.1 Clear and Specific Goals:

- **Define Your Objectives:** Before approaching a potential mentor, clarify your goal. Whether it's career guidance, industry-specific advice, or skill development, having clear goals will help you and your mentor focus your efforts.

- **Personalized Request:** When reaching out, personalize your message. Mention why you admire their work, how their expertise aligns with your goals, and what you hope to gain from the mentorship.

2.2 Respecting Their Time:

- **Be Concise and Respectful:** Keep your initial request brief and to the point. Acknowledge their busy schedule and propose a time for a brief meeting or conversation that works for them.

- **Flexible Commitment:** To accommodate their availability, offer a flexible mentorship arrangement, such as monthly check-ins or informal advice sessions.

3. Building a Productive Mentor-Mentee Relationship

3.1 Establishing Expectations:

- **Define the Relationship:** Discuss and agree on your interactions' format, frequency, and duration. This ensures both parties are on the same page and can commit to the arrangement.

- **Set Mutual Goals:** Work together to set achievable goals for your mentorship. This could include specific career milestones, learning new skills, or navigating complex projects.

3.2 Active Participation:

- **Be Prepared:** Come to meetings with specific questions, challenges, or topics for discussion. This shows your mentor that you value their time and are committed to your growth.

- **Seek Feedback:** Regularly ask for constructive feedback on your progress and how you can improve. This will help you grow faster and demonstrate your willingness to learn.

4. Leveraging the Mentorship for Career Growth

4.1 Applying Insights:

- **Implement Advice:** Act on your mentor's advice and guidance. Whether you're learning a new skill, pursuing additional education, or refining your approach to work, applying their insights can accelerate your development.

- **Document Progress:** Record your goals, actions taken, and the outcomes achieved through mentorship. This will help you track your growth and make the most of the relationship.

4.2 Expanding Your Network:

- **Introductions and Referrals:** Your mentor can introduce you to other professionals in their network, opening doors to new opportunities, collaborations, or even job prospects.

- **Visibility in the Industry:** Participating in industry events or contributing to projects alongside your mentor can increase your visibility and reputation in the field.

5. Giving Back and Maintaining the Relationship

5.1 Showing Appreciation:

- **Express Gratitude:** Regularly thank your mentor for their time and guidance. A simple thank-you note or acknowledging their support in your achievements goes a long way in maintaining a positive relationship.

- **Reciprocate When Possible:** Offer to help your mentor in ways you can, such as assisting with projects, sharing useful resources, or providing feedback on areas where you have expertise.

5.2 Long-Term Connection:

- **Stay in Touch:** Contact your mentor even after formal mentorship ends. Update them on your progress, celebrate milestones together, and continue seeking occasional advice.

- **Become a Mentor:** As you grow in your career, consider mentoring others. This will help you solidify your knowledge and leadership skills and contribute to the professional community.

Conclusion

Finding and leveraging mentorship opportunities requires intentional effort and a proactive approach. By identifying the right mentors, building strong relationships, and actively applying their guidance, you can significantly enhance your career trajectory in pharmaceutical product management. Remember, mentorship is a two-way street that thrives on mutual respect, clear communication, and a shared commitment to growth.

32.3 Continuous Learning and Professional Development

In the ever-evolving field of pharmaceutical product management, continuous learning and professional development are crucial for staying current with industry advancements, maintaining expertise, and driving career growth. Here's a comprehensive overview of what continuous learning and professional development entail:

1. **Importance of Continuous Learning**

   1.1 Keeping Up with Industry Trends:

   - **Emerging Technologies:** Stay informed about new technologies, such as AI, digital health, and telemedicine, that are reshaping the pharmaceutical landscape.

   - **Regulatory Changes:** Understanding regulations and compliance standards updates that impact product development and marketing.

   1.2 Enhancing Skills and Knowledge:

   - **Skill Enhancement:** Regularly updating skills to remain competitive and effectively manage pharmaceutical products.

   - **Knowledge Expansion:** Gaining insights into new therapeutic areas, scientific advancements, and market dynamics.

   1.3 Career Advancement:

   - **Professional Growth:** Leveraging new skills and knowledge to advance your career, take on more complex roles, and achieve career goals.

   - **Leadership Development:** Developing leadership qualities to effectively manage teams and projects in the evolving digital landscape.

2. **Methods of Continuous Learning**

   2.1 Formal Education and Certification:

   - **Advanced Degrees:** Pursuing additional degrees, such as an MBA or a Master's in Pharmaceutical Sciences, to deepen knowledge and enhance career prospects.

- **Certifications:** Earning certifications related to pharmaceutical management, project management (e.g., PMP), digital marketing, and regulatory affairs to validate expertise.

2.2 Professional Development Programs:

- **Workshops and Seminars:** Attending industry workshops, seminars, and conferences to gain insights from experts and network with peers.

- **Online Courses and Webinars:** Participating in online courses and webinars offered by professional organizations and educational institutions.

2.3 On-the-Job Learning:

- **Cross-Functional Projects:** Engaging in projects across different departments to gain a broader perspective and develop diverse skills.

- **Mentorship and Coaching:** Seeking guidance from experienced professionals and mentors to gain insights and advice on career development.

2.4 Industry Publications and Research:

- **Reading Journals:** Keeping up with industry journals, research papers, and publications to stay informed about the latest scientific and market developments.

- **Case Studies:** Analyzing case studies to understand successful strategies and learn from industry experiences.

3. Strategies for Effective Professional Development

3.1 Setting Clear Goals:

- **Career Objectives:** Defining short-term and long-term career goals to guide learning and development efforts.

- **Skill Development:** Identifying key skills and knowledge areas to focus on for career advancement.

3.2 Creating a Learning Plan:

- **Structured Learning:** Developing a structured learning plan that includes formal education, self-study, and practical experience.

- **Time Management:** Allocating regular time for learning and development activities to ensure consistent progress.

3.3 Networking and Collaboration:

- **Industry Networking:** Building a network of industry professionals to exchange knowledge, share experiences, and explore new opportunities.

- **Professional Associations:** Joining professional associations and organizations related to pharmaceutical management to access resources and support.

3.4 Reflecting and Adapting:

- **Self-Assessment:** Regularly assessing your progress, strengths, and areas for improvement to adapt your learning approach.

- **Feedback:** Seeking feedback from peers, mentors, and supervisors to identify areas for growth and development.

4. Leveraging Technology for Learning

4.1 E-Learning Platforms:

- **Online Learning Platforms:** Utilizing platforms like Coursera, LinkedIn Learning, and Udemy to access a wide range of courses and training materials.

- **Interactive Tools:** Engaging with interactive tools and simulations to enhance learning experiences and apply knowledge in practical scenarios.

4.2 Digital Communities:

- **Online Forums and Groups:** Participating in online forums, social media groups, and professional networks to stay updated and engage in discussions.

- **Virtual Events:** Attending virtual conferences and webinars to access content and connect with experts worldwide.

5. Embracing Lifelong Learning

5.1 Adapting to Change:

- **Continuous Improvement:** Embracing a mindset of continuous improvement and adaptability to navigate the rapidly changing pharmaceutical landscape.

- **Innovative Thinking:** Encouraging innovative thinking and problem-solving to address emerging challenges and opportunities.

5.2 Personal Development:

- **Soft Skills:** Developing skills such as communication, leadership, and emotional intelligence to complement technical expertise.

- **Work-Life Balance:** Balancing professional development with personal well-being to maintain productivity and satisfaction.

By committing to continuous learning and professional development, pharmaceutical product managers can stay ahead of industry trends, enhance their skills, and drive successful product management strategies in an increasingly complex and dynamic environment.

Part IX.   The Future of Product Management in Pharmaceutical Industry

Pharmaceutical product management is at the cusp of a transformative era driven by rapid technological advancements, evolving healthcare landscapes, and shifting market dynamics. As the industry navigates an increasingly complex ecosystem, product managers must adapt to new challenges and opportunities that were unimaginable just a few decades ago.

This chapter will explore the future of pharmaceutical product management, focusing on the emerging trends, skills, and technologies reshaping the role. Product managers' tools and strategies are evolving from digital health innovations and personalized medicine to artificial intelligence (AI) and data analytics. Success in the future will depend on their ability to integrate these innovations into product development, marketing, and lifecycle management.

As healthcare moves towards a more patient-centric and data-driven model, pharmaceutical product managers' responsibilities will extend beyond traditional marketing and sales strategies. They must collaborate and strategize more with cross-functional teams, regulatory bodies, healthcare professionals (HCPs), patients, and payers. Understanding global health trends, regulatory changes, and ethical considerations will be crucial for navigating the competitive landscape.

This chapter will provide insights into how the role of the pharmaceutical product manager is changing, the essential skills and competencies needed for future success, and the key trends that will shape the industry over the coming years.

The Future of Pharmaceutical Product Management

The future of pharmaceutical product management is poised for transformation as the industry evolves with technological advancements, regulatory changes, and shifting market dynamics. Here's an in-depth look at how these factors are shaping the future of pharma product management:

1. Integration of Advanced Technologies

A. Artificial Intelligence and Machine Learning:

- **Enhanced Decision-Making:** AI and ML will continue to improve data analysis, helping product managers make more informed decisions about drug development, market strategies, and patient needs.

- **Personalized Medicine:** AI will facilitate the creation of tailored treatment plans based on genetic, behavioral, and environmental data.

B. Digital Therapeutics and Telemedicine:

- **Digital Health Solutions:** Integrating digital therapeutics will offer innovative treatment options and improve patient adherence.

- **Remote Monitoring:** Telemedicine and remote patient monitoring will become standard practices, providing real-time data and improving patient engagement.

2. Increased Focus on Patient-Centricity

A. Personalized Treatment:

- **Precision Medicine:** Future product management will emphasize developing treatments tailored to individual genetic profiles and health conditions.

- **Patient Experience:** The core focus will be to enhance patient experience through digital tools, patient education, and support systems.

B. Patient Data Utilization:

- **Data-Driven Insights:** Leveraging patient data to understand treatment outcomes, preferences, and adherence patterns will guide product development and marketing strategies.

3. Emphasis on Real-World Evidence

A. Real-World Data (RWD):

- **Market Access and Reimbursement:** RWD will be crucial in demonstrating the value of new therapies to payers and regulators, influencing market access and reimbursement decisions.

- **Post-Market Surveillance:** Ongoing collection and analysis of RWD will help assess long-term safety and effectiveness, informing lifecycle management strategies.

B. Health Economics and Outcomes Research (HEOR):

- **Value Demonstration:** HEOR will be key in showing new treatments' economic and clinical value and shaping pricing and reimbursement strategies.

4. Regulatory and Compliance Adaptations

A. Dynamic Regulatory Environment:

- **Evolving Regulations:** Product managers must navigate an increasingly complex regulatory landscape, including new guidelines for digital health solutions and personalized medicine.

- **Global Market Considerations:** Compliance with international regulations will be essential for global product launches and market expansions.

B. Agile Compliance Strategies:

- **Adaptive Approaches:** Agile compliance strategies will help manage regulatory changes and streamline approval processes.

5. **Enhanced Cross-Functional Collaboration**

   A. Integrated Teams:

   - **Cross-functional integration:** Future product management will require seamless collaboration across R&D, marketing, sales, and regulatory teams to ensure alignment and efficiency.

   - **External Partnerships:** Collaboration with technology firms, healthcare providers, and patient advocacy groups will be essential for developing and launching innovative solutions.

   B. Effective Communication:

   - **Stakeholder Engagement:** Strong communication skills will be crucial for managing stakeholder relationships, including healthcare professionals, patients, and regulators.

6. **Data-Driven Strategies and Analytics**

   A. Big Data and Analytics:

   - **Strategic Insights:** Advanced data analytics will enable product managers to gain deeper insights into market trends, competitive landscape, and customer behavior.

   - **Predictive Analytics:** Utilizing predictive models will help anticipate market needs and optimize product development and marketing strategies.

   B. Real-Time Monitoring:

   - **Performance Tracking:** Real-time data monitoring will allow timely adjustments in marketing strategies, product positioning, and customer engagement efforts.

7. **Digital Transformation and Innovation**

   A. Digital Marketing:

   - **Omnichannel Strategies:** Embracing digital marketing channels, including social media, content marketing, and digital advertising, will be critical for engaging healthcare professionals and patients.

- **Interactive Platforms:** Leveraging interactive digital platforms and tools for education, engagement, and support will enhance product visibility and adoption.

B. Innovation in Drug Development:

- **Accelerated R&D:** Advancements in technology will shorten the drug development timeline, enabling faster delivery of new therapies to the market.

8. **Sustainability and Corporate Responsibility**

A. Environmental Impact:

- **Sustainable Practices:** Product managers must consider the environmental impact of drug manufacturing and packaging, implementing sustainable practices to reduce the industry's carbon footprint.

B. Ethical Considerations:

- **Corporate Responsibility:** Emphasizing ethical practices and corporate social responsibility will maintain public trust and meet regulatory requirements.

Conclusion

The future of pharmaceutical product management will be characterized by increased technological integration, a focus on patient-centric approaches, and the need for agile and data-driven strategies. As the industry evolves, product managers will be crucial in navigating these changes, driving innovation, and delivering value to patients and stakeholders. Adapting to emerging trends and leveraging new technologies will be key to thriving in the dynamic landscape of pharma product management.

33.1 Staying Ahead: Lifelong Learning in Pharma

In the rapidly evolving pharmaceutical industry, staying ahead requires a commitment to lifelong learning. Continuous education is a professional necessity and a strategic advantage for pharma product managers. Here's how it impacts career development and success:

1. **Adapting to Industry Changes**

   - **Evolving Regulations:** Regulatory environments are constantly changing. Keeping up with new regulations, such as GDPR or evolving FDA guidelines, is critical to ensure compliance and effective market strategies.

   - **Technological Advancements:** As AI, big data, and digital tools continue to transform the industry, product managers must stay informed about the latest technologies to effectively use them in their roles.

   - **Market Dynamics:** It is crucial to understand shifts in healthcare delivery, patient needs, and competitive landscapes. Ongoing learning helps product managers anticipate and respond to these changes.

2. **Enhancing Core Competencies**

   - **Scientific Knowledge:** Deepening your understanding of pharmacology, clinical research, and drug development processes will ensure that you can communicate effectively with cross-functional teams and make informed decisions.

   - **Business Acumen:** Mastering financial management, marketing strategies, and project management skills enables product managers to drive business success and manage resources effectively.

   - **Leadership Skills:** Lifelong learning in leadership development is vital for managing cross-functional teams, driving collaboration, and leading strategic initiatives.

3. **Embracing Digital Transformation**

   - **Digital Marketing:** As digital marketing becomes more central to pharma strategies, staying updated on best practices and emerging trends is key to executing successful campaigns.

- **Data-Driven Decision-Making:** Continuous learning in data analytics empowers product managers to make informed, evidence-based decisions that can enhance product performance and market positioning.

- **Telemedicine and Digital Health:** Understanding the implications of telemedicine and digital health on product management can help you innovate and adapt strategies to meet evolving patient and healthcare provider needs.

4. Networking and Professional Development

- **Industry Conferences:** Attending conferences, seminars, and webinars provides opportunities to learn from industry leaders, stay updated on trends, and expand your professional network.

- **Certifications and Courses:** Certifications in regulatory affairs, digital marketing, or project management can enhance your expertise and credibility.

- **Mentorship and Peer Learning:** Engaging in mentorship programs or peer learning groups allows for knowledge exchange and offers insights into best practices and innovative approaches in product management.

5. Personal Growth and Career Advancement

- **Continuous Improvement:** Lifelong learning fosters a mindset of continuous improvement, driving personal growth and career advancement. It positions you as a proactive, knowledgeable leader in the industry.

- **Adaptability:** Learning and adapting quickly is a significant asset in a fast-paced industry. It ensures you remain relevant and capable of navigating the challenges and opportunities in pharmaceutical product management.

6. Preparing for the Future

- **Emerging Trends:** Staying ahead of trends such as personalized medicine, AI, and digital therapeutics will be crucial for future success. Continuous learning helps you anticipate and integrate these trends into your product strategies.

- **Career Longevity:** Lifelong learning supports career longevity by ensuring your skills and knowledge remain current, making you a valuable asset to your organization and the industry.

Conclusion

Lifelong learning is a pathway to career development and a cornerstone of effective pharmaceutical product management. By committing to continuous education, product managers can enhance their competencies, stay ahead of industry changes, and position themselves as leaders in the digital frontier of pharma.

33.2 The Role of the Product Manager in Pharma's Future

The role of the pharmaceutical product manager is set to evolve significantly as the industry continues to adapt to new technological advancements, regulatory changes, and shifts in consumer expectations. As the intersection between science, business, and technology becomes more complex, product managers will find themselves at the forefront of driving innovation and ensuring the success of their products. Here are the key aspects of the product manager's role in the future of pharma:

1. **Navigating Digital Transformation**

   - **Adapting to Emerging Technologies**: With the increasing integration of digital health tools, AI, and data analytics into pharmaceutical operations, product managers must be proficient in these technologies. They must leverage these tools to optimize product development, marketing strategies, and patient engagement.

   - **Leading Digital Initiatives**: Product managers will be crucial in leading digital transformation projects within their organizations. This includes implementing digital marketing strategies, enhancing digital therapeutics, and utilizing real-world evidence to inform product decisions.

2. **Driving Innovation**

   - **Fostering Cross-Disciplinary Collaboration**: The future of pharmaceutical product management will require collaboration across various disciplines, including R&D, regulatory affairs, marketing, and sales. Product managers must bridge these areas to drive innovation and efficiently bring new products to market.

   - **Championing Patient-Centric Approaches**: As the industry moves towards more personalized medicine, product managers must ensure that their strategies are aligned with patient needs. This includes developing products that offer better outcomes, improved adherence, and enhanced patient experiences.

3. **Strategic Leadership**

   - **Managing Complex Product Lifecycles**: Pharmaceutical products are becoming increasingly complex, especially in areas like biologics and gene therapies. Product managers must be adept at managing these products throughout their entire lifecycle. This includes strategic planning, market entry, lifecycle extension strategies, and competition management.

   - **Enhancing Regulatory Acumen**: As regulations evolve, especially concerning digital health and AI in healthcare, product managers must stay informed and ensure their products comply with the latest requirements. This will be crucial for successful product approvals and maintaining market access.

4. **Building Resilience in Uncertain Markets**

   - **Scenario Planning and Risk Management**: The pharmaceutical industry faces numerous uncertainties, including regulatory changes, market access challenges, and global health crises. Product managers must develop strong scenario planning and risk management skills to navigate these uncertainties effectively.

   - **Sustainability and Ethical Considerations**: With a growing emphasis on sustainability and ethical business practices, product managers will need to incorporate these aspects into their strategies. This includes developing products that align with sustainability goals and ensuring ethical marketing practices.

5. **Career Development and Lifelong Learning**

   - **Continuous Skill Enhancement**: The rapid pace of change in the pharmaceutical industry will require product managers to update their skills continually. This includes staying current with industry trends, acquiring new technical competencies, and developing leadership capabilities.

   - **Networking and Professional Growth**: To stay ahead, product managers must engage in industry networks, attend conferences, and participate in professional development opportunities. This will also be critical for career advancement and transitioning into more senior roles.

6. Shaping the Future of Healthcare

- **Influencing Industry Standards**: As key decision-makers, product managers can influence industry standards and best practices. By participating in industry groups, contributing to thought leadership, and collaborating with regulatory bodies, they can shape the future of pharmaceutical product management.

- **Advancing Global Health**: Product managers will play a pivotal role in advancing global health by ensuring that innovative and effective treatments reach patients worldwide. This will involve navigating global regulatory environments, understanding diverse market needs, and addressing health disparities.

In conclusion, the role of the pharmaceutical product manager will become increasingly strategic and multifaceted in the future. Those who can adapt to the changing landscape, embrace new technologies, and lead with a patient-centric focus will be well-positioned to drive success in the evolving pharmaceutical industry.

33.3 Preparing for the Future: Staying Resilient and Innovative

In the rapidly evolving pharmaceutical industry, staying resilient and innovative is critical for long-term success. As a pharma product manager, you must be prepared to navigate a landscape characterized by constant technological advancements, shifting market dynamics, and evolving regulatory requirements. This section will focus on developing and maintaining resilience while fostering a culture of innovation within your organization.

1. **Understanding Resilience in Pharma**

   - **Definition and Importance:** Resilience refers to the ability to recover from setbacks, adapt to change, and keep going in the face of adversity. Resilience is essential in the pharma industry, where market conditions, regulatory landscapes, and technological environments are in constant flux.

   - **Building Organizational Resilience:** Strategies to create a resilient organization, including fostering a culture of adaptability, investing in employee development, and maintaining a strong focus on patient and market needs.

2. **Embracing Change and Uncertainty**

   - **The Role of Agility:** How agility—both in thought and action—enables pharma product managers to respond quickly to new challenges and opportunities.

   - **Scenario Planning:** Techniques for anticipating future scenarios and preparing for various outcomes, ensuring that your product strategy is robust and flexible.

   - **Learning from Failures:** Encouraging a mindset where failures are seen as learning opportunities helps drive continuous improvement.

3. **Fostering a Culture of Innovation**

   - **Innovation as a Competitive Advantage:** The importance of innovation in staying ahead of the competition, particularly in the digital age where new technologies can disrupt traditional business models.

- **Encouraging Creative Thinking:** Use design thinking and other innovation methodologies to encourage creative problem-solving within your team.

- **Collaboration and Cross-Functional Teams:** Leveraging diverse perspectives and expertise from different departments to drive innovative solutions.

4. Leveraging Technology for Innovation

- **Digital Tools and Platforms:** Exploring the latest digital tools and platforms that can support innovative product management practices, from AI and machine learning to big data analytics.

- **Emerging Technologies:** Staying informed about emerging technologies such as blockchain, AR/VR, and telemedicine and understanding their potential impact on your products and strategies.

- **Digital Transformation:** How to lead and manage digital transformation initiatives within your organization, ensuring that innovation is integrated into all aspects of product management.

5. Continuous Learning and Development

- **Investing in Education and Training:** Continuous learning is essential for staying current with industry trends, technological advancements, and new marketing strategies.

- **Networking and Mentorship:** Building a strong professional network and seeking mentorship opportunities to gain insights and advice from industry leaders.

- **Staying Informed:** Strategies for keeping abreast of the latest research, industry reports, and market developments, ensuring that your knowledge remains current and relevant.

6. Anticipating Future Trends

- **Industry Predictions:** An overview of key trends likely to shape the future of pharma product management, including personalized medicine, digital therapeutics, and patient-centric care.

- **Adapting to Regulatory Changes:** Prepare for potential shifts in regulatory requirements and ensure that your product strategies remain compliant while embracing innovation.

- **Sustainability and Ethical Considerations:** Understanding the importance of sustainability and ethical practices in pharma and how these factors influence product management strategies.

33.4 The Importance of Adaptability in a Changing Industry

Adaptability is crucial in the constantly evolving pharmaceutical industry. As the sector undergoes rapid transformations due to technological advancements, regulatory changes, and shifting market dynamics, the ability to adapt becomes a defining trait for success. Here's why adaptability is vital for pharma product managers and how it impacts their career development:

1. **Responding to Technological Advancements**

   The pharmaceutical industry is increasingly influenced by innovations such as artificial intelligence (AI), big data, and digital health technologies. Product managers must learn and integrate these technologies into their strategies to stay competitive quickly. Adaptability allows them to embrace new tools, optimize product lifecycles, and enhance decision-making processes.

2. **Navigating Regulatory Changes**

   Pharmaceutical regulations are continually evolving to ensure safety, efficacy, and transparency. Product managers must be able to adapt to new regulations promptly. This might involve adjusting marketing strategies, modifying product formulations, or ensuring compliance with new guidelines. Adaptability ensures that these changes are handled efficiently, reducing non-compliance risk.

3. **Managing Market Dynamics**

   The pharmaceutical market is characterized by intense competition, pricing pressures, and changing consumer demands. Adaptable product managers can pivot their strategies to respond to competitors, capitalize on new opportunities, and address shifts in patient or healthcare provider needs. This agility helps in maintaining product relevance and market share.

4. Leading Cross-Functional Teams

Product managers often lead teams from various departments, including R&D, marketing, and sales. Each department may have different perspectives and goals. An adaptable product manager can manage these diverse teams effectively, ensuring alignment and collaboration toward common objectives.

5. Continuous Learning and Skill Development

As the industry evolves, so must the skills of those within it. Adaptability drives a commitment to continuous learning, whether mastering new marketing techniques, understanding the latest scientific research, or developing new leadership skills. This continuous improvement is essential for career progression.

6. Strategic Thinking in Uncertain Times

Thinking strategically during uncertain times—such as a public health crisis or economic downturn—requires adaptability. Product managers must be able to reassess priorities, develop contingency plans, and make decisions that balance short-term needs with long-term goals.

7. Career Advancement

In an industry that values innovation and responsiveness, adaptability is often a key factor in career advancement. Product managers who can thrive in changing environments will likely be recognized for leadership roles and other growth opportunities.

8. Fostering a Culture of Innovation

Adaptable product managers contribute to a culture of innovation within their organizations. By encouraging flexibility and creative problem-solving, they help their teams anticipate and react to changes, and and drive them. This proactive approach is critical in a fast-paced industry like pharmaceuticals.

Conclusion

Adaptability is not just a beneficial trait for pharmaceutical product managers—it's a necessity. The ability to swiftly adjust to new technologies, regulations, market conditions, and organizational changes ensures that product managers can effectively lead their products to success, maintain compliance, and grow professionally in a dynamic industry. As the pharmaceutical landscape continues to evolve, those who embrace adaptability will be better equipped to navigate the challenges and capitalize on the opportunities.

Part X. Final Thoughts and Conclusion

The role of a pharmaceutical product manager has evolved significantly, particularly with the rise of digital technologies and data-driven strategies. As the healthcare landscape changes, pharma product managers are at the forefront, bridging the gap between scientific innovation, regulatory compliance, and market needs. They play a pivotal role in driving the success of pharmaceutical products, from development through to launch and lifecycle management.

Product managers must continuously adapt and refine their skills in this rapidly evolving environment. Integrating digital marketing, data analytics, AI, and emerging technologies into their workflows is not optional but essential for staying competitive. Moreover, their ability to lead cross-functional teams, manage stakeholder relationships, and make informed, strategic decisions is crucial to navigating the complexities of the pharmaceutical industry.

As we look to the future, the demand for skilled pharmaceutical product managers who can thrive in this dynamic environment will only increase. Those who can effectively harness the power of technology while maintaining a strong foundation in their role's scientific and business aspects will be well-positioned to lead their organizations to success.

In conclusion, the future of pharmaceutical product management is bright and filled with opportunities for those willing to embrace change and drive innovation. As a product manager, your ability to adapt, learn, and lead will be key to your career success and the success of the products and patients you serve.

Recap of Key Points

1. **Introduction to Pharmaceutical Product Management**: Understanding the evolution, importance, and impact of pharmaceutical product management in the digital age, where digital tools and strategies are transforming the industry.

2. **Role and Responsibilities of a Pharmaceutical Product Manager**: Core competencies, such as scientific knowledge, business acumen, financial management, project management, and regulatory expertise, are essential for success. The role significantly impacts business outcomes and requires effective leadership and communication skills.

3. **Transformation of Pharmaceutical Product Management**: The shift from traditional to digital product management has introduced new opportunities and challenges. Product managers must adapt to digital tools, emerging technologies, and changing market dynamics to stay competitive.

4. **Product Launch and Lifecycle Management**: Effective planning, execution, and monitoring of product launches and lifecycle management are critical for sustaining product success. This includes pre-launch planning, launch execution, post-launch monitoring, and strategies for lifecycle extension.

5. **Strategic Planning and Brand Management**: Building strategic marketing plans, aligning them with business objectives, conducting competitive analysis, and scenario planning are crucial for navigating the complex pharmaceutical market.

6. **Digital Marketing in Pharma**: Understanding digital marketing channels, data-driven decision-making, and leveraging AI and emerging technologies is key to thriving in the digital era.

7. **Leadership and Communication**: Leading cross-functional teams, managing conflicts, and engaging stakeholders are vital for effective product management. Clear communication and strong relationships with stakeholders are essential for success.

8. **Emerging Technologies and Future Trends**: The role of AI, digital health, and telemedicine is growing in importance. Pharma product managers must prepare for future trends by developing the necessary skills and competencies to navigate these changes.

9. **Career Development and Advancement**: Continuous learning, skill development, and staying updated with industry trends are essential for career growth and advancement in pharmaceutical product management.

This recap highlights the critical aspects of pharmaceutical product management in the digital age, emphasizing the need for adaptability, strategic thinking, and continuous development to succeed in this evolving field.

34.1 Call to Action

As the pharmaceutical industry continues to evolve, particularly in the digital age, the role of the pharmaceutical product manager is becoming more critical and multifaceted than ever before. This book has provided a comprehensive guide to navigating the complexities of modern pharmaceutical product management, emphasizing the importance of adapting to new technologies, embracing data-driven strategies, and leading cross-functional teams with a clear, strategic vision.

Pharmaceutical product managers are at the intersection of science, business, and technology. They must understand the intricacies of drug development and regulatory landscapes and have the foresight to anticipate market trends and patient needs. As explored throughout this book, the successful product manager is agile, continuously learning, and capable of making informed decisions that drive product success and improve patient outcomes.

The digital frontier offers unparalleled opportunities but also presents significant challenges. As a product manager, your ability to leverage digital tools, understand emerging technologies, and apply strategic thinking will be key to your success. Whether managing a product launch, navigating lifecycle management, or engaging with stakeholders, the skills and knowledge outlined in this book will serve as your toolkit for effective leadership and innovation.

Call to Action

Now is the time to take action. As you conclude this book, consider how you can apply the insights and strategies discussed to your daily work. Reflect on the areas where you can improve, whether it's deepening your understanding of digital marketing, enhancing your leadership skills, or becoming more proficient in data analysis.

Embrace digital transformation confidently, knowing you have the tools and knowledge to succeed. Stay curious, stay informed, and always be ready to adapt to the pharmaceutical industry's ever-changing landscape.

Your journey as a pharmaceutical product manager is ongoing. The steps you take today will shape the future of your career and your impact on the industry. Take charge, lead purposefully, and continue striving for excellence in all you do.

Glossary of Terms

Adherence: The extent to which patients take their medications as prescribed, including the correct dosage and timing.

Behavioral Economics: A field of study that examines how psychological, cognitive, emotional, cultural, and social factors affect economic decisions and behaviors.

Big Data: Large and complex datasets that traditional data-processing software cannot manage effectively. Big data can analyze patient outcomes, treatment efficacy, and other key metrics in healthcare.

Blockchain Technology: A decentralized digital ledger that records transactions across multiple computers securely and transparently. In pharma, it can be used for secure supply chain management and data integrity.

Brand Metrics: Quantitative measures used to evaluate the success of a brand, including brand awareness, perception, loyalty, and market share.

Brand Positioning is the strategy used to differentiate a brand in the target audience's mind, highlighting its unique value and benefits compared to competitors.

Clinical Trials: Research studies performed on patients to evaluate the efficacy and safety of medical treatments or interventions.

Compliance: Adherence to regulatory guidelines, laws, and ethical standards governing pharmaceutical marketing and operations.

Competitive Analysis: The process of assessing competitors' strengths, weaknesses, and strategies to identify opportunities and threats in the market.

Conflict Resolution: Techniques and strategies to resolve disagreements and conflicts within teams or organizations.

Cultural Competency: Understanding, communicating, and effectively interacting with people across different cultures.

Decision-Making Processes: Behavioral economics often influences the steps and considerations involved in healthcare decisions.

Digital Marketing: Using digital channels such as social media, email, and search engines to promote products and engage with customers.

Drug Discovery: Identifying new medication candidates involves target identification, compound screening, and lead optimization.

Economic Evaluations: Assessments of the cost-effectiveness of drug therapies, comparing their costs and health outcomes.

Epidemiology studies population disease patterns, causes, and effects, which inform public health decisions and strategies.

Genomics: The study of an organism's complete set of DNA, including all its genes, used in personalized medicine to tailor treatments to individual genetic profiles.

Healthcare Professionals (HCPs) are medical practitioners, such as doctors, nurses, and pharmacists, involved in patient care and treatment.

Internet of Things (IoT): A network of interconnected devices that collect and exchange data, used in healthcare for remote patient monitoring and data collection.

Market Share: The percentage of total sales in a market captured by a particular brand or product.

Market Research: The process of gathering, analyzing, and interpreting information about a market, including customer needs, preferences, and behaviors.

Nudging: Subtle interventions designed to influence behavior and decision-making, often used to encourage healthier choices among patients.

Pharmacoeconomics: The study of the economic impact of drugs and therapies, including cost-effectiveness and value.

Pharmacology: The science of drugs, including their composition, uses, and effects on the body.

Profit and Loss (P&L) Management: Managing a company's revenues, costs, and expenses to ensure profitability.

Project Management: Applying processes, methods, skills, and knowledge to achieve specific project objectives within defined parameters.

Regulatory Affairs: The field focused on ensuring compliance with regulations and guidelines governing pharmaceutical development, approval, and marketing.

Resource Allocation: The process of distributing resources, such as time, money, and personnel, effectively to achieve project goals.

Risk Management: Identifying, assessing, and mitigating potential risks impacting a project or business.

Scenario Planning: A strategic planning method used to envision and prepare for different future scenarios and their potential impacts.

Sentiment Analysis: Using natural language processing and other techniques to analyze opinions, attitudes, and emotions expressed in text.

Social Media: Online platforms that enable users to create and share content or participate in social networking.

Strategic Marketing Planning: Defining marketing strategies and tactics to achieve business goals and build a strong brand.

SWOT Analysis: A strategic planning tool used to identify strengths, weaknesses, opportunities, and threats related to a business or project.

Telemedicine: The remote diagnosis and treatment of patients through telecommunications technology.

Therapeutics: The branch of medicine concerned with the treatment of disease and the action of remedial agents.

Webinars: Online seminars or presentations conducted online, often used for educational or promotional purposes.

List of Resources and Further Reading

Here are some resources that pharmaceutical product managers can use to deepen their knowledge and stay updated on the industry's latest trends and best practices.

Books

1. **Pharmaceutical Marketing by Mickey Smith**

   - Provides an in-depth overview of pharmaceutical marketing, including strategies, regulatory issues, and case studies.

2. **Marketing Management by Philip Kotler**

   - A comprehensive guide to marketing principles applicable to the pharmaceutical industry with adaptations.

3. **Pharmaceutical Market Access in Developed Markets by GÔvenc Kockaya and Albert Wertheimer**

   - Explores market access strategies and challenges in developed markets.

4. **Digital Marketing for Dummies by Ryan Deiss and Russ Henneberry**

   - A practical guide to digital marketing, covering strategies and tools relevant to pharma marketing.

5. **Blue Ocean Strategy by W. Chan Kim and Renée Mauborgne**

   - Offers innovative strategies for creating new market spaces and making the competition irrelevant.

Articles and Journals

1. **Journal of Pharmaceutical Marketing and Management**

   - Publishes research on the latest pharmaceutical marketing trends, strategies, and practices.

2. **The Impact of Digital Transformation on the Pharmaceutical Industry - McKinsey & Company**

   - Analyzes how digital technologies are reshaping the pharmaceutical industry and the implications for marketing.

3. **Patient Engagement in the Digital Age - Deloitte Insights**
   - Discusses strategies for engaging patients through digital platforms and technologies.
4. **Pharmacoeconomics: From Theory to Practice - Value in Health Journal**
   - Provides insights into the economic evaluation of drug therapies and their impact on healthcare decision-making.

Online Resources

1. **Pharma Marketing Network** - www.pharma-mkting.com
2. **Digital Pharma** - www.digitalpharma.com
3. **PhRMA (Pharmaceutical Research and Manufacturers of America)** - www.phrma.org
4. **FDA (U.S. Food and Drug Administration)** - www.fda.gov

Professional Organizations

1. **Healthcare Businesswomen's Association (HBA)**
   - Offers networking, education, and leadership opportunities for women in healthcare, including pharma marketing.
2. **Pharmaceutical Marketing Society (PM Society)**
   - Provides events, training, and resources for professionals in pharmaceutical marketing.
3. **American Association of Pharmaceutical Scientists (AAPS)**
   - Offers resources, conferences, and publications related to pharmaceutical science and marketing.

Online Courses and Certifications

1. **Coursera**: Digital Marketing Specialization
2. **edX**: Digital Transformation in Healthcare
3. **LinkedIn Learning**: Pharmaceutical Marketing
4. **Harvard Online**: Healthcare Economics

Case Studies and Reports

1. **Pharma 2020**: Marketing the Future: PwC Report
2. **The Digital Imperative in Pharma** - BCG Report
3. **Building Effective Patient Support Programs** - Accenture Case Study

Templates and Tools for Product Management

Strategic Planning Templates

1. SWOT Analysis Template

- **Description**: A tool to identify strengths, weaknesses, opportunities, and threats.
- **Components**:
  - **Strengths**: Internal factors that are advantageous.
  - **Weaknesses**: Internal factors that could hinder progress.
  - **Opportunities:** External factors that can be capitalized on.
  - **Threats**: External factors that could pose challenges.
  - **Use Case**: Analyzing the competitive position of a new diabetes medication like Glucotrol.
  - **Tools**: Excel/Google Sheets templates, online tools like Lucidchart or Miro.

2. Scenario Planning Template

- **Description**: A method for envisioning and preparing for different future scenarios.
- **Components**:
  - Best-case scenario
  - Worst-case scenario
  - Most likely scenario
  - Contingency plans for each scenario
- **Use Case**: Preparing for market responses to a new drug launch.
- **Tools**: PowerPoint/Google Slides, dedicated software like Scenario Planner.

3. Marketing Plan Template

- **Description**: A detailed plan outlining marketing strategies and tactics.
- **Components**:
  - Market analysis

- Target audience
- Marketing objectives
- Marketing strategies and tactics
- Budget and resources
- Key performance indicators (KPIs)
- **Use Case**: Annual marketing plan for Glucotrol.
- **Tools**: Word/Google Docs, marketing software like HubSpot or Marketo.

4. **Project Management Tools**

  1. **Gantt Chart Template**
     - **Description**: A visual timeline for planning and tracking project schedules.
     - **Components**:
       - Project tasks
       - Start and end dates
       - Dependencies
       - Milestones
     - **Use Case**: Planning the development and launch phases of Glucotrol
     - **Tools**: Excel/Google Sheets, project management software like Microsoft Project or Asana.

  2. **Risk Management Plan Template**
     - **Description**: A plan for identifying, assessing, and mitigating risks.
     - **Components**:
       - Risk identification
       - Risk assessment (probability and impact)
       - Mitigation strategies
       - Monitoring and review
     - **Use Case**: Identifying potential regulatory and market risks for Glucotrol.

- **Tools**: Word/Google Docs, risk management software like RiskWatch or LogicManager.

5. **Financial Management Tools**

   1. **Budgeting Template**

      - **Description**: A detailed financial plan for managing expenses and revenues.
      - **Components**:
        - Revenue forecasts
        - Expense categories
        - Variance analysis
        - Profit margins
      - **Use Case**: Annual budget for marketing and operational expenses for Glucotrol.
      - **Tools**: Excel/Google Sheets, financial management software like QuickBooks or Xero.

   2. **Profit and Loss Statement Template**

      - **Description**: A financial report summarizing revenues, costs, and expenses.
      - **Components**:
        - Total revenue
        - Cost of goods sold (COGS)
        - Gross Profit
        - Operating expenses
        - Net profit
      - **Use Case**: Quarterly financial performance review of Glucotrol
      - **Tools**: Excel/Google Sheets, accounting software like Freshbooks or Sage.

6. **Marketing Research Analysis Tools**

   1. **Market Research Report Template**

      - **Description**: A comprehensive report on market conditions and consumer insights.

- **Components**:
  - Market size and growth
  - Competitive landscape
  - Customer demographics and preferences
  - SWOT analysis
- **Use Case**: Market research for positioning Glucotrol against competitors.
- **Tools**: Word/Google Docs, market research platforms like Qualitrics or SurveyMonkey.

2. **Sentiment Analysis Dashboard**

- **Description**: A tool for analyzing patient and HCP sentiment from various data sources.
- **Components**:
  - Data sources (social media, forums, surveys)
  - Sentiment scores
  - Trend analysis:
  - Key insights:
- **Use Case**: Tracking patient sentiment towards Glucotrol post-launch.
- **Tools**: Data visualization software like Tableau or Power BI, sentiment analysis tools like BrandWatch or Lexalytics.

7. **Digital Marketing Tools**

1. **Email Marketing Campaign Template**

- **Description**: A structured plan for executing email marketing campaigns.
- **Components**:
  - Campaign objectives
  - Target audience
  - Email content and design
  - Scheduling
  - Metrics and KPIs

- **Use Case**: Email campaign to promote Glucotrol to healthcare professionals.
- **Tools**: Word/Google Docs, email marketing platforms like MailChimp or ConstantContact.

2. Social Media Content Calendar

- Description: A calendar for planning and scheduling social media posts.
- Components:
    - Content themes
    - Post dates and times
    - Platforms
    - Engagement metrics
- Use Case: Social media strategy for building awareness of Glucotrol.
- Tools: Excel/Google Sheets, social media management tools like Hootsuite or Buffer.

8. Professional Development Tools

1. Personal Development Plan (PDP) Template

- **Description**: A plan for setting and tracking personal career goals and development activities.
- **Components**:
    - Career goals
    - Skills assessment
    - Development activities
    - Progress tracking
- **Use Case**: Developing a career advancement plan for a pharma product manager.
- **Tools**: Word/Google Docs, professional development platforms like LinkedIn Learning or Coursera.

2. Networking and Mentorship Plan

- **Description**: A strategy for building professional relationships and finding mentors.

- **Components**:
  - Networking goals
  - Potential mentors
  - Engagement activities
  - Follow-up actions
- **Use Case**: Establishing a mentorship relationship for career growth in pharma.
- **Tools**: Word/Google Docs, professional networking platforms like LinkedIn.

Pharmaceutical product managers can use these templates and tools to plan, execute, and manage their projects, ensuring successful product launches and sustained market performance.

Index

T

U

V

W

X

Z

ABOUT THE AUTHOR

Subba Rao Chaganti He has a master's in business administration and over fifty-seven years of experience in pharmaceutical marketing, covering all facets of the industry, from selling to sales management, product management, and heading the total marketing activity. His experience covers domestic and international marketing and Indian and multinational sectors.

He also taught a course on Advertising and Brand Management at Gitam Institute of Foreign Trade (now part of Gitam University) at Visakhapatnam for a few years as an adjunct professor and a course on Marketing at Jawaharlal Nehru Technological University (JNTU), Hyderabad, as a visiting faculty member.

He lives in Farmington, Connecticut, USA, and can be reached at subbarao.chaganti@gmail.com

Here is a list of his publications:

Books Published:

1. Pharmaceutical Marketing in India: Concepts, Cases, Strategy
2. Game Plans for Post-Gatt Era: Action Agenda of Indian Pharmaceutical Industry
3. Compete or Forfeit: Strategies for Sustainable Competitive Edge in Pharma Product Patent Era
4. Pharmaceutical Marketing in India for Today and Tomorrow - 25th Anniversary Edition
5. Bullseyes and Blunders: Lessons from 100 Cases in Pharmaceutical Marketing
6. Digital Pharma Marketing Playbook: Winning With the New Rules of Engagement
7. Cracking the Generics Code: Your Single-Source Success Manual for Winning in Multi-Source Product Markets
8. Reimagine Pharma Marketing: Make It Future-Proof!
9. Brand Positioning in Pharma

www.ingramcontent.com/pod-product-compliance
Lightning Source LLC
LaVergne TN
LVHW050734200726
843507LV00001B/1